When Deep Learning Meets Geometry for Air-to-Ground Perception on Drones

When Deep Learning Meets Geometry for Air-to-Ground Perception on Drones

Editors

Dongdong Li
Gongjian Wen
Yangliu Kuai
Runmin Cong

MDPI

Basel • Beijing • Wuhan • Barcelona • Belgrade • Novi Sad • Cluj • Manchester

Editors
Dongdong Li
National University of
Defense Technology
Changsha
China

Gongjian Wen
National University of
Defense Technology
Changsha
China

Yangliu Kuai
National University of
Defense Technology
Changsha
China

Runmin Cong
Shandong University
Jinan
China

Editorial Office
MDPI AG
Grosspeteranlage 5
4052 Basel, Switzerland

This is a reprint of articles from the Special Issue published online in the open access journal *Drones* (ISSN 2504-446X) (available at: https://www.mdpi.com/journal/drones/special_issues/3080158YOJ).

For citation purposes, cite each article independently as indicated on the article page online and as indicated below:

Lastname, A.A.; Lastname, B.B. Article Title. *Journal Name* **Year**, *Volume Number*, Page Range.

ISBN 978-3-7258-2507-3 (Hbk)
ISBN 978-3-7258-2508-0 (PDF)
doi.org/10.3390/books978-3-7258-2508-0

Contents

drones

Article

Drone-Based Visible–Thermal Object Detection with Transformers and Prompt Tuning

Rui Chen [1], Dongdong Li [1,*], Zhinan Gao [1], Yangliu Kuai [2] and Chengyuan Wang [3]

1 College of Electronic Science and Technology, National University of Defense Technology, Changsha 410073, China; chenrui23@nudt.edu.cn (R.C.); gaozhinan22@nudt.edu.cn (Z.G.)
2 College of Intelligent Science and Technology, National University of Defense Technology, Changsha 410073, China; kuaiyangliu09@nudt.edu.cn
3 Information and Communication College, National University of Defense Technology, Wuhan 430010, China; wangchengyuan@nudt.edu.cn
* Correspondence: lidongdong12@nudt.edu.cn

Abstract: The use of unmanned aerial vehicles (UAVs) for visible–thermal object detection has emerged as a powerful technique to improve accuracy and resilience in challenging contexts, including dim lighting and severe weather conditions. However, most existing research relies on Convolutional Neural Network (CNN) frameworks, limiting the application of the Transformer's attention mechanism to mere fusion modules and neglecting its potential for comprehensive global feature modeling. In response to this limitation, this study introduces an innovative dual-modal object detection framework called **Vi**sual **P**rompt multi-modal **Det**ection (**VIP-Det**) that harnesses the Transformer architecture as the primary feature extractor and integrates vision prompts for refined feature fusion. Our approach begins with the training of a single-modal baseline model to solidify robust model representations, which is then refined through fine-tuning that incorporates additional modal data and prompts. Tests on the DroneVehicle dataset show that our algorithm achieves remarkable accuracy, outperforming comparable Transformer-based methods. These findings indicate that our proposed methodology marks a significant advancement in the realm of UAV-based object detection, holding significant promise for enhancing autonomous surveillance and monitoring capabilities in varied and challenging environments.

Keywords: drone-based object detection; visible–thermal object detection; vision transformer; vision prompt tuning

Citation: Chen, R.; Li, D.; Gao, Z.; Kuai, Y.; Wang, C. Drone-Based Visible–Thermal Object Detection with Transformers and Prompt Tuning. *Drones* **2024**, *8*, 451. https://doi.org/10.3390/drones8090451

Academic Editor: Oleg Yakimenko

Received: 31 July 2024
Revised: 21 August 2024
Accepted: 29 August 2024
Published: 1 September 2024

1. Introduction

Object detection, a central challenge in computer vision, necessitates algorithms that possess robust classification capabilities and precise spatial localization for the identification and location of various targets, such as humans, animals, and vehicles, in images and videos. The performance of detection has been markedly improved by the rapid advancement of deep learning, particularly Convolutional Neural Networks (CNNs) [1], fueling progress in the field and spurring interest in downstream tasks [2–4]. The rise of unmanned aerial vehicles (UAVs), with their agility and efficient data collection capabilities, has given birth to the task of drone-based object detection [5]. However, the significant scale variations and variable angles in UAV imagery pose challenges to object detection. Existing algorithms for rotated object detection [6–10], often designed for remote sensing images, struggle to meet these demands.

In the field of drone-based object detection, current algorithms primarily depend on visible light imagery, which inherently limits their effectiveness in complex environments such as nighttime, rainy conditions, dense fog, and instances of occlusion (See Figure 1). With the advancement of sensor technology, modern drones are equipped with a variety of sensors, including infrared payloads, vastly expanding their range of applications and making dual-modal object detection a hot topic of research in the drone sector. The distinct imaging mechanism of infrared, which captures thermal energy, complements visible light imagery, markedly improving the precision and robustness of object detection. However, existing dual-modal detection algorithms often employ dual-stream backbone networks to process each modality separately, neglecting the issue of information imbalance between the two modalities. This leads to a substantial amount of parameter redundancy, highlighting the need for research into more efficient fusion strategies.

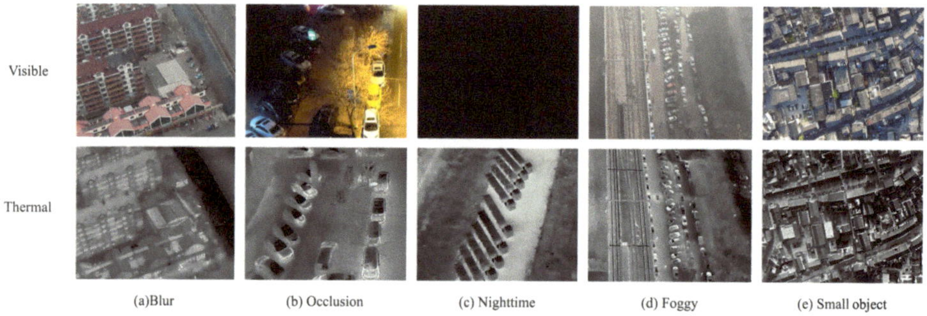

Figure 1. Difficulties and challenges faced by dual-modal object detection of UAV. (**a**) Low resolution of infrared images, and unclear target textures. (**b**) Under occlusion conditions, parts of target in visible light images are covered by trees. (**c**) Under night conditions, visible light imaging completely fails. (**d**) Under heavy fog conditions, visibility of targets in visible light is obstructed. (**e**) Flight height of drones is unstable, resulting in uneven target scales.

The Transformer architecture has achieved great success in the field of natural language processing [11,12] and has since been adopted by researchers in the realm of computer vision [13–15]. Its efficiency in processing long-range dependencies and parallelization capabilities have established it as a new paradigm. Dealing with issues such as lighting variations and target occlusions in dual-modal object detection poses challenges for CNNs. CNNs excel at local feature extraction via convolutions but struggle with lighting changes that alter pixel values and occlusion that disrupts these local patterns. Their limited global context understanding and multi-modal interaction hinder performance. Transformers, on the other hand, leverage global context capture capabilities, enabling better generalization across different lighting conditions. For occlusions, Transformers utilize pre-trained masking mechanisms to handle obscured regions, and their self-attention mechanism tracks information changes before and after occlusion, facilitating robust multi-modal global information interaction. However, in the domain of dual-modal object detection, the application of Transformers is limited, with their attention mechanisms often confined to the fusion module [16], not fully harnessing their potential for understanding global context. Additionally, with the emergence of efficient self-supervised learning methods like Masked Autoencoders (MAEs) [17], the Vision Transformer (VIT) [13] architecture can leverage a wealth of pre-trained weights, offering superior feature extraction and generalization for downstream tasks. Therefore, employing a VIT for dual-modal object detection is a promising and innovative approach.

Considering the fine-tuning of models pre-trained on extensive datasets, visual prompt tuning has emerged as a dominant approach. It significantly lightens the computational load and storage requirements of model fine-tuning by introducing only a few parameters. The VPT [18] integrates prompts into pre-trained networks through embeddings, yielding favorable results across 24 downstream tasks in fine-grained classification. The ViPT [19] creatively employs prompts as a dual-modal fusion tool, expanding visible light object tracking to include infrared, depth, and event-based image tracking. Drawing from this insight, we can conceptualize dual-modal object detection as a fine-tuning task. By refining single-modal benchmark models with prompts, we can transition them to dual-modal detection, thereby improving their versatility and robustness for complex scenarios.

In conclusion, we develop a Transformer-based algorithm for visible–infrared object detection, named **Vi**sual **P**rompt multi-modal **Det**ection **(VIP-Det)**. To fully exploit the capabilities of Transformer-based dual-modal algorithms, the Vision Transformer is utilized as the backbone for feature extraction, leveraging its strength in capturing long-range dependencies and global context. To simplify the complex architecture of dual-modal object detection, a prompt-based fusion module is devised that introduces prompts for fusion within a single-stream network, significantly reducing the number of parameters. To optimize pre-trained models and balance modal information, a stage-wise optimization strategy is introduced that commences with training single-modal benchmark models and subsequently refines features with additional modalities, fostering more effective modal integration and refined feature extraction. Our algorithm is tested on the DroneVehicle dataset, and the results demonstrate that it achieves high precision and adeptly accommodates the demands of object detection in intricate settings.

In summary, the contributions of this paper are as follows:

- We propose a novel Transformer-based framework for dual-modal object detection, which incorporates the Vision Transformer (VIT) as a backbone, capable of efficiently extracting features and enhancing the precision of object detection;
- We introduce a prompt-based fusion module and a stage-wise optimization strategy, utilizing prompts to guide feature fusion and enhance the aggregation capabilities of dual-modal information. Additionally, we employ a phased fine-tuning approach to guide parameter optimization, thereby better transferring the feature representation capabilities of the original model;
- We assess the performance of our proposed framework on the DroneVehicle dataset and showcase its superior accuracy when compared to other comparable Transformer-based methods.

2. Related Work

2.1. Visible–Thermal Object Detection

Visible–thermal object detection algorithms stand as prime examples of image fusion technology, overcoming the limitations of single-modality images in complex environments by integrating complementary data from visible and thermal imagery. This synergy significantly enhances the precision and robustness of object detection. Researchers have not only compiled diverse datasets such as KAIST [20], DVTOD [21], and DroneVehicle [22] but have also proposed various cutting-edge algorithmic frameworks. Halfway Fusion [23] excels in merging visible and thermal information at the midlevel feature stage through a unique ConvNet architecture. UA-CMDet [22] introduces an uncertainty-aware mechanism that dynamically assesses the uncertainty of each modality and proposes a novel light-aware cross-modal non-maximum suppression algorithm to further improve detection. C2Former [16] focuses on cross-modal attention learning, facilitating interaction between RGB and thermal data via the ICA module while enhancing computational efficiency with the AFS module. TSFADet [24] offers the TSRA module for precise alignment of features.

However, current visible–thermal object detection algorithms, often based on dual-backbone networks, grapple with high complexity and a large number of parameters. The unequal significance of visible and thermal information under different environmental conditions challenges the assumption of equal importance, necessitating the development of more efficient, lightweight fusion strategies and intelligent mechanisms for adjusting modal weights as a critical research focus.

2.2. Vision Transformer for Object Detection

Inspired by the way humans process information, attention mechanisms in deep learning models dynamically adjust the weights of different parts to enable the models to focus on the critical portions of the input data, thereby enhancing their performance [25,26]. The

Transformer is one of the best examples that showcases the power of attention mechanisms. The Transformer model, renowned for its global modeling and parallel processing capabilities in NLP [11,12], has intrigued the field of computer vision. The Vision Transformer (ViT) [13] revolutionized image processing by treating image patches as core processing elements. In the realm of object detection, DETR [15] introduced a new approach by discarding conventional anchor boxes and non-maximum suppression, utilizing an attention-based encoder–decoder framework for direct bounding box and category prediction. The Swin Transformer [14,27] significantly accelerates its computation speed through the use of a sliding window mechanism and hierarchical structure. The innovation of Masked Autoencoders (MAE) [17] for ViT pre-training advanced the field, facilitating self-supervised learning through the prediction of masked pixels, leading to the emergence of models like ViTDet [28], MIMDet [29], and ImTed [30] that enhance detection with an MAE's pre-trained weights. For remote sensing, RVSA [31,32] tailored a ViT for detecting rotating objects by adjusting attention mechanisms, while STD [33] employed separate network branches to predict bounding box attributes, harnessing ViT's spatial transformation abilities. In the field of drone-based object detection, a Hybrid Convolutional–Transformer framework [34] was proposed to address the challenge of weak supervision in drone-view imagery.

Nevertheless, the full capacity of Transformers in visible–thermal object detection remains untapped. Currently, Transformers are predominantly used as fusion components alongside CNNs [16], rather than independently harnessing their global modeling and spatial transformation strengths. Future research should focus on the explicit and customized application of Transformer models to visible–thermal object detection. This necessitates developing Transformer architectures that cater to the unique aspects of visible and infrared imagery, propelling advancements in this domain.

2.3. Vision Prompt Tuning

Fine-tuning large-scale pre-trained models on downstream tasks has become a prevalent training strategy for numerous NLP and CV tasks. The essence of this approach is to perform a comprehensive update of the model parameters on a specific dataset. However, this method is less efficient in terms of parameter utilization, as it necessitates creating unique model replicas for each new task and requires storing the enormous pre-trained models. In contrast to past conventional methods, Prompt has emerged as a novel training paradigm and is increasingly becoming the dominant approach for fine-tuning in computer vision. This involves training a large foundational visual model with extensive data and then using different prompts to accomplish various tasks. The Image Inpainting [35] algorithm has trained a model with the objective function, allowing it to rely on visual prompts to perform various tasks. The SAM [36] algorithm uses repeated prompts to direct the model's output, with prompt formats such as points, bounding boxes, masks, and text, which describe target objects for segmentation. VPT [18] outperforms fine-tuning in classification tasks by embedding prompt parameters before input. ViPT [19] learns modality-specific prompts to adapt frozen pre-trained foundational models to a range of downstream dual-modal tracking tasks, including RGB + Depth, RGB + Thermal, and RGB + Event tracking.

3. Models and Methods

In this section, we introduce **VIP-Det** (**Vi**sual **P**rompt dual-modal **Det**ection), an innovative algorithm for drone-based visible–thermal object detection that leverages the Vision Transformer architecture. This section commences with an exposition of the motivations that drove the development of the algorithm and an elucidation of its overarching framework. Subsequently, it delves into the technical nuances of the implementation of the prompt-based fusion module. This section concludes with an elucidation of the algorithm's stage-wise training optimization strategy.

3.1. Overview

Traditional drone-based object detection algorithms are often limited to visible light imagery and may fail under complex environmental conditions such as nighttime, rainy weather, fog, and occlusions. Existing visible–thermal object detection algorithms typically rely on dual-stream backbone networks for feature extraction, which significantly increases the number of parameters and is hindered by the imbalance between the two modalities, thereby limiting the efficiency of their fusion. Vision Transformers (ViTs) have demonstrated impressive performance across a wide range of visual tasks; however, in the domain of visible–thermal object detection, their attention mechanisms are often confined to the fusion module, and the potential of their feature modeling has not been fully exploited.

To address these limitations, our VIP-Det, designed for visible–thermal object detection, introduces the Vision Transformer as its backbone. The algorithm adopts a single-stream network architecture to concurrently extract features from visible and infrared images. A novel prompt mechanism is employed to introduce a small set of learnable parameters for feature-level integration. During training, the algorithm first establishes a baseline model on single-modal data and then refines the model parameters using dual-modal data. The overall network architecture is designed to efficiently integrate the information from both modalities, aiming to enhance the algorithm's capability in object detection. The overall architecture is illustrated in the accompanying Figure 2.

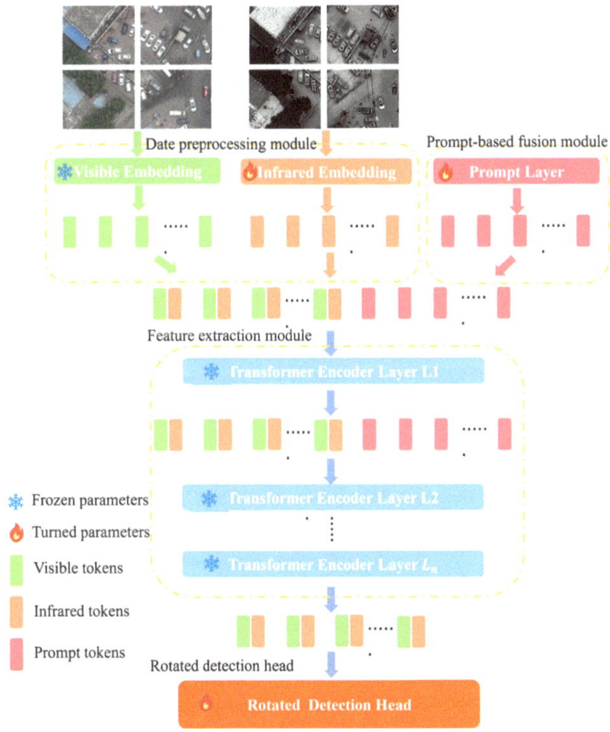

Figure 2. The overarching architecture of VIP-Det encompasses several principal components: a data preprocessing module, a prompt-based fusion module, a feature extraction module, and a rotation detection head. Firstly, the dual-modal images are input in the data preprocessing module to generate visible light tokens and infrared tokens separately. Then, the prompt-based fusion module initializes and generates prompt tokens, which are merged with the tokens from both modalities and jointly input into the feature extraction module. The feature extraction module, comprising multiple Transformer layers, performs feature extraction on the merged tokens. Finally, the extracted feature maps are fed into the rotated detection head to obtain results.

Our VIP-Det algorithm is composed of four main components: a data preprocessing module, a prompt-based fusion module, a feature extraction module, and a rotated detection head. The data preprocessing module processes both visible light and infrared imagery by patching them into tokenized form and stacking them. The prompt-based fusion module introduces prompts as learnable parameters, guiding feature fusion through training iterations. These prompt-embedded tokens are then inputted into the feature extraction module. This module employs an MAE pre-trained Vision Transformer model, which features 12 layers of Transformer blocks, as its backbone network instead of ResNet-50. The extracted features are then fed into the rotated detection head for classification and regression. In our experiments, we utilized the rotated detection head of STD [33] to achieve the most precise detection results. During the training process, we initially selected one modality as the baseline model and trained it to establish a foundation. Subsequently, we integrated the other modality to facilitate fusion, achieving dual-modal object detection with minimal parameter adjustments for efficient fine-tuning.

3.2. *Vision Transformer Architecture*

The Vision Transformer (ViT) represents a significant advancement in computer vision, reshaping the traditional approach of Convolutional Neural Networks (CNNs). Instead of sliding convolutional kernels across an image to extract features, the ViT divides the input image into a grid of nonoverlapping patches. Each of these patches is then flattened and converted into a vector, effectively transforming the two-dimensional image data into a sequence of one-dimensional vectors. To encode the spatial relationships between these patches, positional encodings are added to the vector representations, ensuring that the model can distinguish and utilize the positional information. These enriched embeddings serve as the input to a stack of Transformer encoder layers, which form the core of ViT's architecture. Each encoder layer leverages self-attention mechanisms to allow each patch to attend to and interact with every other patch in the sequence, capturing long-range dependencies and contextual information. This is complemented by feedforward neural networks, which introduce non-linearities and enable the model to learn complex feature representations. As the embeddings traverse through the stacked encoder layers, they are progressively transformed and enriched, ultimately encoding a rich semantic understanding of the input image. In the context of this specific task, the output of the final encoder layer, now enriched with features extracted from both visible and infrared modalities, serves as the foundation for subsequent dual-modal target detection. These features, reflecting the unique properties of both spectra, empower the model to detect and identify objects with unprecedented accuracy and robustness, demonstrating the versatility and power of the Vision Transformer framework in addressing complex computer vision challenges.

3.3. *Prompt-Based Fusion*

3.3.1. Overview

To fine-tune a single-modal object detection model based on prompts, we first need a pre-trained baseline model for the specific modality, where the embedding layer and Vision Transformer layers are already equipped with relevant parameters. During subsequent fine-tuning, these layers are frozen to preserve their feature extraction capabilities. When pre-training the baseline model, the embedding layer and Transformer layers for feature extraction of that modality, along with the detection head, are trained, while the prompt layer remains untrained.

For simplicity, let us assume there is a pre-trained baseline model for the visible light modality; hence, the visible embedding layer is frozen, and a certain number of Vision Transformer layers are also frozen as per the requirement. Since the infrared modality has not been trained, the infrared embedding layer requires fine-tuning.

For input images, visible light and infrared images are separately fed into their corresponding embedding layers, where they undergo patch partitioning and encoding to obtain visible tokens and infrared tokens. This step is performed in the data preprocessing module. Subsequently, the prompt layer is initialized to generate a certain number of prompt tokens, which are then combined with the previously extracted visible and infrared tokens to form fused tokens. As training progresses, the parameters of the prompt layer are iteratively fine-tuned to produce prompt tokens with lower losses.

These fused tokens are then fed into the Transformer layers of the feature extraction module, where the output from the previous layer, including the prompt tokens, serves as input for the next layer. This process continues through all Transformer layers, from which the visible and infrared tokens are extracted to obtain feature maps. These feature maps are then input into the rotated detection head. For the specific code process, refer to Appendix A.

3.3.2. Introduction

Given a pair of prealigned and coregistered visible and thermal images, denoted $x_v \in \mathbb{R}^{3\times H\times W}$ (where v stands for visible) and $x_t \in \mathbb{R}^{3\times H\times W}$ (where t stands for thermal), respectively, with H and W being the height and width of the images, and assuming a batch size of 1 for simplicity, we explore the integration of these modalities within a Vision Transformer framework for object detection.

3.3.3. Image Patch Embedding

A typical ViT with N layers divides the input images into m fixed-size patches $I_v^j \in \mathbb{R}^{3\times h\times w}$ and $I_t^j \in \mathbb{R}^{3\times h\times w}$ for $j \in \mathbb{N}, 1 \leq j \leq m$, where h and w are the height and width of each patch. These patches are then embedded into a d-dimensional latent space and position encodings are added:

$$\begin{aligned} e_{v0}^{j} &= Embed(I_v{}^j) \quad e_{v0}^{j} \in \mathbb{R}^d, j = 1,2,\ldots,m \\ e_{t0}^{j} &= Embed(I_t{}^j) \quad e_{t0}^{j} \in \mathbb{R}^d, j = 1,2,\ldots,m \end{aligned} \tag{1}$$

The sets of patch tokens at layer i are represented as:

$$\begin{aligned} E_{vi} &= \{e_{v_i}^{j} \in \mathbb{R}^d \mid j \in \mathbb{N}, 1 \leq j \leq m\} \\ E_{ti} &= \{e_{t_i}^{j} \in \mathbb{R}^d \mid j \in \mathbb{N}, 1 \leq j \leq m\} \end{aligned} \tag{2}$$

3.3.4. Prompt-Based Feature Fusion

To facilitate dual-modal feature fusion, we introduce a set of continuous prompt tokens $P = \{p^k \in \mathbb{R}^d \mid k \in \mathbb{N}, 1 \leq k \leq p\}$ initialized randomly and inserted before the first encoder layer L_1 of the pre-trained Transformer. During fine-tuning, only the task-relevant prompts are updated, while the main Transformer parameters are frozen. This leads to:

$$[E_{v1}, E_{t1}, Z_1] = L_1([E_{v0}, E_{t0}, P]) \tag{3}$$

where Z represents the prompt parameters after iteration within the network. The forward pass through the Transformer layers can be expressed as:

$$[E_{vi}, E_{vi}, Z_i] = L_i([E_{vi-1}, E_{ti-1}, Z_{i-1}]) \quad i = 1,2,\ldots,N \tag{4}$$

Each layer L_i consists of multihead self-attention (MSA) and a feedforward network (FFN), accompanied by Layer Normalization (LayerNorm) and residual connections.

3.3.5. Detection Head

Finally, a rotated object detection head denoted as Head processes the fused features from the last layer to predict rotated bounding boxes and categories:

$$y = \text{Head}(E_{vN}, E_{tN}) \tag{5}$$

3.4. Stage-Wise Training Optimization

The typical approach to dual-modal object detection adheres to a standardized process: Initially, two separate backbone networks are employed to extract features from paired visible and infrared images independently. Subsequently, these dual-modal features are fed into a feature fusion module, integrating information from the two distinct modalities. Ultimately, an object detection head is leveraged for regression prediction, enabling the localization and classification of objects within the images.

However, a notable issue arises from this methodology: During a single training cycle, all network architecture parameters must be learned, resulting in a substantial parameter count and sluggish training speed. To address this challenge, we propose a stage-wise training optimization strategy. See Figure 3.

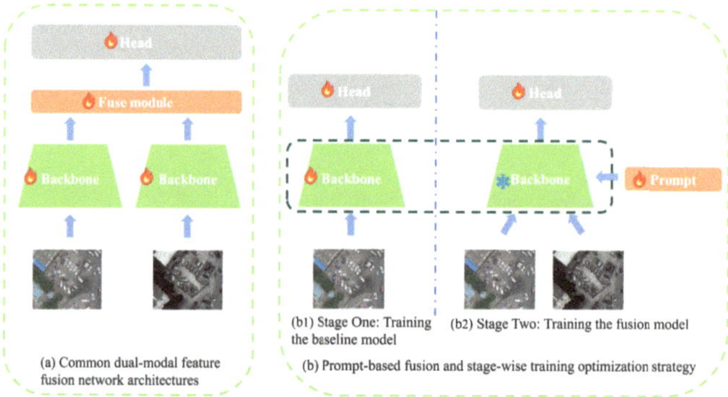

Figure 3. A comparison between the stage-wise training optimization strategy and common dual-modal object detection algorithms. (**a**) shows the common dual-modal object detection framework. (**b**) represents the prompt-based fusion and stage-wise training optimization strategy. It is divided into two stages. (**b1**) shows the process of training the baseline model. (**b2**) illustrates the process of training the fusion model.

First, we individually train the mono-modal visible and infrared images using a Vision Transformer backbone network, aiming to develop the capacity to extract fundamental and generic features. This phase targets the establishment of benchmark mono-modal models. Next, we proceed with dual-modal image inputs for modal fine-tuning, freezing partial weights within the backbone networks and introducing prompt parameters for fine-tuning. This promotes efficient dual-modal feature fusion.

By adopting this training paradigm, we not only drastically reduce the parameter count but also simplify the overall model architecture. As there is no need for a separate feature fusion module, our approach relies solely on a minimal set of prompt parameters to achieve dual-modal feature fusion. This not only decreases model complexity but also renders the model more concise and interpretable.

Furthermore, our method harnesses the power of pre-trained models, facilitating seamless migration to dual-modal object detection tasks. By maintaining the invariance of selected weights from the pre-trained models during the fine-tuning phase, our approach effectively leverages the rich feature representations already learned, further enhancing the performance of dual-modal object detection.

4. Results

In this section, we commence by detailing the datasets and evaluation metrics employed in our experiments. Subsequently, we provide the pertinent setup and configuration details. We proceed with a series of ablation studies to validate the efficacy of our algorithm. Finally, we conduct comparative experiments against related algorithms.

4.1. Datasets and Evaluation Metric

The DroneVehicle [22] dataset is a comprehensive and diverse collection of RGB–infrared (RGB-IR) images captured by drones. This dataset encompasses a wide range of scenarios, including urban roads, residential areas, parking lots, and other environments, spanning various times of the day and night. The dataset consists of 28,439 image pairs, each pair containing corresponding RGB and infrared images that have been precisely aligned to ensure accurate representation of the scene. The annotations provided by the dataset authors are extensive and include oriented bounding boxes for five distinct vehicle categories: cars, buses, trucks, vans, and freight cars. The dataset is organized into a training set and a test set, with the training set comprising 17,990 image pairs and the test set consisting of 1469 image pairs. Our experiments were conducted on this DroneVehicle dataset, leveraging its richness and variability to test and refine our vehicle detection algorithm.

We utilized the mean average precision (mAP) as the primary evaluation metric for our detection algorithm, applied to the validation set. To ensure accurate detections, we employed an Intersection over Union (IoU) threshold of 0.5, which helped filter out false positives and contributed to a reliable assessment of the algorithm's performance.

4.2. Implementation Details

Utilizing pre-trained weights initialized from the MAE, we embarked on training our network specifically for the DroneVehicle dataset, leveraging the computational prowess of an NVIDIA RTX 4090 GPU. Our training strategy commenced with initializing a single-modal base model through 12 epochs, subsequently transitioning into a 12-epoch fusion model training phase, where prompts were integrated to enhance performance. The optimization process employed stochastic gradient descent (SGD) equipped with a momentum factor of 0.9 and a weight decay rate of 0.0001.

During each training iteration, we processed batches containing two images apiece, initiating with a learning rate of 0.001. This learning rate underwent a strategic halving at epochs 8 and 11 to facilitate a smoother convergence. To augment the training data and bolster the model's generalization capabilities, we applied various image transformations such as flipping, cropping, and splicing.

Post-training, during the inference phase, we utilized non-maximum suppression (NMS) with an Intersection over Union (IoU) threshold set at 0.3 to effectively eliminate redundant bounding boxes, ensuring the precision of our detections. Throughout these endeavors, we leveraged customized versions of the MMRotate and MMDetection frameworks.

4.3. Ablation Experiment

4.3.1. Ablation on Prompt-Based Fusion

To validate the efficacy of our prompt-based fusion module in enhancing the quality of fusion outcomes, we conducted a rigorous ablation study. This investigation entailed a comparative analysis between two experimental setups: the baseline approach, which directly stacked modalities without utilizing the fusion module, and our algorithm augmented with the integrated prompt-based fusion module.

The outcomes of this study, tabulated in Table 1, reveal a notable improvement. Specifically, the inclusion of the prompt-based fusion module resulted in a marked 1.3% increase in mean average precision (mAP). This substantial gain underscores the pivotal role played by our fusion module in bolstering the overall performance of the algorithm, highlighting its effectiveness in fostering seamless and effective modality integration.

Table 1. Ablation on prompt-based fusion. Compared to direct feature stacking and fusion, the introduction of prompts simplifies task complexity by minimizing direct modifications to model parameters. This approach mitigates the risk of overfitting and, through the incorporation of additional parameters, enables the model to adapt more flexibly to feature transformations and weight adjustments. Consequently, it enhances the model's generalization capabilities, making it more robust and versatile across diverse scenarios. The red mark indicates an increase in the number of parameters.

Method	Car	Truck	Freight Car	Bus	Van	mAP	Param
baseline	90.3	68.1	62.2	90.0	56.3	73.4	70.02 M
baseline + prompt	90.4	78.5	61.4	89.8	57.5	75.5	70.10 M (+0.08 M)

4.3.2. Ablation on the Number of Frozen Layers

In the two-stage optimization training strategy, we strategically froze the Transformer encoder layers to safeguard the model's foundational representation and generalization capabilities. The experiment compared freezing the first 6 layers, fine-tuning the last 6 layers, and fine-tuning all 12 layers, assessing the trade-off between accuracy and efficiency.

The results in Table 2, tabulated, reveal that freezing the first six layers results in a minimal 0.4% decrease in mAP compared to fine-tuning of the full layer, showcasing the effectiveness of partial freezing in reducing training parameters without compromising accuracy. This approach accelerates training and reduces computational demands, facilitating large-scale deployments and iterations.

Table 2. Ablation on the number of frozen layers. Due to the discrepancy between the pre-training task and the new task, some features in the pre-trained model may not be suitable for the new task. Freezing certain layers can potentially limit the model's ability to represent features tailored to the new task, resulting in a certain degree of accuracy degradation. However, this approach also brings benefits such as reducing the number of parameters and accelerating the model training speed. The red mark indicates an increase in the number of parameters.

Method	Car	Truck	Freight Car	Bus	Van	mAP	Param
frozen 6 layers	90.3	71.2	63.2	90.1	57.8	74.5	59.45 M
fine-tune 12 layers	90.4	78.5	61.4	89.8	57.5	75.5	70.10 M (+10.65 M)

In conclusion, partial freezing of the backbone network layers in two-stage optimization training is an efficient and practical method, allowing us to balance speed and accuracy by adjusting the number of frozen layers. This discovery offers an innovative approach to optimizing deep learning model training workflows.

4.3.3. Ablation on Stage-Wise Optimization

We conducted experiments to train both a single-stage mono-modal baseline model and a two-stage dual-modal model to validate the effectiveness of the staged training optimization for the detection of dual-modal objects. In our setup, we initially trained mono-modal models for visible light and infrared data, and then we introduced the data of the other modality to fine-tune the corresponding models. The results, as tabulated in Table 3, show that the fine-tuned models exhibited impressive improvements in mAP: the visible light model saw a remarkable 15.6% increase, and the infrared model experienced a 3% increase. The dual-modal object detection algorithms outperformed their mono-modal counterparts on the dataset, which contains challenging environments such as nighttime. The introduction of infrared data mitigates the limitations of using only visible light for object detection, enhancing performance in complex scenarios.

Table 3. Ablation on stage-wise optimization. With the addition of information from another modality, the fine-tuned model can fully leverage the complementary nature of the data, achieving higher performance. Meanwhile, since the ground truth is uniformly adopted from infrared annotations, the infrared detection performance tends to be better than that of visible light. The red mark indicates an improvement in accuracy.

Method	Car	Truck	Freight Car	Bus	Van	mAP	Modality
visible baseline	78.3	54.9	38.8	83.8	43.8	59.9	RGB
thermal baseline	90.3	72.5	57.8	88.8	52.9	72.5	T
visible + fine-tune	90.4	78.5	61.4	89.8	57.5	75.5 (+15.6)	RGB + T
thermal + fine-tune	90.4	78.5	59.8	89.6	56.9	75.0 (+2.50)	

4.4. Performance Comparison

For the purpose of comparison, our experiment involved the implementation of a dual-modal object detection algorithm that underwent fine-tuning on the visible light base model, referred to as VIP-Det. This was set against a range of baseline mono-modal object detection algorithms, including the single-stage R3Det [8], the two-stage Oriented R-CNN [9], and the anchor-free SASM [10]. To ensure a fair comparison, these baseline algorithms were trained separately on either the visible light or infrared datasets. Furthermore, we conducted a meticulous re-implementation of four established RGB + T multispectral methodologies—Halfway Fusion [23], UA-CMDet [22], TSFADet [24], and C2Former [16]—with the objective of rigorously assessing their efficacy in the realm of RGB-IR object detection.

4.4.1. Comparison with Single-Modal Algorithm

The experimental results, as presented in Table 4, offer profound insights upon analysis. Specifically, in the comparison of single-modality algorithms for visible light, Oriented R-CNN emerges as the top performer, surpassing the single-modality benchmark VIP-Det algorithm in terms of precision. This underscores the advanced detection framework and optimization strategies employed by Oriented R-CNN in handling complex scenes and recognizing intricate features.

Table 4. Performance comparison of single-modal and dual-modal object detection algorithms. This concise table comprehensively evaluates the performance of diverse object detection algorithms across single-modal (RGB or thermal IR) and dual-modal (RGB + thermal IR) setups. By analyzing average precision (AP) in detecting cars, trucks, freight cars, buses, and vans and calculating mean average precision (mAP), it underscores VIP-Det's excellence in harnessing dual-modal information. The table highlights the modality used, revealing how modality choice impacts detection accuracy, offering valuable insights. The red mark indicates the maximum precision value in the column.

Method	Car	Truck	Freight Car	Bus	Van	mAP	Modality
R3Det [8]	87.8	35.0	16.1	75.9	16.2	46.20	RGB
Oriented R-CNN [9]	88.9	61.7	39.7	87.9	40.5	63.74	
SASM [10]	88.6	52.4	31.6	82.0	32.0	57.32	
VIP-Det (V)	78.3	54.9	38.8	83.8	43.8	59.90	
R3Det [8]	89.5	29.5	22.3	73.1	16.2	46.12	T
Oriented R-CNN [9]	90.1	61.7	48.2	88.6	39.7	65.66	
SASM [10]	89.6	46.8	36.2	80.7	28.8	56.42	
VIP-Det (T)	90.3	57.8	61.4	88.8	52.9	72.50	
VIP-Det (ours)	90.4	78.5	61.4	89.8	57.5	75.50	RGB + T

However, when shifting our focus to infrared data, the narrative shifts. VIP-Det, when trained solely on infrared datasets, demonstrates a remarkable superiority over other single-modality object detection algorithms, including the formidable Oriented R-CNN in visible light. Its precision advantage over Oriented R-CNN reaches a significant 6.84%, highlighting VIP-Det's unique strengths in processing infrared imagery, possibly attributed to its sensitivity and adaptability to spectral characteristics.

Moreover, in the head-to-head comparison between single-modality and dual-modality algorithms, VIP-Det claims the highest precision level. This achievement not only validates the inherent superiority of the VIP-Det algorithm but also underscores the profound impact of multi-modality information fusion on enhancing object detection performance. By integrating information from both visible and infrared spectra, VIP-Det is able to comprehensively capture target features, mitigating information loss and interference inherent in single-modality approaches. Consequently, it achieves more precise and robust target detection in complex environments.

4.4.2. Comparison with Dual-Modal Algorithm

As shown in Table 5, when compared with the current state-of-the-art dual-modal object detection algorithms, our algorithm has demonstrated remarkable performance, achieving a significant mAP of 75.5%. This achievement not only surpasses other relevant dual-modal algorithms but also validates the effectiveness of our algorithmic innovations in multi-source information fusion and efficient feature extraction.

Table 5. A comparison with dual-modal algorithms. This refined table introduces a comparative analysis of VIP-Det against leading dual-modal (RGB + thermal) object detection algorithms. By assessing their performance in detecting various vehicle types and calculating the mean average precision (mAP), it offers valuable insights into how VIP-Det fares against the most advanced techniques in the field, further elucidating its strengths and positioning within the current state of the art. The red mark indicates the maximum precision value in the column.

Method	Car	Truck	Freight Car	Bus	Van	mAP	Modality
Halfway Fusion [23]	89.85	60.34	55.51	88.97	46.28	68.19	
UA-CMDet [22]	87.51	60.70	46.80	87.00	38.00	64.00	
TSFADet [24]	90.01	69.15	65.45	89.70	55.19	73.90	RGB + T
C2Former [16]	90.20	68.30	64.40	89.80	58.50	74.20	
VIP-Det (ours)	90.40	78.50	61.40	89.80	57.50	75.50	

4.4.3. Comparison of Visual Detection Results

In this experiment, we aimed to validate the robustness of our algorithm and explore the efficacy of multi-modal object detection in complex environments. To this end, we selected Oriented R-CNN, the top-performing algorithm under single-modality conditions, as a benchmark for comparison. Our objective was to demonstrate the advantages of dual-modal object detection in the same environments where Oriented R-CNN is typically applied.

To comprehensively assess performance, we chose four distinct scenarios: daylight, nighttime, rainy/foggy conditions, and scenes with occlusion. Each of these scenarios poses unique challenges to object detection systems, requiring robust algorithms that can overcome factors such as illumination variations, poor visibility, and partial visibility of targets. The results are shown in Figure 4.

In daylight conditions, visible light predominates. Regarding the section highlighted in the red frame, the single-modality Oriented R-CNN misclassifies it in infrared imagery, whereas VIP-Det accurately determines the target category, effectively addressing the issues of low resolution and lack of clarity in infrared images.

Under nighttime conditions, the Oriented R-CNN fails to detect the red-framed area entirely in the single visible light modality due to insufficient information. Conversely, VIP-Det supplements visible light information with infrared imagery, mitigating the inability of visible light-based target detection at night.

In rainy or foggy environments, visible light images tend to blur due to light reflection by raindrops or fog particles, whereas infrared imaging, relying on thermal conduction, is less affected. For the red-framed section, Oriented R-CNN, operating solely on visible light, misses the detection, while VIP-Det, leveraging infrared imagery as an aid, adeptly resolves the issue of unclear textures and blurred contours in visible light images under adverse weather conditions.

In cases of occlusion, such as by trees or other objects, visible light imaging suffers from information loss due to reflection. However, thermal radiation from targets can penetrate certain obstructions. For the red-framed region, Oriented R-CNN, using only visible light, experiences missed detections, whereas VIP-Det, by successfully fusing infrared and visible light information, is capable of detecting occluded targets. For a more extensive showcase of visual results, please refer to Appendix B.

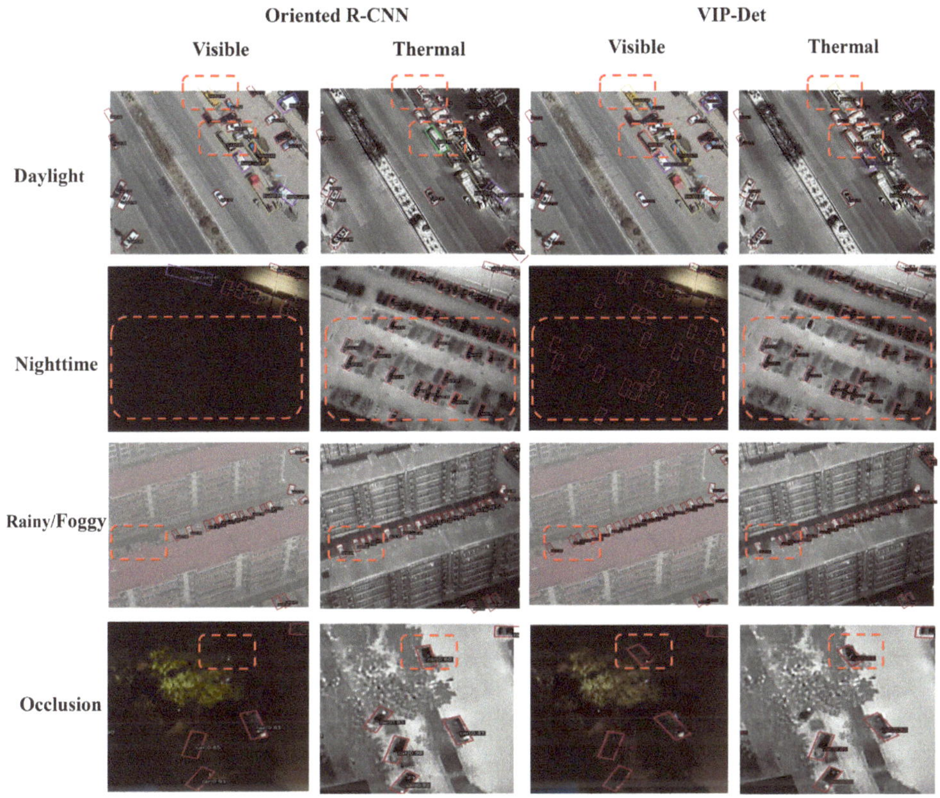

Figure 4. A comparison of visual detection results. In this table, we present a visual comparison of the detection results between the single-modal algorithm, Oriented R-CNN, and our dual-modal algorithm, VIP-Det, across different scene environments. Each set of images encapsulates the detection outcomes from the same pair of visible and infrared images within a given scene. The red bounding boxes highlight the performance differences demonstrated by the algorithms under those specific scenarios, providing a clear visualization of their respective strengths and capabilities.

5. Discussion

Our research is primarily focused on the dual-modal object detection task within the UAV field. We have conducted a comprehensive set of ablation studies to validate the reliability of our proposed modules, and we have compared our algorithm with relevant state-of-the-art methods, showcasing its superior detection accuracy and impressive visual results. In contrast to single-modal object detection algorithms, our approach ingeniously fuses features through the use of prompts, endowing it with the capability of dual-modal complementarity and heightened robustness. Compared to existing dual-modal detection algorithms, our method fully exploits the representation and modeling power of Vision Transformers, achieving even better dual-modal feature extraction.

Looking ahead, we envision numerous avenues for further exploration to enhance the performance and practicality of our algorithm in real-world applications. Beyond developing more effective fusion modules and simplifying network architectures, we aim to optimize our model for seamless integration with UAV edge devices, enabling real-time, accurate detections under diverse environmental conditions. Additionally, we will investigate the potential of leveraging both visual and thermal data for battlefield reconnaissance and target identification, paving the way for safer and more efficient drone operations in the field.

6. Conclusions

In this work, our main contribution lies in the introduction of a Transformer-based algorithm for visible–thermal object detection tailored for applications of unmanned aerial vehicles (UAVs), named VIP-Det (**Vi**sual **P**rompt dual-modal **Det**ection). VIP-Det employs a Vision Transformer as its backbone network, innovatively incorporates a prompt-based fusion module for refined feature integration, and adopts a stage-wise optimization strategy for efficient fine-tuning. Through a series of quantitative and qualitative experiments conducted on the DroneVehicle dataset, we demonstrate that VIP-Det surpasses existing dual-modal object detection algorithms, effectively tackling complex UAV-to-ground target detection scenarios, including rainy conditions, nighttime environments, and occlusion, with remarkable performance. This underscores the significant advancement of our proposed methodology in the realm of UAV-based object detection, which has immense potential to improve autonomous surveillance and monitoring capabilities in diverse and challenging environments.

Author Contributions: Conceptualization, R.C. and D.L.; methodology, R.C.; software, R.C.; validation, R.C., D.L. and Z.G.; formal analysis, D.L.; investigation, R.C.; resources, R.C.; data curation, R.C.; writing—original draft preparation, R.C.; writing—review and editing, Y.K. and D.L.; visualization, R.C.; supervision, C.W. and Y.K.; project administration, D.L.; funding acquisition, D.L. All authors have read and agreed to the published version of the manuscript.

Funding: This work is supported by the National Natural Science Foundation of China (NSFC) (No. 62102426) and the scientific research project of National University of Defense Technology (No. ZK21-29).

Data Availability Statement: All data utilized in this study originate from the publicly available "DroneVehicle" dataset, released by the VisDrone project team and hosted on the GitHub platform at the following link: https://github.com/VisDrone/DroneVehicle. This comprehensive dataset encompasses images and video footage specifically designed for tasks such as drone and vehicle detection, tracking, and beyond, thereby forming a solid foundation for our experimental endeavors. Throughout our experiments, we directly leveraged the image and video frames within this dataset, engaging in data preprocessing, model training, and result validation. All reported experimental outcomes are solely based on this dataset, adhering strictly to the data usage protocols and terms of the VisDrone project. It is worth noting that as the dataset has undergone public processing and sharing, we are exempted from concerns related to data privacy or ethical implications. Furthermore, we wholeheartedly encourage fellow researchers to harness this valuable resource, fostering collaborative efforts towards advancing the field of drone and vehicle detection technologies.

Conflicts of Interest: The authors declare no conflicts of interest.

Appendix A

To better elucidate the prompt-based fusion module, we offer a streamlined pseudocode flow of the algorithm, facilitating comprehension. Our algorithmic module encompasses two pivotal steps: pre-training and fine-tuning. During pre-training, the focus lies in provisioning initial embedding weights and relevant Transformer layer parameters. The fine-tuning phase, on the other hand, introduces prompt parameters for optimization. By integrating these two steps, our prompt-based fusion module efficiently leverages pre-trained knowledge while flexibly adapting to various tasks through optimized prompts, enhancing overall performance and versatility.

Algorithm A1 Prompt-based fusion

procedure PRE-TRAIN($\mathbf{x_v}$)
    Divide $\mathbf{x_v}$ into patches $\mathbf{I_v}^j \in \mathbb{R}^{3\times h\times w}$,$1 \leq j \leq m$
    for $j = 1$ to m **do**
        $e_{v0}^j = Embed_v(I_v{}^j), e_{v0}^j \in \mathbb{R}^d$
        $E_{v0} = Concatenate(e_{v0}^j)$
    end for
    for $i = 1$ to N **do**
        $\mathbf{E_v}_i = Transformer_i([\mathbf{E_v}_{i-1})$
    end for
    $\mathbf{y} = Head(\mathbf{E_v}_N)$
    $Loss(\mathbf{y})$
    $update(parameters)$
end procedure
Retrieve weights of visible light embedding and Transformer layers

procedure FINE-TUNING($\mathbf{x_v}$,$\mathbf{x_i}$)
    Frozen weights of visible light embedding and Transformer layers
    Divide $\mathbf{x_v}$ and $\mathbf{x_t}$ into patches $\mathbf{I_v}^j, \mathbf{I_t}^j \in \mathbb{R}^{3\times h\times w}$
    for $j = 1$ to m **do**
        $e_{v0}^j = Embed_v(I_v{}^j), e_{v0}^j \in \mathbb{R}^d$
        $e_{t0}^j = Embed_t(I_t{}^j), e_{t0}^j \in \mathbb{R}^d$
    end for
    Initialize prompt tokens $\mathbf{P} = \{\mathbf{p}^k \in \mathbb{R}^d \mid k = 1, 2, \ldots, p\}$
    $[\mathbf{E_v}_0, \mathbf{E_t}_0, \mathbf{Z}_0] = Concatenate(\mathbf{E_v}_0, \mathbf{E_t}_0, \mathbf{P})$
    for $i = 1$ to N **do**
        $[\mathbf{E_v}_i, \mathbf{E_t}_i, \mathbf{Z}_i] = Transformer_i([\mathbf{E_v}_{i-1}, \mathbf{E_t}_{i-1}, \mathbf{Z}_{i-1}])$
    end for
    $\mathbf{y} = Head(\mathbf{E_v}_N, \mathbf{E_t}_N)$
    $Loss(\mathbf{y})$
    update(parameters)(excluding parameters of visible light embedding and Transformer layers)
end procedure

Appendix B

In this supplementary section, we incorporate four comprehensive sets of visual comparison graphs to showcase the detection outcomes of our proposed method under diverse and challenging environmental conditions. These include scenarios of daytime, nighttime, foggy weather, and occlusion, providing a more prominent demonstration of the superiority of our approach. Within each set of images, we include four pairs of images specific to that environmental scenario. Each image pair comprises a visible light image on the left and its corresponding infrared image on the right. With the left image depicting the visible light influence and the right image corresponding to the infrared imagery, we can more effectively demonstrate the inherent differences between these two modalities and underscore the algorithm's adept utilization of their complementary information.

Figure A1. The additional visualization results obtained using VIP-Det in the daytime scenarios. This set showcases the baseline performance under optimal lighting conditions.

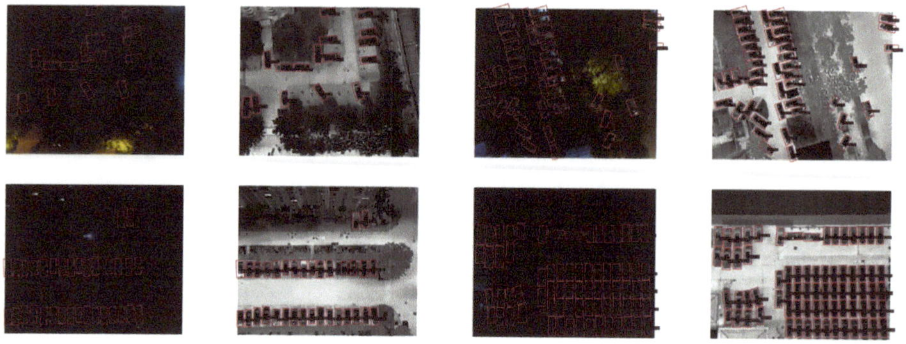

Figure A2. The additional visualization results obtained using VIP-Det in the nighttime scenarios. The nighttime set reveals the effectiveness of our algorithm in low-light environments.

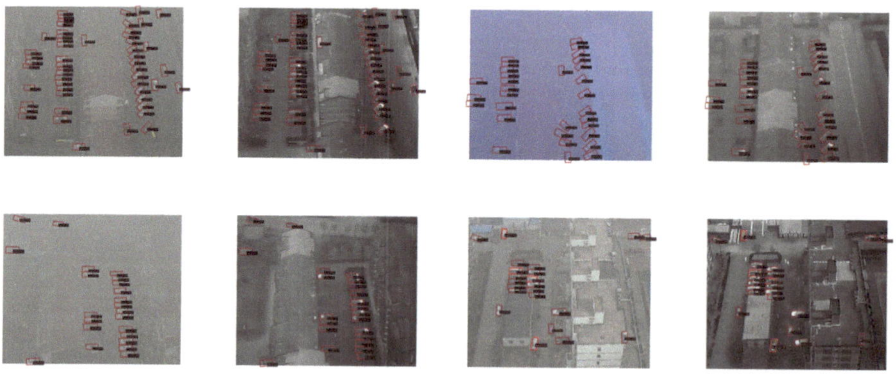

Figure A3. The additional visualization results obtained using VIP-Det in the foggy scenarios. This set highlights the ability of our method to penetrate visual obscurities and accurately detect objects, demonstrating its resilience against atmospheric disturbances.

Figure A4. The additional visualization results obtained using VIP-Det in the occlusion scenarios. This set underscores the capability of our approach to recognize objects even when partially hidden or obstructed, illustrating its robustness against occlusion challenges.

References

1. He, K.; Zhang, X.; Ren, S.; Sun, J. Deep Residual Learning for Image Recognition. In Proceedings of the 2016 IEEE Conference on Computer Vision and Pattern Recognition (CVPR), Las Vegas, NV, USA, 27–30 June 2016.
2. Girshick, R. Fast r-cnn. In Proceedings of the IEEE International Conference on Computer Vision, Santiago, Chile, 7–13 December 2015; pp. 1440–1448.
3. Ren, S.; He, K.; Girshick, R.; Sun, J. Faster r-cnn: Towards real-time object detection with region proposal networks. *IEEE Trans. Pattern Anal. Mach. Intell.* **2017**, *29*, 1137–1149. [CrossRef] [PubMed]
4. Redmon, J.; Divvala, S.; Girshick, R.; Farhadi, A. You only look once: Unified, real-time object detection. In Proceedings of the IEEE Conference on Computer Vision and Pattern Recognition, Las Vegas, NV, USA, 27–30 June 2016; pp. 779–788.
5. Du, D.; Zhu, P.; Wen, L.; Bian, X.; Liu, Z.M. VisDrone-DET2019: The Vision Meets Drone Object Detection in Image Challenge Results. In Proceedings of the ICCV Visdrone Workshop, Seoul, Republic of Korea, 27–28 October 2019.
6. Xia, G.S.; Bai, X.; Ding, J.; Zhu, Z.; Belongie, S.; Luo, J.; Datcu, M.; Pelillo, M.; Zhang, L. DOTA: A Large-scale Dataset for Object Detection in Aerial Images. In Proceedings of the 2018 IEEE/CVF Conference on Computer Vision and Pattern Recognition (CVPR), Salt Lake City, UT, USA, 18–23 June 2018.
7. Ding, J.; Xue, N.; Long, Y.; Xia, G.S.; Lu, Q. Learning RoI Transformer for Oriented Object Detection in Aerial Images. In Proceedings of the 2019 IEEE/CVF Conference on Computer Vision and Pattern Recognition (CVPR), Long Beach, CA, USA, 15–20 June 2019.
8. Yang, X.; Yan, J.; Feng, Z.; He, T. R3det: Refined single-stage detector with feature refinement for rotating object. In Proceedings of the AAAI Conference on Artificial Intelligence, Vancouver, BC, Canada, 2–9 February 2021; Volume 35, pp. 3163–3171.
9. Xie, X.; Cheng, G.; Wang, J.; Yao, X.; Han, J. Oriented R-CNN for object detection. In Proceedings of the IEEE/CVF International Conference on Computer Vision, Montreal, QC, Canada, 10–17 October 2021; pp. 3520–3529.
10. Hou, L.; Lu, K.; Xue, J.; Li, Y. Shape-adaptive selection and measurement for oriented object detection. In Proceedings of the AAAI Conference on Artificial Intelligence, Online, 22 February–1 March 2022; Volume 36, pp. 923–932.
11. Vaswani, A.; Shazeer, N.; Parmar, N.; Uszkoreit, J.; Jones, L.; Gomez, A.N.; Kaiser, Ł.; Polosukhin, I. Attention is all you need. *arXiv* **2017**, arXiv:1706.0376.
12. Devlin, J.; Chang, M.W.; Lee, K.; Toutanova, K. Bert: Pre-training of deep bidirectional transformers for language understanding. *arXiv* **2018**, arXiv:1810.04805.
13. Dosovitskiy, A.; Beyer, L.; Kolesnikov, A.; Weissenborn, D.; Zhai, X.; Unterthiner, T.; Dehghani, M.; Minderer, M.; Heigold, G.; Gelly, S.; et al. An image is worth 16x16 words: Transformers for image recognition at scale. *arXiv* **2020**, arXiv:2010.11929.
14. Liu, Z.; Lin, Y.; Cao, Y.; Hu, H.; Wei, Y.; Zhang, Z.; Lin, S.; Guo, B. Swin transformer: Hierarchical vision transformer using shifted windows. In Proceedings of the IEEE/CVF International Conference on Computer Vision, Montreal, BC, Canada, 11–17 October 2021; pp. 10012–10022.
15. Carion, N.; Massa, F.; Synnaeve, G.; Usunier, N.; Kirillov, A.; Zagoruyko, S. End-to-end object detection with transformers. In Proceedings of the European Conference on Computer Vision, Glasgow, UK, 23–28 August 2020; pp. 213–229.
16. Yuan, M.; Wei, X. C 2 Former: Calibrated and Complementary Transformer for RGB-Infrared Object Detection. *IEEE Trans. Geosci. Remote Sens.* **2024**, *62*, 5403712. [CrossRef]
17. He, K.; Chen, X.; Xie, S.; Li, Y.; Dollár, P.; Girshick, R. Masked autoencoders are scalable vision learners. In Proceedings of the IEEE/CVF Conference on Computer Vision and Pattern Recognition, New Orleans, LA, USA, 18–24 June 2022; pp. 16000–16009.
18. Jia, M.; Tang, L.; Chen, B.C.; Cardie, C.; Belongie, S.; Hariharan, B.; Lim, S.N. Visual prompt tuning. In Proceedings of the European Conference on Computer Vision, Tel Aviv, Israel, 23–27 October 2022; pp. 709–727.

19. Zhu, J.; Lai, S.; Chen, X.; Wang, D.; Lu, H. Visual prompt multi-modal tracking. In Proceedings of the IEEE/CVF Conference on Computer Vision and Pattern Recognition, Vancouver, BC, Canada, 18–22 June 2023; pp. 9516–9526.
20. Hwang, S.; Park, J.; Kim, N.; Choi, Y.; So Kweon, I. Multispectral pedestrian detection: Benchmark dataset and baseline. In Proceedings of the IEEE Conference on Computer Vision and Pattern Recognition, Boston, MA, USA, 7–12 June 2015; pp. 1037–1045.
21. Song, K.; Xue, X.; Wen, H.; Ji, Y.; Yan, Y.; Meng, Q. Misaligned Visible-Thermal Object Detection: A Drone-based Benchmark and Baseline. *IEEE Trans. Intell. Veh.* **2024**, *early access*. [CrossRef]
22. Sun, Y.; Cao, B.; Zhu, P.; Hu, Q. Drone-based RGB-infrared cross-modality vehicle detection via uncertainty-aware learning. *IEEE Trans. Circuits Syst. Video Technol.* **2022**, *32*, 6700–6713. [CrossRef]
23. Liu, J.; Zhang, S.; Wang, S.; Metaxas, D.N. Multispectral deep neural networks for pedestrian detection. *arXiv* **2016**, arXiv:1611.02644.
24. Yuan, M.; Wang, Y.; Wei, X. Translation, scale and rotation: Cross-modal alignment meets RGB-infrared vehicle detection. In Proceedings of the European Conference on Computer Vision, Tel Aviv, Israel, 23–27 October 2022; pp. 509–525.
25. Zhang, W.; Zhao, W.; Li, J.; Zhuang, P.; Sun, H.; Xu, Y.; Li, C. CVANet: Cascaded visual attention network for single image super-resolution. *Neural Netw.* **2024**, *170*, 622–634. [CrossRef] [PubMed]
26. Zhang, W.; Li, Z.; Li, G.; Zhuang, P.; Hou, G.; Zhang, Q.; Li, C. Gacnet: Generate adversarial-driven cross-aware network for hyperspectral wheat variety identification. *IEEE Trans. Geosci. Remote Sens.* **2023**, *62*, 5503314. [CrossRef]
27. Cui, L.; Jing, X.; Wang, Y.; Huan, Y.; Xu, Y.; Zhang, Q. Improved swin transformer-based semantic segmentation of postearthquake dense buildings in urban areas using remote sensing images. *IEEE J. Sel. Top. Appl. Earth Obs. Remote Sens.* **2022**, *16*, 369–385. [CrossRef]
28. Li, Y.; Mao, H.; Girshick, R.; He, K. Exploring plain vision transformer backbones for object detection. In Proceedings of the European Conference on Computer Vision, Tel Aviv, Israel, 23–27 October 2022; pp. 280–296.
29. Fang, Y.; Yang, S.; Wang, S.; Ge, Y.; Shan, Y.; Wang, X. Unleashing vanilla vision transformer with masked image modeling for object detection. In Proceedings of the IEEE/CVF International Conference on Computer Vision, Paris, France, 1–6 October 2023; pp. 6244–6253.
30. Liu, F.; Zhang, X.; Peng, Z.; Guo, Z.; Wan, F.; Ji, X.; Ye, Q. Integrally migrating pre-trained transformer encoder-decoders for visual object detection. In Proceedings of the IEEE/CVF International Conference on Computer Vision, Paris, France, 1–6 October 2023; pp. 6825–6834.
31. Wang, D.; Zhang, Q.; Xu, Y.; Zhang, J.; Du, B.; Tao, D.; Zhang, L. Advancing plain vision transformer toward remote sensing foundation model. *IEEE Trans. Geosci. Remote Sens.* **2022**, *61*, 1–15. [CrossRef]
32. Zhang, Q.; Xu, Y.; Zhang, J.; Tao, D. Vsa: Learning varied-size window attention in vision transformers. In Proceedings of the European Conference on Computer Vision, Tel Aviv, Israel, 23–27 October 2022; pp. 466–483.
33. Yu, H.; Tian, Y.; Ye, Q.; Liu, Y. Spatial transform decoupling for oriented object detection. In Proceedings of the AAAI Conference on Artificial Intelligence, Vancouver, BC, Canada, 20–27 February 2024; Volume 38, pp. 6782–6790.
34. Li, S.; Xue, L.; Feng, L.; Yao, C.; Wang, D. Hybrid Convolutional-Transformer framework for drone-based few-shot weakly supervised object detection. *Comput. Electr. Eng.* **2022**, *102*, 108154. [CrossRef]
35. Bar, A.; Gandelsman, Y.; Darrell, T.; Globerson, A.; Efros, A. Visual prompting via image inpainting. *Adv. Neural Inf. Process. Syst.* **2022**, *35*, 25005–25017.
36. Kirillov, A.; Mintun, E.; Ravi, N.; Mao, H.; Rolland, C.; Gustafson, L.; Xiao, T.; Whitehead, S.; Berg, A.C.; Lo, W.Y.; et al. Segment anything. In Proceedings of the IEEE/CVF International Conference on Computer Vision, Paris, France, 1–6 October 2023; pp. 4015–4026.

drones

MDPI

Article

A Large Scale Benchmark of Person Re-Identification

Qingze Yin [1,†] and Guodong Ding [2,*,†]

1 School of Computer and Information Engineering, Institute for Artificial Intelligence, Shanghai Polytechnic University, Shanghai 201209, China; qzyin@sspu.edu.cn
2 School of Computing, National University of Singapore, Singapore 117416, Singapore
* Correspondence: dinggd@comp.nus.edu.sg
† These authors contributed equally to this work.

Abstract: Unmanned aerial vehicles (UAVs)-based Person Re-Identification (ReID) is a novel field. Person ReID is the task of identifying individuals across different frames or views, often in surveillance or security contexts. At the same time, UAVs enhance person ReID through their mobility, real-time monitoring, and ability to access challenging areas despite privacy, legal, and technical challenges.To facilitate the advancement and adaptation of existing person ReID approach to the UAV scenarios, this paper introduces a baseline along with two datasets, i.e., LSMS and LSMS-UAV. Both datasets have the following key features: (1) LSMS: Raw videos captured by a network of 29 cameras deployed across complex outdoor environments. LSMS-UAV: captured by 1 UAV. (2) LSMS: Videos span both winter and spring seasons, encompassing diverse weather conditions and various lighting conditions throughout different times of the day. (3) LSMS: Including the largest number of annotated identities, comprising 7730 identities and 286,695 bounding boxes. LSMS-UAV: comprising 500 identities and 2000 bounding boxes. Comprehensive experiments demonstrate LSMS's excellent capability in addressing the domain gap issue when facing complex and unknown environments. The LSMS-UAV dataset verifies that UAV data has strong transferability to traditional camera-based data.

Keywords: Person Re-Identification; UAVs-based Person Re-Identification; large scale dataset; Multi-Scene; multi-time; multi-camera

Citation: Yin, Q.; Ding, G. A Large Scale Benchmark of Person Re-Identification. *Drones* **2024**, *8*, 279. https://doi.org/10.3390/drones8070279

Academic Editor: Anastasios Dimou

Received: 5 June 2024
Revised: 19 June 2024
Accepted: 19 June 2024
Published: 21 June 2024

1. Introduction

Unmanned aerial vehicles (UAVs), commonly known as drones, have seen a rise in accessibility and have significantly impacted various domains such as photography [1], transportation [2], and search operations [3], providing substantial benefits to the public. Among them, utilizing drones for person ReID tasks in urban settings is a relatively novel direction. Compared to traditional person ReID systems based on camera setups, UAV-based person ReID offers faster response times. This is because it eliminates the need for complex camera retrieval with multiple different parameters and allows for direct video transmission on a single drone.

Traditional Person Re-Identification (ReID) aims to match and retrieve images of a specific individual from a vast gallery dataset captured by camera networks. Due to its significance in surveillance and security applications, ReID has garnered considerable attention from both industrial and academic sectors [4–6]. With the advancements in deep learning techniques and the various public datasets, the performance of ReID has witnessed remarkable improvements. For instance, on the Market1501 [7] dataset, the Rank-1 accuracy of a single query has increased from 43.8% [8] to 96.1% [9]. Similarly, on the CUHK03 [10], the Rank-1 accuracy has risen from 19.9% [10] to 88.5% [11]. Furthermore, on the MSMT17 [12] dataset, the Rank-1 accuracy has risen from 47.6% [12] to 89.7% [13]. A comprehensive review of current methodologies will be provided in Section 2.

Although the current ReID algorithms have a good effect on the existing datasets, there are still some unresolved problems that impede its applications in reality. One major issue is the disparity between existing public datasets and real-world data. Many current datasets are limited in scope, either containing only a limited number of identities or are captured under controlled environments. For instance, even the largest dataset, MSMT17 [12], comprises less than 4101 identities and features simplistic lighting variations. However, in real-world scenes, ReID typically operates within camera networks that are set up across diverse environments, processing videos captured over extended periods of time. As a result, real-world applications must contend with challenges such as a huge number of person identities, and complex variations in lighting, view, and weather conditions, which current methods may struggle to adequately settle.

Another significant issue that has been identified is the domain gap between various person ReID datasets. This refers to the phenomenon where ReID models trained on one dataset while tested with another often experience a significant performance drop. For instance, a model with a classic person ReID algorithm, Bag of Tricks (BoT) [14], trained on Market1501 [7] achieves only a Rank-1 accuracy of 28.6% when tested on MSMT17 [12]. As illustrated in Figure 1, the domain gap can be attributed to various factors such as differences in lighting conditions, viewpoints, resolutions, seasons, weather, backgrounds, etc. For example, most pedestrians from Market1501 are captured during summer, wearing bright-colored short sleeves and shorts. Conversely, the DukeMTMC-ReID dataset was collected during winter, so pedestrians are mostly dressed in dark-colored and thick clothing. The MSMT17 dataset has provided more variations in lighting, but pedestrians still predominantly wear thick clothing, which, to some extent, limits the diversity of dataset styles. This challenge poses a significant obstacle to the practical applications of person ReID, as the data from the existing training set cannot be efficiently applied to the new test set.

Figure 1. An illustration of the domain gaps across MSMT17, Market1501, and DukeMTMC-ReID reveals distinct styles, including variations in lighting, resolution, human demographics, seasonal conditions, and backgrounds. These discrepancies pose challenges in achieving high accuracy when using any one of them as the training set and the others as the test set.

To advance research efforts toward real-world applications, this paper presents a curated large-scale dataset named Large-Scale Multi-Scene (LSMS). Distinguished from existing datasets, LSMS offers several novel features. Firstly, the raw videos were captured by a network of 29 cameras deployed across complex outdoor environments on campus, including academic and residential sectors. Consequently, the dataset showcases intricate scene transformations and diverse backgrounds. For example, it includes images of pedestrians, such as elderly people, children, and teenagers. It also features diversity with images of both cyclists and walking pedestrians. Secondly, the videos span a considerable duration of time, covering nine days within three months under different weather conditions across winter and spring seasons. In addition, it features footage captured during the morning, noon, and afternoon hours. This results in a dataset with complex variations in lighting conditions and person clothes styles. Lastly, LSMS provides the largest number of labeled bounding boxes and identities to date, comprising 286,695 bounding boxes and 7730 identities. To the best of our knowledge, LSMS stands as the most challenging and the largest open dataset available for ReID research. We'll elaborate on the dataset in Section 3.

In order to address the person ReID under drone surveillance, we also propose a dataset collected using drones, namely, LSMS-UAV. It has the following features: Firstly, the raw videos were captured by a drone on a different road from LSMS, including complex outdoor environments such as academic and residential areas on campus. Secondly, the videos span two days, with each day capturing 20 min of footage during various periods, including morning, noon, and afternoon, showcasing different lighting conditions and variations. Lastly, the dataset comprises 500 identities and 2000 bounding boxes, which is sufficient for the test set.

Our contributions can be delineated into four key aspects. (1) A challenging large-scale dataset LSMS is curated, available at https://github.com/QingzeYin/LSMS, for realistic person ReID tasks, advancing research in the field. (2) A UAV-based person ReID dataset is proposed, with tests conducted on several other classic camera-based ReID datasets. Experimental results demonstrate that models trained on traditional person ReID datasets perform well on UAV-based datasets. This provides a benchmark for subsequent research on ReID based on UAVs. (3) The comparison and analysis of the most typical person ReID algorithms were conducted on four public classic camera-based ReID datasets and one LSMS-UAV dataset. LSMS demonstrated significant advantages in complexity, authenticity, and robustness. (4) This paper comprehensively analyzes the issues hindering practical applications of person ReID, such as monotonous backgrounds in training data, uniform clothing, and limited variation in person samples. It also highlights the potential of LSMS to drive future research in the field.

2. Related Work

This research is closely related to the standard ReID datasets, descriptor learning in person ReID and UAV applications. We provide a brief summary of these three categories of research as follows.

2.1. Standard ReID Datasets

To improve the performance of person ReID gradually, researchers have proposed most of the related datasets. Earlier, Cheng et al. [15] introduced a novel dataset named CAVIAR which includes 72 identities with 610 bounding boxes captured from two cameras. Then, Hirzer et al. [16] proposed a novel dataset named PRID which includes 934 identities with 1134 bounding boxes captured from two cameras. Recently, Li et al. [10] proposed a novel dataset named CUHK03 which includes 1467 identities with 28,192 bounding boxes captured from two cameras. Zheng et al. [7] introduced the Market1501 dataset for person ReID, addressing limitations of existing datasets by offering over 1501 identities with 32,668 annotated bounding boxes across 6 cameras. Images are produced using the Deformable Part Model (DPM) as a pedestrian detector, and the dataset features multiple images for each identity under each camera. Ristani et al. [17] introduced new precision-recall measures and the largest fully-annotated dataset named DukeMTMC-ReID for multi-target, multi-camera tracking, which includes 1812 identities with 36,411 bounding boxes captured from 8 cameras. Wei et al. [12] introduced the MSMT17 dataset with features captured from a 15-camera network and 4101 annotated identities with 126,441 bounding boxes, aiming to address challenges in person ReID.

2.2. Descriptor Learning in ReID

Descriptors based on deep learning have demonstrated significant superiority over hand-crafted features in the majority of ReID datasets. Some studies [18,19] employ deep descriptors learned from entire images using classification models, treating each person ID as a distinct category. Others [20,21] combine classification and verification models to train descriptors. In [22], Hermans et al.have shown that triplet loss efficiently enhances person ReID accuracy, while Chen et al. [23] have proposed quadruplet networks for representation learning.

However, the aforementioned approaches focus on learning global descriptors and overlook detailed cues that may be crucial for distinguishing individuals. To explicitly leverage local cues, Yin et al. [24] introduce a multi-view part-based network for discriminative descriptor learning. Wu et al. [25] discovered that hand-crafted features could complement deep features by dividing the global picture into five fixed-length areas and extracting histogram descriptors for each region concatenated with the global deep descriptor. Despite their effectiveness, these methods overlook misalignment issues that stem from the rigid division of body parts. Addressing this concern, Wei et al. [26] detected three coarse body regions by utilizing Deepercut [27] and subsequently learned a global-local-alignment descriptor. Zhao et al. [28] localized the fine-grained part areas and input them into the raised Spindle Net to learn descriptors. Similarly, in [29], Li et al. detected latent part regions by employing Spatial Transform Networks (STN) [30] and then training descriptors on those regions.

2.3. UAV Detection, Classification, and 3D Tracking Techniques

The integration of deep learning methods across diverse sensor modalities has significantly advanced UAV detection [31,32] and classification techniques [33]. Vision-based detection systems, leveraging neural networks for processing visual data from cameras, have demonstrated notable success. Notably, models from the YOLO series [34] have exhibited remarkable accuracy in bounding box classification and regression tasks. Liu et al. [35] proposed an enhanced detection and classification approach utilizing clustering support vector machines, yielding improved performance. Additionally, segmentation methods [36] have been employed to augment detection capabilities.

In real-world applications, UAV 3D tracking finds extensive utility across various domains such as military, transportation [37], and security [38]. Techniques leveraging learning-based methodologies have been pivotal in enhancing tracking accuracy. For instance, Lan et al. [39] utilized a sparse learning approach for RGB-T tracking, effectively mitigating cross-modality discrepancies. Moreover, transformer-based algorithms for multi-object tracking [40] hold promise for UAV detection scenarios, demonstrating potential effectiveness in handling complex data associations.

3. LSMS and LSMS-UAV Dataset

3.1. Overview of Previous ReID Datasets

The current landscape of person ReID datasets has significantly propelled research in this field. Notably, as shown in Table 1, datasets such as MSMT17 [12], DukeMTMC-ReID [17], CUHK03 [10], and Market1501 [7] exhibit larger scales in terms of the number of cameras and identities compared to predecessors like VIPeR [41], CAVIAR [15], and PRID [16]. This abundance of training data enables the development of deep models that showcase their discriminative prowess in person ReID tasks. Despite the high accuracy achieved by current algorithms on these datasets, the practical application of person ReID in real-world scenarios remains a challenge. Therefore, it is imperative to conduct a thorough analysis of the limitations present in existing datasets.

Table 1. Comparison between LSMS and other person ReID datasets.

Dataset	LSMS	MSMT17 [12]	DukeMTMC-ReID [17]	Market-1501 [7]	CUHK03 [10]	VIPeR [41]	PRID [16]	CAVIAR [15]
BBoxes	286,695	126,441	36,411	32,668	28,192	1264	1134	610
Identities	7730	4101	1812	1501	1467	632	934	72
Cameras	29	15	8	6	2	2	2	2
Detector	Faster RCNN	Faster RCNN	hand	DPM	DPM, hand	hand	hand	hand

Current datasets, in contrast to those gathered in real-world scenes, exhibit limitations across four key dimensions: (1) The number of bounding boxes and identities is insufficient, particularly when compared to authentic surveillance video data. For instance, the largest dataset comprises only 126,441 bounding boxes and less than 4101 identities, as indicated in Table 1. (2) Most of the existing datasets contain fewer cameras, such as the largest dataset MSMT17 only utilizes 15 cameras. A deficient number of cameras would lead to a weak performance of person ReID because of image conditions of pedestrians are changeless, which is reflected in the resolution, viewpoints, background, and occlusion. (3) Many datasets originate from short-duration surveillance system videos that lack distinct variations in lighting conditions, limiting their applicability to real-world scenarios. (4) The consistent weather conditions lead to uniform pedestrian attire, consequently reducing pedestrian attribute features, such as umbrellas, among others. Unlike real-world weather conditions, this scenario does not favor the robustness of training models. These constraints underscore the need for larger and more representative datasets to advance person ReID research.

3.2. Description to LSMS and LSMS-UAV

3.2.1. Description to LSMS

To mitigate the aforementioned constraints, we have curated a novel person ReID dataset named LSMS, which aimed at emulating real-world scenarios as closely as feasible. Leveraging a network of 29 cameras stationed across three major thoroughfares spanning over a dozen intersections within the campus, encompassing both academic and residential sectors. We meticulously selected nine days over three months to capture varying weather conditions, with each day featuring 4-h video segments captured during the morning, forenoon, noon, and afternoon periods, facilitating pedestrian detection and annotation. The resultant dataset comprises 486 h of final raw video footage across 29 outdoor cameras, spanning 36 distinct time slots. Pedestrian bounding box detection was performed using Faster Region-based Convolutional Neural Networks (Faster RCNN) [42], with 13 labelers assigned to annotate ID labels over a two-month period, yielding a total of 286,695 bounding boxes corresponding to 7730 unique identities.

Figure 2 showcases and compares sample images from this dataset. It is evident that the LSMS dataset poses a more challenging and realistic ReID challenge. Figure 3 provides statistical insights into LSMS. In Figure 3a, the distribution of person bounding box numbers across different training and test sets based on various seasons is shown. It can be observed that the training set contains the highest proportion of bounding box numbers, which is intended to train a more robust model. Additionally, the number of bounding boxes in spring is higher than in winter because people's clothing styles are more varied in spring compared to winter. Figure 3b,c, respectively, show the comparison of person identities and the number of bounding boxes captured by different cameras in different seasons. Firstly, both figures indicate that the number of images in spring is greater than in winter. Secondly, it can be seen that the cameras positioned in the front, middle, and end captured more images. This is because these cameras are located at intersections where person traffic is higher, making it easier to collect more images. Finally, Figure 3d shows the distribution of the number of bounding boxes across different time periods in different seasons. It can be observed that the bounding box collection in spring is evenly distributed across various time periods, whereas in winter, there are more bounding boxes collected at noon and fewer in the morning and evening. This is due to the insufficient sunlight in the morning and evening during winter, making it harder to capture suitable person images for model training.

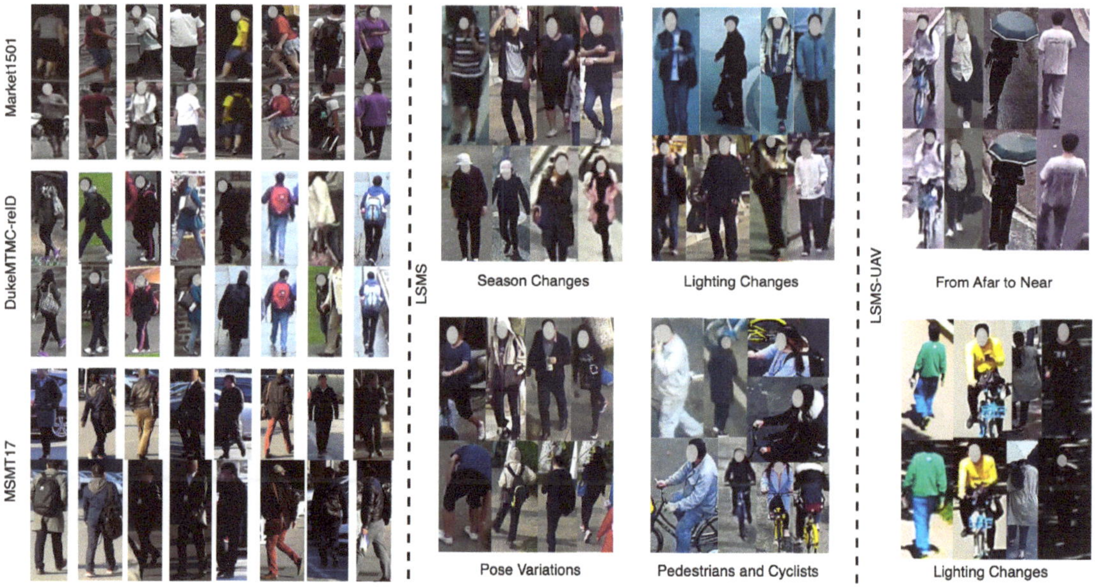

Figure 2. Comparing person images across Market1501, MSMT17, DukeMTMC-ReID, LSMS, and LSMS-UAV. Each column contains paired pictures of the same individual, except for the LSMS 'season changes', where each row represents a different season.

In comparison to existing datasets, the novel features of LSMS are delineated as follows:

(1) *Larger number of identities and bounding boxes.* As far as we know, LSMS currently stands as the largest person ReID dataset. As demonstrated in Table 1, LSMS encompasses 286,695 bounding boxes and includes 7730 identities, representing a significant increase compared to previous datasets.

(2) *Complex viewpoints and backgrounds.* LSMS boasts the highest camera count among existing datasets, with a total of 29 cameras strategically positioned in various locations. The distribution of cameras takes into account the activity patterns of pedestrians. For instance, the academic area mainly comprises young students dressed uniformly, whereas the residential area encompasses a broader demographic, including brightly dressed children and elderly individuals. This inclusion contributes to the dataset's complexity by introducing diverse backgrounds and viewpoints variations, rendering LSMS more captivating and demanding for research purposes.

(3) *Multiple time slots introduce variations in lighting conditions.* LSMS comprises 36 time slots, encompassing morning, forenoon, noon, and afternoon over nine days. Although this setup better mirrors real-world scenarios compared to previous datasets, it also introduces substantial variations in lighting conditions.

(4) *More reliable individuals outfits.* Compared with existing datasets, LSMS captures pedestrian clothing styles from both winter and spring seasons, enhancing the realism and complexity of the dataset's appearance features. Additionally, it includes various weather conditions, adding additional attributes to pedestrians, such as umbrellas.

In addition, for better comparison and analysis of the influence of pedestrian attire across different seasons on person ReID, we also calibrated the distribution of data in the LSMS dataset according to different seasons. As shown in Table 2, it can be observed that the proportion of pedestrian images in the spring season in the LSMS dataset exceeds a large portion, spanning more cameras than in the winter season. The aforementioned advantages illustrate that LSMS possesses broader applicability and robustness, which can better drive the advancement of person ReID solutions in real-world scenarios.

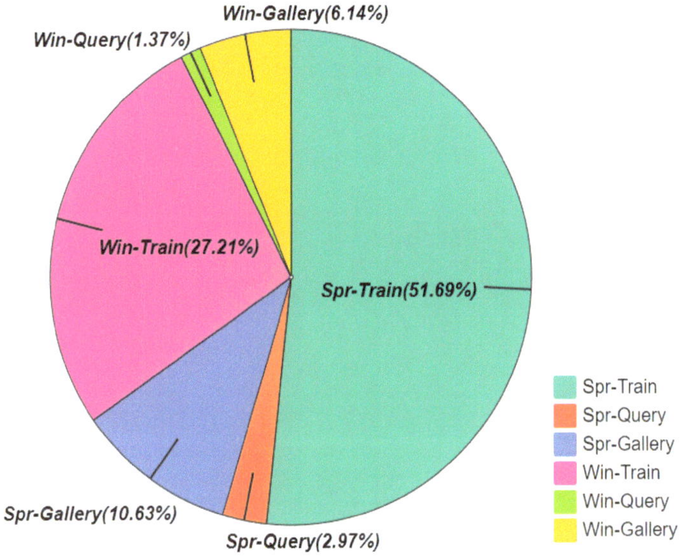

(**a**)

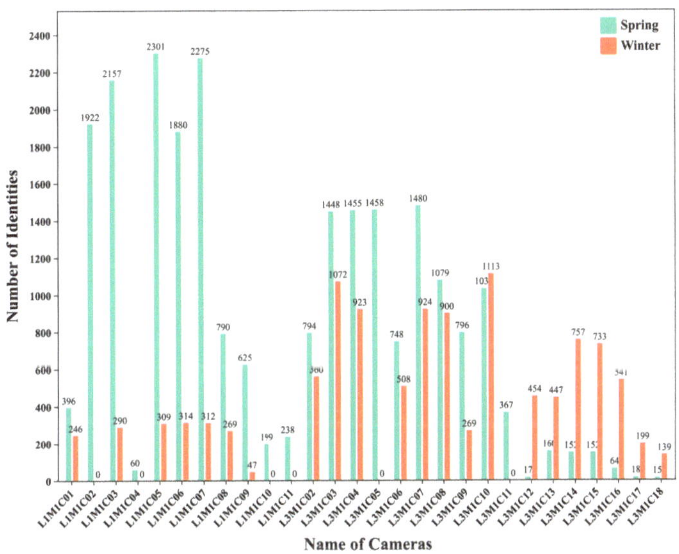

(**b**)

Figure 3. *Cont.*

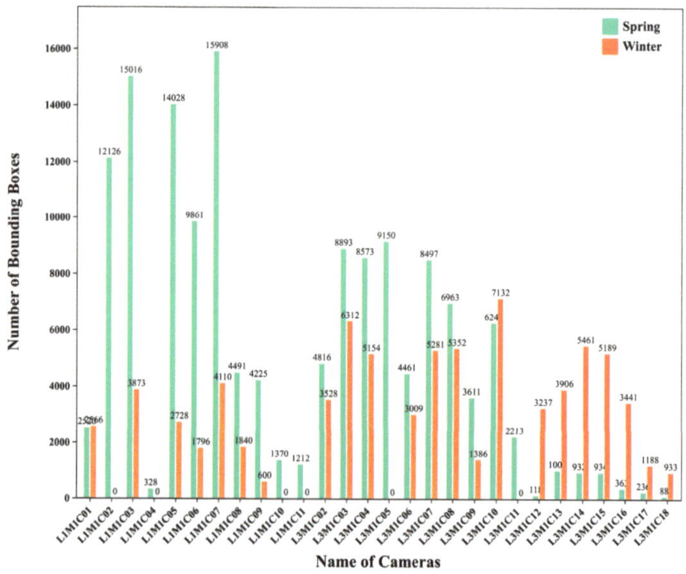

(c)

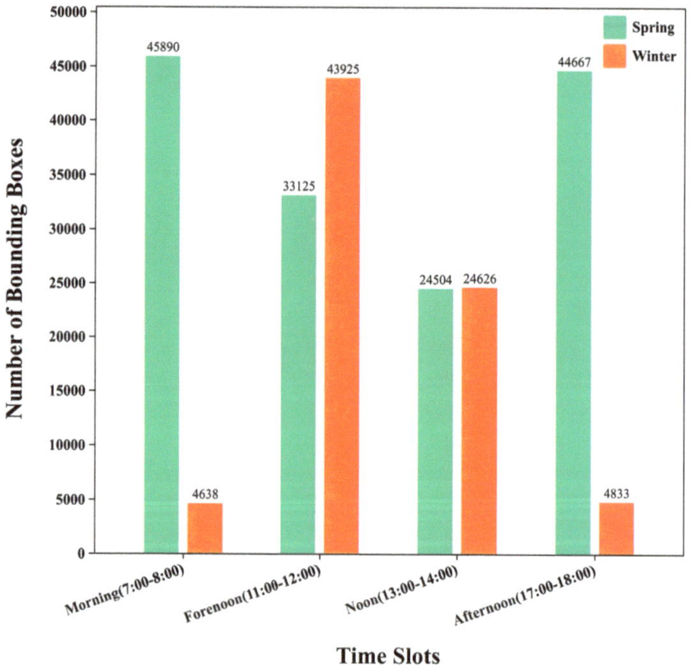

(d)

Figure 3. Statistics of LSMS. (**a**) Distribution of Bounding Box numbers across two seasons. (**b**) Comparison of the distribution of identity numbers based on two seasons for each camera in the training set. (**c**) Comparison of the distribution of Bounding Box numbers based on two seasons for each camera in the training set. (**d**) Comparison of the distribution of Bounding Box numbers based on two seasons for each time slot in the training set.

Table 2. Detailed distribution of LSMS across spring and winter seasons.

Seasons	Spring			Winter		
	Training	Testing		Training	Testing	
		Query	Gallery		Query	Gallery
Bboxes	148,186	8511	30,466	78,022	3915	17,595
Identities	3869	1217	1217	1905	739	739
Cameras	28	27	27	22	22	23

3.2.2. Description to LSMS-UAV

Current person ReID algorithms primarily train and test models on person ReID datasets. To enhance the application of UAVs in the ReID field, it is crucial to train a more effective model. For this purpose, a large-scale ReID dataset, LSMS, is introduced for training ReID models. Additionally, for testing purposes, a novel UAV-based dataset, LSMS-UAV, is proposed for transfer learning comparisons to assess the performance of UAVs in the person ReID domain. Here, the LSMS-UAV dataset is used as the test set for the ReID model, while the LSMS dataset serves as the training set.

During the image collection process, both datasets were gathered within the same campus, encompassing both academic and residential areas. The difference lies in the fact that they were collected on different days and on different streets, ensuring no overlap in person identities.

The LSMS-UAV dataset has the following characteristics: (1) It includes 500 identities and 2000 bounding boxes. (2) Data was collected using a single UAV. (3) Data collection spanned 2 days, with 20-min sessions each in the morning, forenoon, noon, and afternoon, totaling 160 min of video. (4) Compared to the LSMS dataset, LSMS-UAV was collected on a different road. Since the LSMS-UAV dataset was collected using a UAV, the images feature varying resolutions due to the nature of capturing from afar to near. The angles are predominantly overhead, and there are variations in lighting conditions. These characteristics can be seen in Figure 2.

3.3. *Evaluation Protocol*

We employ a random division approach to partition our LSMS dataset into training and test sets. Unlike previous datasets, where the two parts are divided equally, we set the training-to-testing ratio as 3:1. Consequently, the training set comprises 226,208 bounding boxes corresponding to 5774 identities, while the test set includes 60,487 bounding boxes representing 1956 identities. Within the test set, 12,426 bounding boxes are stochastically chosen as query images, with the remaining 48,061 bounding boxes serving as gallery images. This is also shown in Table 3.

Table 3. Detailed distribution of LSMS.

LSMS	Bounding Boxes	Identities
Training set	226,208	5774
Query set	12,426	1956
Gallery set	48,061	1956

Similarly, as shown in Table 4, the LSMS-UAV dataset serves as the test set, comprising a total of 2000 bounding boxes and 500 identities. Due to the smaller data size, the query set contains 500 bounding boxes, while the gallery set includes 1500 bounding boxes.

Table 4. Detailed distribution of LSMS-UAV.

LSMS-UAV	Bounding Boxes	Identities
Query set	500	500
Gallery set	1500	500

Consistent with the majority of existing datasets, the Cumulative Matching Characteristics (CMC) curve is employed to assess the accuracy of ReID. This evaluation method considers that each query bounding box may yield multiple true positives. Consequently, we treat ReID as a searching task. In addition to the CMC curve, the mean Average Precision (mAP) is also used as an evaluation metric.

4. Classic ReID Algorithms

To better evaluate the advantages of the LSMS dataset, validation is performed against three classic person ReID algorithms. Below, introductions to each of these algorithms are provided.

4.1. *Bag of Tricks (BoT)*

In recent years, deep neural networks have propelled person ReID to high-performance levels, but many state-of-the-art methods employ complex network architectures and feature concatenation. Luo et al. [14] collect and assess effective training tricks in person ReID, achieving notable performance improvements with ResNet50 [43] reaching 94.5% rank-1 accuracy and 85.9% mAP on Market1501 using global features. However, a survey of articles from high-quality journals reveals that most works build upon weak baselines. This paper addresses this by enhancing the standard baseline with training tricks to establish a robust baseline, emphasizing the importance of considering these tricks in method comparisons. Additionally, the industry's preference is for simple and efficient models, hence focusing on leveraging global features to attain high accuracy while minimizing computational overhead. The contributions of this paper include identifying and evaluating six effective training tricks, introducing a new neck structure named BNNeck, and providing a strong ReID baseline, achieving exceptional performance on Market1501 with global features from ResNet50.

4.2. *Part-Based Convolutional Baseline (PCB)*

Deeply-learned representations, especially when aggregated from part features, demonstrate high discriminative ability. State-of-the-art results on ReID benchmarks are achieved using part-informed deep features. However, accurately locating parts remains crucial for learning discriminative features.

Recent methods for partitioning vary in their strategies. Some leverage external cues, such as human pose estimation, while others abandon semantic cues and achieve competitive accuracy. In this context, a network called Part-based Convolutional Baseline (PCB) [44] is proposed, which conducts uniform partitioning on the convolutional layer for learning part-level features. PCB does not explicitly partition images but outputs a convolutional feature, demonstrating higher discriminative ability compared to fully connected descriptors. Additionally, an adaptive pooling method named Refined Part Pooling (RPP) is introduced to improve uniform partitioning. RPP relocates outliers within each part to reinforce within-part consistency without requiring part labels for training.

4.3. *Pose-Driven Deep Convolutional (PDC)*

To address the challenges posed by pose variations, Su et al. [11] propose a Pose-driven Deep Convolutional (PDC) model for ReID. This model simultaneously learns global representations of the whole body and local representations of body parts. The global representation is trained using Softmax Loss [11], while a Feature Embedding sub-Net (FEN) automatically adjusts and relocates body parts for improved recognition across different cameras. A Pose Transformation Network (PTN) further eliminates pose variations, enabling the learning of local representations on transformed regions. Additionally, a Feature Weighting sub-Net (FWN) was introduced to learn weights for global and local representations, facilitating more effective feature fusion for similarity measurement.

Detailed illustrations of the local representation generation process are provided, demonstrating how key body joints are located, body parts are extracted and normalized, and pose variations are eliminated using PTN. These normalized and transformed part regions are then used to train a deep neural network for learning local representations. This then emphasizes the importance of considering human pose cues and weights of representations on different parts, which are jointly learned end-to-end.

5. Experiments

5.1. Typical Datasets

Except as LSMS and LSMS-UAV, our experiments utilize three widely used person ReID datasets.

DukeMTMC-ReID [17] comprises 36,411 bounding boxes and 1812 identities. In the training set, it has 702 identities and 16,522 bounding boxes of that. The remaining identities are reserved for the test set.

Market1501 [7] is composed of 32,668 bounding boxes and 1501 identities. In the training set, it encompasses 751 identities and 12,936 bounding boxes of that, while the remaining 750 identities constitute the test set. Market1501 is abbreviated as Market.

MSMT17 [12] includes 4101 identities and 126,441 bounding boxes generated by Faster RCNN. Here, 32,621 bounding boxes of 1041 identities are designated for training, while 93,820 bounding boxes of 3060 identities are reserved for testing. Out of the test set, 11,659 bounding boxes are chosen at random for query images, with the remaining 82,161 bounding boxes allocated for use as gallery images.

5.2. Implementation Details

Based on the approach outlined in [45], the batch size is configured to 64, with an input image size of 256×128. Training epochs are 120, starting with an initial learning rate of 3.5×10^{-4}, which is reduced to $0.1\times$ after 40 epochs and further to $0.01\times$ after 70 epochs. A warm-up period of 10 epochs is implemented.

5.3. Performance on LSMS and LSMS-UAV across Different Datasets

In order to demonstrate our dataset LSMS can achieve outstanding performance on person ReID and the LSMS-UAV dataset enjoys excellent transferability, we compare the domain transfer learning by using the classic ReID method BoT [14] across three widely used ReID datasets, including DukeMTMC-ReID, MSMT17, and Market1501, also with LSMS and LSMS-UAV datasets. The compared results are reported in Figure 4.

In summary, as Figure 4a shows, when the training set and test set are Market1501, the results of BoT are the best which are 99.8% Rank-1 and 99.7% mAP. While the test set is DukeMTMC-ReID, the model achieves 58.1% Rank-1 and 38.1% mAP, which are the sub-optimal results. This is because the Market1501 dataset and the DukeMTMC-ReID dataset enjoy a similar distribution of data scales, hence yielding relatively good results. On the contrary, the results are much weaker when MSMT17 and LSMS are used as the test set. This is because MSMT17 and LSMS, serving as the test set, encompass many scenarios not present in the training set. These include a larger number of bounding boxes, more complex lighting conditions, and richer variations in human body poses. Consequently, models trained on Market1501 and tested on MSMT17 and LSMS exhibit poorer performance. Additionally, due to the fact that LSMS contains a more diverse range of pedestrian images and background conditions compared to MSMT17, the performance of the model tested on LSMS is weaker than those tested on MSMT17.

The same pattern is also observed when DukeMTMC-ReID is used as the training set, which can be observed in Figure 4b. When the test sets are Market1501 and DukeMTMC-ReID, both mAP (36.4%, 98.6%) and Rank-1 (67.0%, 99.3%) are relatively high. However, when the test sets are MSMT17 and LSMS, their mAP (13.2%, 12.1%) and Rank-1 (31.4%, 30.6%) are comparatively low. Similarly, this is also because the data included in DukeMTMC-ReID as the training set is weaker in terms of both quantity and complexity compared to MSMT17 and LSMS.

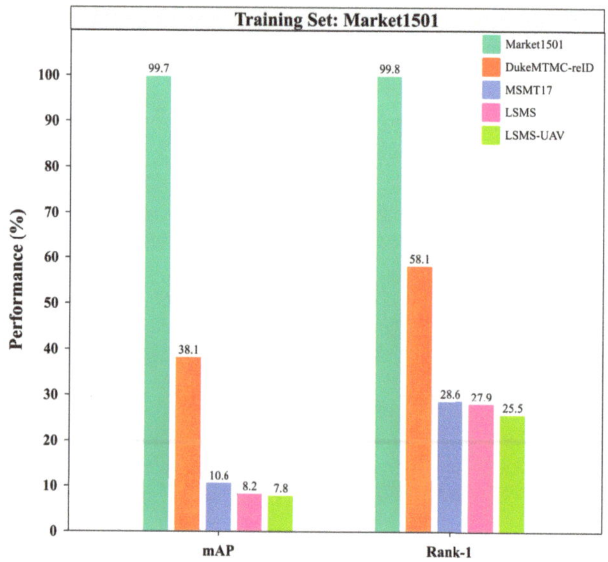

(a)

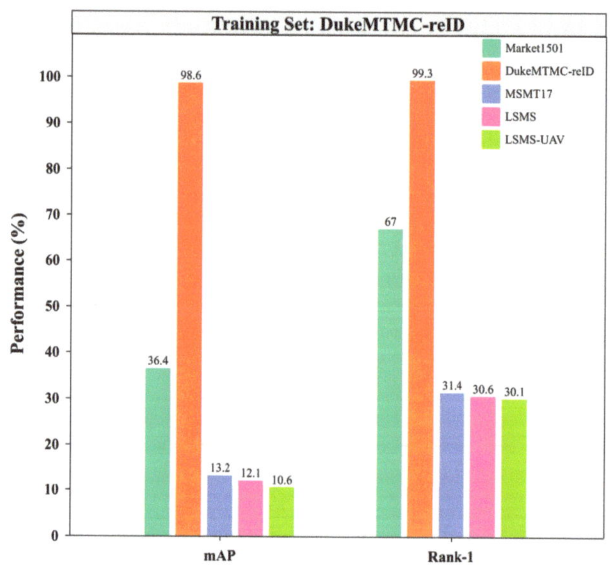

(b)

Figure 4. *Cont.*

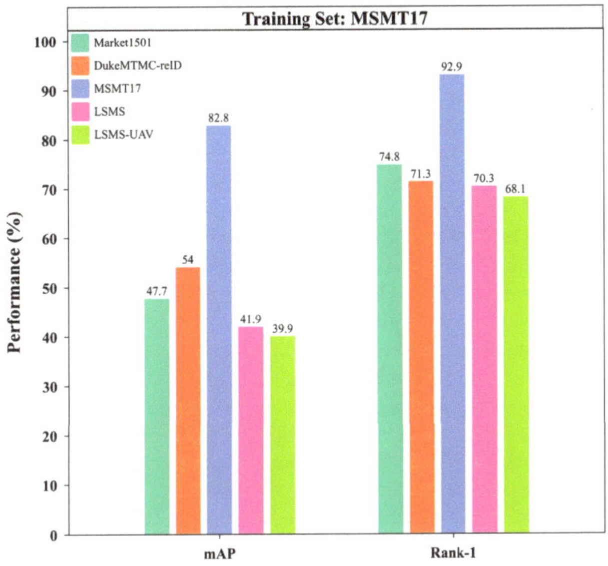

(c)

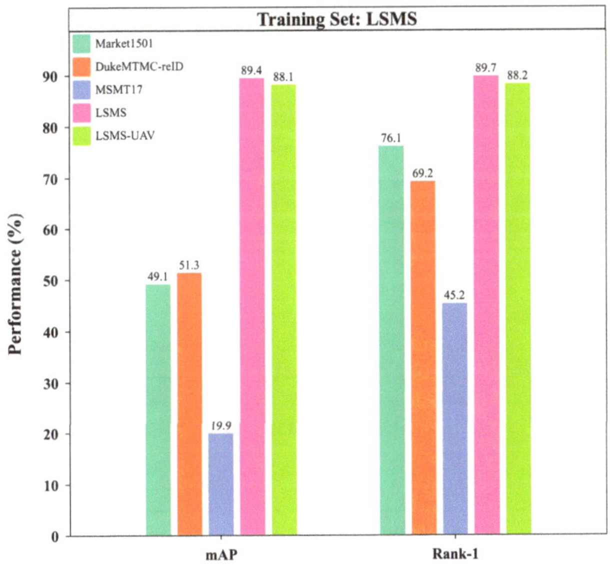

(d)

Figure 4. The performance of the BoT algorithm is compared across different datasets using transfer learning. (**a**) The transfer learning performance across different datasets, with Market1501 as the source domain and Market1501, DukeMTMC-ReID, MSMT17, LSMS, and LSMS-UAV as the target domain, separately. (**b**) The transfer learning performance across different datasets, with DukeMTMC-ReID as the source domain and Market1501, DukeMTMC-ReID, MSMT17, LSMS, and LSMS-UAV as the target domain, separately. (**c**) The transfer learning performance across different datasets, with MSMT17 as the source domain and Market1501, DukeMTMC-ReID, MSMT17, LSMS, and LSMS-UAV as the target domain, separately. (**d**) The transfer learning performance across different datasets, with LSMS as the source domain and Market1501, DukeMTMC-ReID, MSMT17, LSMS, and LSMS-UAV as the target domain, separately.

When MSMT17 and LSMS are used, respectively, as their own training and test sets, the results show that MSMT17 outperforms LSMS. As shown in Figure 4c,d, when LSMS is both the training and test set, its mAP and Rank-1 are 89.4 and 89.7%, respectively. When MSMT17 is both the training and test set, its mAP and Rank-1 are 82.8% and 92.9%, respectively. This is because our dataset LSMS is more challenging, as it contains a greater variety of complex variations in person images, such as variations in seasons, person pose, lighting, viewpoint, background, etc. In addition, when LSMS is used as the training set and MSMT17 as the test set, the mAP (19.9%) and Rank-1 (45.2%) are lower compared to when MSMT17 is the training set and LSMS is the test set (mAP: 41.9% and Rank-1: 70.3%). This is because LSMS contains many images of pedestrians riding bicycles, which introduces more complex noise features during model training. However, this can also be considered a characteristic of the LSMS dataset: unlike the traditional person ReID datasets, LSMS contains images of both pedestrians and cyclists, making it more representative of real-world person ReID scenarios.

Additionally, when Market1501, DukeMTMC-ReID, MSMT17, and LSMS are used as the training sets, and LSMS-UAV is used as the test set, the resulting mAP are 7.8%, 10.6%, 39.9%, 88.1% and its Rank-1 accuracy are 25.5%, 30.1%, 68.1%, 88.2%, respectively. As a conclusion, the lower performance of the LSMS-UAV dataset as a test set compared to LSMS can be attributed to the fact that LSMS-UAV data consists mainly of overhead angle images. This strong bias towards specific features may result in lower performance when facing diverse training features. However, despite this bias, the performance is still close to that of LSMS.

5.4. *Performance on LSMS across Different Methods*

This subsection aims to validate the assertion made in Section 3 regarding the challenging yet realistic nature of LSMS. This is achieved through the examination of existing algorithms on the LSMS dataset.

We review the classic advancements in the field. Notably, BoT, introduced by Luo et al. [14], demonstrated superior performance on most ReID datasets. While PDC, introduced by Su et al. [11], showcased the best results on CUHK03 [10]. Additionally, as a common practice in person ReID research, PCB proposed by Sun et al. [44] also served as our comparison method.

The experimental findings are summarized in Table 5. The baseline model PDC [11] achieves a Rank-1 and mAP are 82.9% and 80.3% on LSMS. Notably, PCB [44] and BoT [14] significantly surpass the baseline by incorporating additional part and regional features. Among them, BoT obtains the best performance, with a Rank-1 of 89.7% and mAP of 89.4%, which notably lags behind its reported results on other datasets, such as Rank-1 of 94.5% on Market [14]. These results underscore the challenges posed by LSMS.

Table 5. The performance of the classic methods on LSMS.

Methods	Rank-1	mAP
PDC [11]	82.9	80.3
PCB [44]	86.7	86.1
BoT [14]	89.7	89.4

We qualitatively present retrieval results in Figure 5, which underscore the realism and challenges encapsulated within the ReID task defined by LSMS. In real-world scenarios, individuals may exhibit similar clothing cues, while images of the same person can vary significantly in terms of lighting, background, and pose. As depicted in Figure 5, false positive samples often bear resemblances to the query person, while true positives exhibit diverse lighting conditions, poses, and backgrounds. Thus, LSMS emerges as a valuable resource for advancing research in ReID.

Figure 5. Sample person ReID outcomes produced by the BoT [14] on LSMS.

6. Conclusions

This paper introduces two novel datasets: LSMS and LSMS-UVA for the person ReID task. The former is a large-scale camera-based dataset for traditional person ReID, while the latter provides UAV-captured person images, facilitating UAV-based person ReID. LSMS offers significant variations in lighting conditions, seasons, backgrounds, human poses, etc. Similarly, the LSMS-UAV dataset exhibits characteristics such as resolution disparities, variations in lighting, and person images captured from an overhead perspective. As the largest dataset for person ReID currently available, LSMS defines a more realistic and challenging task compared to existing datasets. In future work, we will focus on exploring more effective and efficient strategies for transferring knowledge between persons in large datasets. Additionally, we will continue to research the transfer learning between persons and cyclists' studies based on UAV datasets.

Author Contributions: Writing—original draft preparation, Q.Y.; writing—review and editing, G.D. All authors have read and agreed to the published version of the manuscript.

Funding: This research was funded by Shanghai Polytechnic University 2024 University-level Research Program for Graduate Student Associate Supervisors to Improve Their Research Abilities OF FUNDER grant number EGD24DS14.

Data Availability Statement: The raw data supporting the conclusions of this article will be made available by the authors on request.

Conflicts of Interest: The authors declare no conflicts of interest.

References

1. Xie, H.; Deng, T.; Wang, J.; Chen, W. Angular Tracking Consistency Guided Fast Feature Association for Visual-Inertial SLAM. *IEEE Trans. Instrum. Meas.* **2024**, *73*, 5006614. [CrossRef]
2. Deng, T.; Liu, S.; Wang, X.; Liu, Y.; Wang, D.; Chen, W. ProSGNeRF: Progressive Dynamic Neural Scene Graph with Frequency Modulated Auto-Encoder in Urban Scenes. *arXiv* **2023**, arXiv:2312.09076.
3. Wang, Y.; Fan, Y.; Wang, J.; Chen, W. Long-term navigation for autonomous robots based on spatio-temporal map prediction. *Robot. Auton. Syst.* **2024**, *179*, 104724. [CrossRef]
4. Ding, G.; Zhang, S.; Khan, S.; Tang, Z.; Zhang, J.; Porikli, F. Feature affinity-based pseudo labeling for semi-supervised person re-identification. *IEEE Trans. Multimed.* **2019**, *21*, 2891–2902. [CrossRef]
5. Ding, G.; Khan, S.; Tang, Z.; Porikli, F. Feature mask network for person re-identification. *Pattern Recognit. Lett.* **2020**, *137*, 91–98. [CrossRef]
6. Yin, Q.; Wang, G.A.; Wu, J.; Luo, H.; Tang, Z. Dynamic re-weighting and cross-camera learning for unsupervised person re-identification. *Mathematics* **2022**, *10*, 1654. [CrossRef]
7. Zheng, L.; Shen, L.; Tian, L.; Wang, S.; Wang, J.; Tian, Q. Scalable person re-identification: A benchmark. In Proceedings of the IEEE International Conference on Computer Vision, Santiago, Chile, 7–13 December 2015; pp. 1116–1124.
8. Liao, S.; Hu, Y.; Zhu, X.; Li, S.Z. Person re-identification by local maximal occurrence representation and metric learning. In Proceedings of the IEEE Conference on Computer Vision and Pattern Recognition, Santiago, Chile, 7–13 December 2015; pp. 2197–2206.
9. Zhang, G.; Zhang, Y.; Zhang, T.; Li, B.; Pu, S. PHA: Patch-wise high-frequency augmentation for transformer-based person re-identification. In Proceedings of the IEEE/CVF Conference on Computer Vision and Pattern Recognition, Vancouver, BC, Canada, 17–24 June 2023; pp. 14133–14142.
10. Li, W.; Zhao, R.; Xiao, T.; Wang, X. Deepreid: Deep filter pairing neural network for person re-identification. In Proceedings of the IEEE Conference on Computer Vision and Pattern Recognition, Zurich, Switzerland, 6–12 September 2014; pp. 152–159.
11. Su, C.; Li, J.; Zhang, S.; Xing, J.; Gao, W.; Tian, Q. Pose-driven deep convolutional model for person re-identification. In Proceedings of the IEEE International Conference on Computer Vision, Venice, Italy, 22–29 October 2017; pp. 3960–3969.
12. Wei, L.; Zhang, S.; Gao, W.; Tian, Q. Person transfer gan to bridge domain gap for person re-identification. In Proceedings of the IEEE Conference on Computer Vision and Pattern Recognition, Salt Lake City, UT, USA, 18–23 June 2018; pp. 79–88.
13. Chen, W.; Xu, X.; Jia, J.; Luo, H.; Wang, Y.; Wang, F.; Sun, X. Beyond appearance: A semantic controllable self-supervised learning framework for human-centric visual tasks. In Proceedings of the IEEE/CVF Conference on Computer Vision and Pattern Recognition, Vancouver, BC, Canada, 17–24 June 2023; pp. 15050–15061.
14. Luo, H.; Gu, Y.; Liao, X.; Lai, S.; Jiang, W. Bag of tricks and a strong baseline for deep person re-identification. In Proceedings of the IEEE/CVF Conference on Computer Vision and Pattern Recognition Workshops, Long Beach, CA, USA, 16–17 June 2019.
15. Cheng, D.S.; Cristani, M.; Stoppa, M.; Bazzani, L.; Murino, V. Custom pictorial structures for re-identification. In Proceedings of the BMVC, Dundee, UK, 29 August–2 September 2011; Volume 1, p. 6.
16. Hirzer, M.; Beleznai, C.; Roth, P.M.; Bischof, H. Person re-identification by descriptive and discriminative classification. In *Image Analysis: Proceedings of the 17th Scandinavian Conference, SCIA 2011, Ystad, Sweden, 1 May 2011*; Springer: Berlin/Heidelberg, Germany, 2011; pp. 91–102.
17. Ristani, E.; Solera, F.; Zou, R.; Cucchiara, R.; Tomasi, C. Performance measures and a data set for multi-target, multi-camera tracking. In *European Conference on Computer Vision*; Springer International Publishing: Cham, Switzerland, 2016; pp. 17–35.
18. Xiao, T.; Li, H.; Ouyang, W.; Wang, X. Learning deep feature representations with domain guided dropout for person re-identification. In Proceedings of the IEEE Conference on Computer Vision and Pattern Recognition, Las Vegas, NV, USA, 27–30 June 2016; pp. 1249–1258.
19. Zheng, Z.; Zheng, L.; Yang, Y. Unlabeled samples generated by gan improve the person re-identification baseline in vitro. In Proceedings of the IEEE International Conference on Computer Vision, Venice, Italy, 22–29 October 2017; pp. 3754–3762.
20. Zheng, Z.; Zheng, L.; Yang, Y. A discriminatively learned cnn embedding for person reidentification. *ACM Trans. Multimed. Comput. Commun. Appl. (TOMM)* **2017**, *14*, 1–20. [CrossRef]
21. Geng, M.; Wang, Y.; Xiang, T.; Tian, Y. Deep transfer learning for person re-identification. *arXiv* **2016**, arXiv:1611.05244.
22. Hermans, A.; Beyer, L.; Leibe, B. In defense of the triplet loss for person re-identification. *arXiv* **2017**, arXiv:1703.07737.
23. Chen, W.; Chen, X.; Zhang, J.; Huang, K. Beyond triplet loss: A deep quadruplet network for person re-identification. In Proceedings of the IEEE Conference on Computer Vision and Pattern Recognition, Honolulu, HI, USA, 21–26 July 2017; pp. 403–412.
24. Yin, Q.; Ding, G.; Gong, S.; Tang, Z. Multi-view label prediction for unsupervised learning person re-identification. *IEEE Signal Process. Lett.* **2021**, *28*, 1390–1394. [CrossRef]
25. Wu, S.; Chen, Y.C.; Li, X.; Wu, A.C.; You, J.J.; Zheng, W.S. An enhanced deep feature representation for person re-identification. In Proceedings of the 2016 IEEE Winter Conference on Applications of Computer Vision (WACV), Lake Placid, NY, USA, 7–10 March 2016; pp. 1–8.
26. Wei, L.; Zhang, S.; Yao, H.; Gao, W.; Tian, Q. Glad: Global-local-alignment descriptor for pedestrian retrieval. In Proceedings of the 25th ACM International Conference on Multimedia, Mountain View, CA, USA, 23–27 October 2017; pp. 420–428.

27. Insafutdinov, E.; Pishchulin, L.; Andres, B.; Andriluka, M.; Schiele, B. Deepercut: A deeper, stronger, and faster multi-person pose estimation model. In Proceedings of the Computer Vision—ECCV 2016: 14th European Conference, Amsterdam, The Netherlands, 11–14 October 2016; Part VI 14; pp. 34–50.
28. Zhao, H.; Tian, M.; Sun, S.; Shao, J.; Yan, J.; Yi, S.; Tang, X. Spindle net: Person re-identification with human body region guided feature decomposition and fusion. In Proceedings of the IEEE Conference on Computer Vision and Pattern Recognition, Honolulu, HI, USA, 21–26 July 2017; pp. 1077–1085.
29. Li, D.; Chen, X.; Zhang, Z.; Huang, K. Learning deep context-aware features over body and latent parts for person re-identification. In Proceedings of the IEEE Conference on Computer Vision and Pattern Recognition, Honolulu, HI, USA, 21–26 July 2017; pp. 384–393.
30. Jaderberg, M.; Simonyan, K.; Zisserman, A. Spatial transformer networks. In Proceedings of the Advances in Neural Information Processing Systems, Montreal, QC, Canada, 7–12 December 2015; Volume 28.
31. Liu, T.; Cai, Q.; Xu, C.; Zhou, Z.; Ni, F.; Qiao, Y.; Yang, T. Rumor Detection with a novel graph neural network approach. *arXiv* **2024**, arXiv:2403.16206.
32. Yao, A.; Jiang, F.; Li, X.; Dong, C.; Xu, J.; Xu, Y.; Liu, X. A novel security framework for edge computing based uav delivery system. In Proceedings of the 2021 IEEE 20th International Conference on Trust, Security and Privacy in Computing and Communications, Shenyang, China, 20–22 October 2021; pp. 1031–1038.
33. Tong, K.W.; Wu, J.; Hou, Y.H. Robust Drogue Positioning System Based on Detection and Tracking for Autonomous Aerial Refueling of UAVs. *IEEE Trans. Autom. Sci. Eng.* **2023**. [CrossRef]
34. Wang, C.Y.; Bochkovskiy, A.; Liao, H.Y.M. YOLOv7: Trainable bag-of-freebies sets new state-of-the-art for real-time object detectors. In Proceedings of the IEEE/CVF Conference on Computer Vision and Pattern Recognition, Vancouver, BC, Canada, 17–24 June 2023; pp. 7464–7475.
35. Liu, R.; Xu, X.; Shen, Y.; Zhu, A.; Yu, C.; Chen, T.; Zhang, Y. Enhanced Detection Classification via Clustering SVM for Various Robot Collaboration Task. *arXiv* **2024**, arXiv:2405.03026.
36. Liu, T.; Xu, C.; Qiao, Y.; Jiang, C.; Yu, J. Particle Filter SLAM for Vehicle Localization. *arXiv* **2024**, arXiv:2402.07429.
37. Ru, J.; Yu, H.; Liu, H.; Liu, J.; Zhang, X.; Xu, H. A Bounded Near-Bottom Cruise Trajectory Planning Algorithm for Underwater Vehicles. *J. Mar. Sci. Eng.* **2022**, *11*, 7. [CrossRef]
38. Weng, Y. Big data and machine learning in defence. *Int. J. Comput. Sci. Inf. Technol.* **2024**, *16*, 25–35. [CrossRef]
39. Lan, X.; Ye, M.; Zhang, S.; Yuen, P. Robust collaborative discriminative learning for RGB-infrared tracking. In Proceedings of the AAAI Conference on Artificial Intelligence, New Orleans, LA, USA, 2–7 February 2018; Volume 32.
40. Liu, J.; Wang, G.; Jiang, C.; Liu, Z.; Wang, H. Translo: A window-based masked point transformer framework for large-scale lidar odometry. In Proceedings of the AAAI Conference on Artificial Intelligence, Washington, DC, USA, 7–14 February 2023; Volume 37, pp. 1683–1691.
41. Gray, D.; Tao, H. Viewpoint invariant pedestrian recognition with an ensemble of localized features. In Proceedings of the Computer Vision—ECCV 2008: 10th European Conference on Computer Vision, Marseille, France, 12–18 October 2008; Part I 10; pp. 262–275.
42. Ren, S.; He, K.; Girshick, R.; Sun, J. Faster R-CNN: Towards real-time object detection with region proposal networks. *IEEE Trans. Pattern Anal. Mach. Intell.* **2016**, *39*, 1137–1149. [CrossRef]
43. He, K.; Zhang, X.; Ren, S.; Sun, J. Deep residual learning for image recognition. In Proceedings of the IEEE Conference on Computer Vision and Pattern Recognition, Las Vegas, NV, USA, 27–30 June 2016; pp. 770–778.
44. Sun, Y.; Zheng, L.; Yang, Y.; Tian, Q.; Wang, S. Beyond part models: Person retrieval with refined part pooling (and a strong convolutional baseline). In Proceedings of the European Conference on Computer Vision, Munich, Germany, 8–14 September 2018; pp. 480–496.
45. Yin, Q.; Wang, G.A.; Ding, G.; Li, Q.; Gong, S.; Tang, Z. Rapid Person Re-Identification via Sub-space Consistency Regularization. *Neural Process. Lett.* **2023**, *55*, 3149–3168. [CrossRef]

drones

MDPI

Article

A Dynamic Visual SLAM System Incorporating Object Tracking for UAVs

Minglei Li [1,*,†], Jia Li [1,†], Yanan Cao [1] and Guangyong Chen [2]

1 College of Electronic and Information Engineering, Nanjing University of Aeronautics and Astronautics, Nanjing 211106, China; lijia0131@nuaa.edu.cn (J.L.); caoyanan@nuaa.edu.cn (Y.C.)
2 Chinese Aeronautical Radio Electronics Research Institute, Shanghai 200241, China; chengy088@avic.com
* Correspondence: minglei_li@nuaa.edu.cn
† These authors contributed equally to this work.

Abstract: The capability of unmanned aerial vehicles (UAVs) to capture and utilize dynamic object information assumes critical significance for decision making and scene understanding. This paper presents a method for UAV relative positioning and target tracking based on a visual simultaneousocalization and mapping (SLAM) framework. By integrating an object detection neural network into the SLAM framework, this method can detect moving objects and effectively reconstruct the 3D map of the environment from image sequences. For multiple object tracking tasks, we combine the region matching of semantic detection boxes and the point matching of the optical flow method to perform dynamic object association. This joint association strategy can prevent trackingoss due to the small proportion of the object in the whole image sequence. To address the problem ofacking scale information in the visual SLAM system, we recover the altitude data based on a RANSAC-based plane estimation approach. The proposed method is tested on both the self-created UAV dataset and the KITTI dataset to evaluate its performance. The results demonstrate the robustness and effectiveness of the solution in facilitating UAV flights.

Keywords: visual SLAM; UAVs; multiple object tracking; dynamic objects

Citation: Li, M.; Li, J.; Cao, Y.; Chen, G. A Dynamic Visual SLAM System Incorporating Object Tracking for UAVs. *Drones* **2024**, *8*, 222. https://doi.org/10.3390/drones8060222

Academic Editors: Dongdong Li, Gongjian Wen, Yangliu Kuai and Runmin Cong

Received: 25 March 2024
Revised: 16 May 2024
Accepted: 23 May 2024
Published: 29 May 2024

1. Introduction

Unmanned aerial vehicles (UAVs) have been used in diverse domains, such asogistics, rescue operations, and wildlife protection. Enhancing the visual perception capabilities of UAVs is essential for robust navigation in some challenging flight scenarios. The UAVs equipped with cameras can capture images of surroundings, which provide motion cues for trajectory estimation and 3D mapping. This visual perception approach primarily relies on the utilization of simultaneousocalization and mapping (SLAM) or visual odometry (VO) technologies [1]. While there are various visual SLAM frameworks available [2–6], their direct application to UAV navigation applications often overlooks the presence of dynamic objects within the environment. Thisimitation hampers their applicability in real-world scenarios, where the detection, tracking, and mapping of dynamic objects are crucial for the safety of UAVs [7]. Therefore, there is a need for specialized algorithms that can effectively handle dynamic objects in visual SLAM systems for UAVs.

Over the past decade, significant attention has been devoted to addressing the challenge of handling dynamic objects in SLAM algorithms. Traditional approaches employ two main strategies: (1) detecting moving regions within the scene and disregarding these regions [8–11]; (2) synthesizing plausible color, texture, and geometry in regions occluded by dynamic objects during image stream processing [12,13]. Both strategies result in the exclusion of information about dynamic objects, leading to the generation of static-only maps. Most recently, certain researchers have adopted a different perspective by integrating dynamic object tracking into the SLAM problem [14]. By taking into account the dynamics of moving objects, these approaches strive to go beyond static mapping andocalization, aiming to improve the overall understanding of the environment.

Despite the numerous efforts aimed to enhance the capabilities of visual SLAM by incorporating the detection and tracking of dynamic objects, there is still a significant gap in the field of UAV navigation. A common problem is that the UAV-borne monocular camera oftenacks the ability to restore real scale information, making it challenging to estimate the actual speed of dynamic objects. In addition, objects captured in UAV-borne images often exhibit sparsity and uneven distribution, which consequently increases the probability of missed detections. In this paper, we present a monocular visual SLAM algorithm explicitly designed for UAVs, which aims to achieve efficient 3D mapping and target tracking and positioning. The scale recovery method enables converting the semantic detection results into meaningful geometric motion results of objects, which can provide target motion parameters with actual physical quantities. Our work was inspired by the VDO-SLAM method. However, one innovation is its ability to estimate object motion models for UAV-borne images, and the proposed method can obtain motion parameters of the targets with real physical scale, which cannot be solved by VDO-SLAM or ORB-SLAM2.

Indeed, we proposed two innovations:

(a) Combining object-wise matching and point-wise matching to track dynamic objects. It solves the problem of tracking instability caused by small target pixel regions and is of great significance for airborne observation systems.
(b) A new trained network model for UAV datasets. It should be noted that the application scenarios of UAVs are different from traditional vehicle scenarios, and simple combinations cannot fully solve such problems. So, we trained a network model suitable for UAV datasets and achieved success through experimental testing.

The proposed SLAM algorithmeverages the random sample consensus (RANSAC) method [15] to estimate and restore scale information by fitting a ground plane. Both object-wise matching and point-wise matching are employed within the algorithm to achieve joint tracking of dynamic objects. Object-wise matching enables efficient and rapid tracking of dynamic objects, while point-wise matching addresses missed detections from the object detection network. Consequently, the final map constructed encompasses both dynamic objects and static environments.

This paper is organized into the following sections. Next, Section 2 provides a comprehensive review of related work in this field. Section 3 outlines the methodology employed in our study. The experimental setup is presented in Section 4, followed by the results and evaluations. Finally, Section 5 summarizes and presents concluding remarks.

2. Related Work

The visual SLAM algorithms applied to UAV flight include several steps, covering a range of research topics. To provide a thorough understanding of the background, we present a review of theiterature in Sections 2.1 and 2.2, covering visual SLAM and dynamic object tracking, respectively. Furthermore, in Section 2.3, we discuss the existing technologies for UAV systems.

2.1. Dynamic Visual SLAM

Early pioneering approaches in visual SLAM are mainly pure feature-based methods. They relied on extracting and matching distinctive features in the images to estimate the cameras' poses relative to the world coordinate system, such as MonoSLAM [16], PTAM [17], ORB-SLAM [4], and ORB-SLAM2 [6]. Inheriting the framework of ORB-SLAM2, subsequent SLAM systems commonly comprise three distinct threads: (1) a tracking thread, responsible for tracking point-wise features (i.e., ORB features [18]) and estimating poses; (2) a mapping thread, which constructs aocal 3D map and eliminates redundant keyframes; and (3) aoop closing thread, which corrects the accumulated drift and performs global optimization. This design enables the algorithms to operate continuously for extended periods inarge-scale scenes with significantoops, ensuring global consistency of the trajectory and the map. Benefiting from the efficiency of this design, many new methodologies [10,11,19–22] are integrated and tested on the widely used ORB-SLAM2 frameworks. The selection of keyframes is important to the system's performance by maintaining good accuracy and robustness [23]. The understanding of dynamic scenes is generally based on keyframes. In many applications, prior knowledge is of great significance for understanding dynamic scenes. However, unlike someiDAR-based SLAM systems [24,25], a purely visual SLAM system cannot directly obtain true physical scale information. Thisack of scale informationimits the use of prior knowledge such as geometric and motion models in the mapping and object tracking algorithms.

With the effectiveness of neural networks, some SLAM algorithms have been proposed to enhance performance in dynamic environments. One such algorithm is Detect-SLAM [19], which incorporates SSD-NET [26] for dynamic object detection within the SLAM pipeline. This algorithm updates the motion probability of feature points in each frame by employing feature matching and neighboring points, thereby capturing the motion of all feature points. Similarly, DynaSLAM [10] leverages Mask R-CNN [27] for semantic segmentation of dynamic objects, and it uses a multi-view geometric method to evaluate the reliability of matched features. Subsequently, Li et al. [21] propose a DP-SLAM algorithm, which integrates the outcomes of geometry constraints and semantic segmentation within a Bayesian probability estimation framework, enabling the tracking of dynamic key points.

The aforementioned SLAM algorithms all utilize the semantic information provided by deepearning to improve system stability. By combining semantic information to detect dynamic objects, these algorithms could differentiate between the static and moving elements in the scene, allowing for more accurate camera pose estimation and map construction. Nevertheless, these methods do not address the challenges related to the positioning of dynamic objects or the restoration of scale in monocular visual mapping.

2.2. Object Tracking in Visual SLAM

The traditional method to solve 3D multi-object tracking is to perform SLAM and multiple object tracking (MOT) separately [28–32]. Notably, Wangsiripitak and Murray [29] present a parallel implementation of monoSLAM with a 3D object tracker, where monocular SLAM supplies the tracker with camera pose information, restoring occluded features and preventing SLAM from utilizing features of dynamic objects. On the other hand, the bearing only tracking (BOT) algorithm [30] aims to reconstruct the motion of dynamic points from a monocular camera and build a 3D dynamic map that encompasses both static structures and the trajectories of moving objects. In a subsequent study [31], a multi-layer dense conditional random field (CRF) is used for motion segmentation and object classabeling. This model incorporates semantic constraints enhancing 3D reconstruction. DYNSLAM [32] is a stereo-based dense mapping algorithm that utilizes sparse scene flow to estimate the 3D motions of detected moving objects. This approach enables the reconstruction of the static background, dynamic objects, and potentially moving but currently stationary objects inarge-scale dynamic urban environments. The limited field of view (FoV) of the camera may cause tracking failure due to sudden changes in perspective or textureless scenes. Fish-eye or panoramic cameras become an alternative [33]. However, these complex camera models increase the tedious work of data calibration and are prone to the calculation error of epipolar geometry.

Recent approaches [20,34–36] try to solve the two problems of SLAM and MOT in a unified framework. Among them, ClusterSLAM [34], as a general SLAM backend, can simultaneously cluster rigid bodies and estimate their motions. Since it is only the backend of the SLAM system, its performance depends on the quality ofandmark tracking and correlation from the front end. Dynamic SLAM [35] exploits semantic segmentation to estimate the motion of rigid objects and generates a map of dynamic and static structures without having any prior knowledge of their 3D models. This method is applied to RGB-D/stereo images, so the authorsater propose a new VDO-SLAM system [36] to explore depth information from a single image. VDO-SLAMeverages semantic information and dense optical flow to achieve accurate motion estimation and tracking of dynamic objects. Similarly, DynaSLAM II [20] utilizes instance semantic segmentation and ORB features for dynamic object tracking. Given these advancements, it is now feasible and applicable to integrate MOT with SLAM for dynamic scene exploration.

2.3. Visual Navigation for UAVs

UAVs equipped with visual navigation systems canocate themselves in GPS-denied areas, which helps them explore unknown environments and avoid obstacles. In general, visual navigation systems can be categorized into map-based navigation and mapless navigation.

Map-based navigation relies on pre-stored maps, which are matched with captured images to determine the UAVs' positions [37–40]. Shan et al. [37] employ a method of the histogram of oriented gradient (HOG) for the registration of UAV-borne images with Google Maps. The method relies on a particle filter to expedite the matching process with an onboard sensor. To tackle the problems ofarge differences in scale and rotation, Zhuo et al. [38] propose an image-matching approach, consisting of a dense feature detection step, a one-to-many matching strategy, and a global geometric verification step. This method requires initial poses from GNSS/IMU to eliminate scale differences in the images. Whenocating a UAV in a wide area, semantic object-based matching [39,40] is sometimes more reliable than feature point-based matching. The algorithms detect the objects in the airborne image by machineearning methods and use the configuration of the objects to find the correspondingocation in the map database.

However, accurate maps are not always available [41], especially in some emergency situations. Consequently, mapless visual navigation approaches, such as SLAM-based algorithms, become more appealing. Qin and Shen [42] present a tightly coupled monocular visual-inertial system (VINS) estimator that enables the autonomous flight of a rotorcraft micro aerial vehicle (MAV) in unknown and unstructured environments. The approach optimizes a fixed history of vehicle states as well as environment features using nonlinear optimization. Subsequently, VINS-Mono [43] is proposed based on this work. The system uses a tightly coupled, nonlinear, optimization-based method to obtain high accuracy visual-inertial odometry by fusing pre-integrated IMU measurements and feature observations. It is successfully applied to medium-scale drone navigation tasks. Fu et al. [44] present a PL-VINS method, which efficiently makes use ofine features to improve the performance of the VINS-Mono. However, these algorithms have not taken into account the presence of moving objects, whichimits the system's wider applicability.

3. Proposed Method

3.1. Overview

The proposed visual SLAM algorithm for UAVs is built upon the ORB-SLAM2 framework [6], which incorporates an object tracking module for avoiding obstacles. It takes images captured by a downward-looking camera on the UAV as input and generates the poses of the camera and dynamic objects along with a map. An overview of the algorithm is presented in Figure 1. Integrating new methodology on the widely used SLAM system, such as ORB-SLAM2, is not a trivial task. In addition to the conventional mapping and positioning steps in a visual SLAM system, our method comprises three main components: image pre-processing, map scale recovery, and object tracking.

The method takes a sequence of images as input. To effectively utilize semantic information, we employ a single-image depth estimation method based on NeW CRFs [45] to derive depth information from the image sequence. Pre-processing of input images involves generating object detection boxes, depth maps, and dense optical flow. We employ two different methods to extract key points for different regions in the images. For static regions, we extract ORB features and calculate depth through a triangulation algorithm. ORB features are also used for the SLAM process, which calculates the camera poses and sparse map points. For regions that potentially contain movable objects (such as pedestrians and vehicles), we directly sample the area at every two points and acquire depth from the depth map. A potential ground plane is fitted, and we calculate the ratio between the distance of the ground plane and the height provided by a barometer to restore the scale of the model. For object motion tracking, we use the Kalman filter [46] to predict the detection boxes of objects and match them with the detection boxes of the target detector to track the detection boxes. Through optical flow, we associate the sampling points in the detection box and estimate the object pose. Finally, the method outputs a static map as well as trajectories of the camera and dynamic objects.

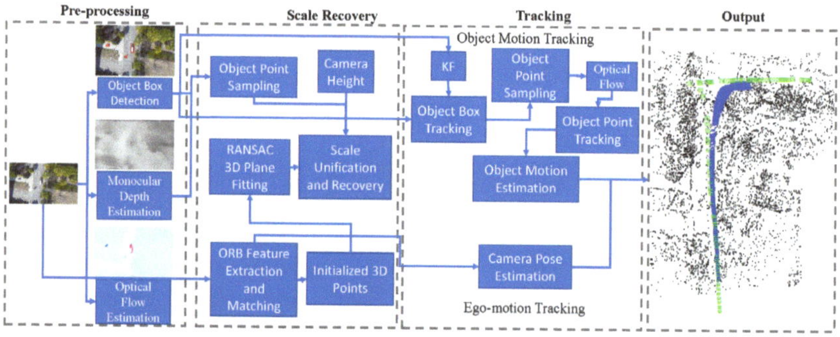

Figure 1. Overview of the visual SLAM method.

3.2. Pre-Processing Module

The pre-processing module faces two challenging problems. Firstly, it needs to effectively distinguish between the static background and the dynamic foreground. Then, it needs to ensure the tracing of dynamic objects over extended periods. When the UAV's camera is used for capturing images, the small proportion of the target within the entire image area poses difficulties in extracting and matching an adequate number of feature points. To overcome this challenge, we utilize recent advancements in computer vision techniques, including monocular depth estimation, object detection, and dense optical flow estimation. These techniques enable accurate dynamic object recognition and stable object tracking. The pre-processing module completes the following three tasks.

(1) Dynamic object detection. Object detection plays a crucial role in identifying dynamic objects within a scene. For instance, buildings and trees are typically static, whereas vehicles may be either stationary or moving. By utilizing object detection results, we can further partition the semantic foreground into distinct areas, thereby facilitating the tracking of individual objects. The dynamic objects in UAV-borne images usually have fewer pixels and are mainly observed from a top-down view.
Compared to pixel-level segmentation, some first-stage object detection networks, such as the YOLO series [47], can offer notable advantages in terms of detection accuracy and speed [48]. Hence, we employ the YOLOv5 network to detect potential dynamic objects and generate object bounding boxes. Our network model used the trained weights from COCO dataset [49] and then fine-tuned them using the VisDrone dataset [50]. A trained deep network model can effectively process UAV-borne images and extract potential dynamic objects.

(2) Monocular depth estimation. Depth estimation facilitates the retrieval of depth information for every pixel in a monocular image, which is crucial for maximizing tracked points on dynamic objects. However, dynamic objects typically occupy only a small portion of UAV-borne images. By employing estimated depth, we can densely sample the monocular images, thereby ensuring stable tracking of moving objects.
We have employed two methods to acquire scene depth. For static regions, we construct sparse maps and calculate the depth map through a triangulation algorithm. For the potential dynamic regions, we derive the depth map from monocular

depth estimation. Specifically, we employ a cutting-edge monocular depth estimation method, i.e., NeW CRFs [45], to calculate the depth map. This method utilizes a novel bottom-up-top-down network architecture and has a significant improvement in the monocular depth estimation. The model is trained on the KITTI Eigen split [51]. The visualization results are shown in Figure 2b.

(3) Optical flow estimation. Dense optical flow provides an alternative approach to establishing feature correspondences by matching sampling points across image sequences, thereby facilitating scene flow estimation. It assists in the consistent tracking of multiple objects, as the optical flow can assign an object recognition marker to each point in the dynamic region and propagate it between frames. This capability becomes particularly valuable in cases where object tracking fails, as dense flow can recover the object area.
We use PWC-Net [52] as the optical flow estimation method. The model is initially trained on the FlyingChairs dataset [53] and subsequently fine-tuned on the Sintel [54] and KITTI training datasets [55]. The visualization results are shown in Figure 2c. The deep network trained in our work can effectively extract the optical flow of targets from drone images. These optical flows form some independent rough contours of objects.

To summarize, in the preprocessing stage, we employed advanced deep network models to achieve some essential tasks, such as depth map estimation, object detection, and optical flow tracking. These network models contribute to extracting valuable information from the input images and enabling subsequent analysis.

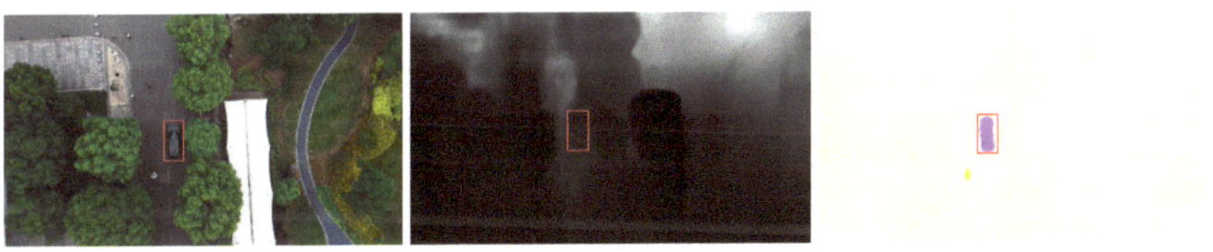

(a) Input image. (b) Depth map. (c) Optical flow.

Figure 2. Visualization results of the pre-processing module.

3.3. Map Scale Restoration

Inheriting the framework of ORB-SLAM2 [6], our SLAM module uses ORB features to reconstruct a sparse environment map. Notably, UAV-borne downward-looking images contain many ground regions, facilitating the fitting of the ground plane from the 3D map points. Assuming the ground is a relatively flat region, the depth values of the ground plane fall within a certain range. To fit the ground plane, our method sorts the sparse map points in ascending order of depth value and selects theowest 40% of points. In practice, we apply the RANSAC-based fitting algorithm to calculate the plane function from the selected points. Then, we use the 2D-pixel positions corresponding to the selected 3D map points to query their depth in the depth map.

The previous calculation can only acquire a reconstructed model scale from monocular images, rather than real physical scale. Therefore, it needs to rely on additional information to restore the true scale. The height h of the camera to the ground plane can be measured using the airborne barometer. It is defined that the camera coordinate system of the first frame is consistent with the world coordinate system. The method computes the ratio between the model distance of the ground plane and the camera's height to restore the scale of the model, as shown in Figure 3.

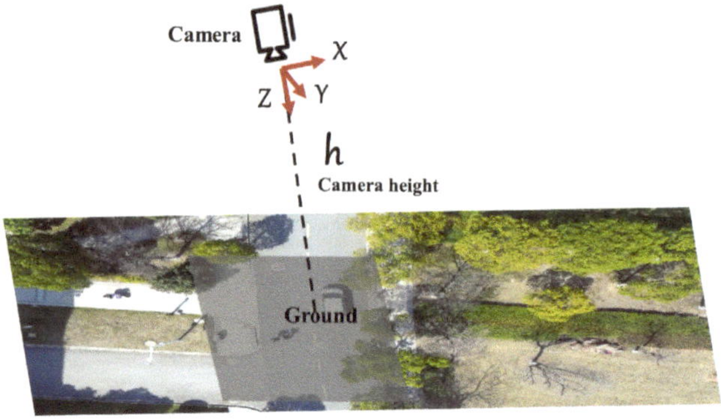

Figure 3. Restore scale information based on height ratio. The grey region in the middle represents a ground plane using a RANSAC-based fitting algorithm.

3.4. Object Tracking and Positioning

In the following, we derive the mathematical calculation process of object tracking. Let $\mathbf{T}^k_{\mathrm{C}_k\mathrm{W}}, \mathbf{T}^k_{\mathrm{O}_k\mathrm{W}} \in \mathbf{SE}(3)$ represent the camera pose and object pose in the world coordinate W at time k, with $k \in \mathcal{T}$ the set of time steps. To distinguish from other symbols, we use calligraphic capitaletters to represent sets of indices. Let $\mathbf{T}^k_{\mathrm{C}_k\mathrm{C}_{k-1}} \in \mathbf{SE}(3)$ be the homogeneous transformation of the camera motion between times $k-1$ and k. In Figure 4, the poses of cameras and objects in the world coordinate are depicted as solid curves, and their relative motion transformations are depicted as dashed curves.

Let $\mathbf{m}^{k,i}_{\mathrm{W}}$ be the homogeneous coordinates of the ith 3D point at time k, with $\mathbf{m}^i_{\mathrm{W}} = \left[m^i_x, m^i_y, m^i_z, 1\right]^{\mathrm{T}} \in \mathbb{R}^4$. The coordinate of a point in camera frame is written as $\mathbf{m}^{k,i}_{\mathrm{C}_k} = \mathbf{T}^k_{\mathrm{C}_k\mathrm{W}} \cdot \mathbf{m}^{k,i}_{\mathrm{W}}$. Define I_k as the image captured by the camera at time k, and let $\mathbf{P}^i_{\mathrm{I}_k} = \left[u^i, v^i, 1\right] \in \mathbb{R}^3$ be the pixelocation on frame I_k corresponding to the homogeneous 3D point $\mathbf{m}^{k,i}_{\mathrm{C}_k}$. The imaging equation is:

$$\mathbf{P}^i_{\mathrm{I}_k} = \lambda \mathbf{K} \cdot \left(\mathbf{T}^k_{\mathrm{C}_k\mathrm{W}} \cdot \mathbf{m}^{k,i}_{\mathrm{W}}\right) = \lambda \mathbf{K} \cdot \mathbf{m}^{k,i}_{\mathrm{C}_k} \tag{1}$$

where $\mathbf{K}$ represents the camera intrinsics. λ indicates that a real physical scale is missing.

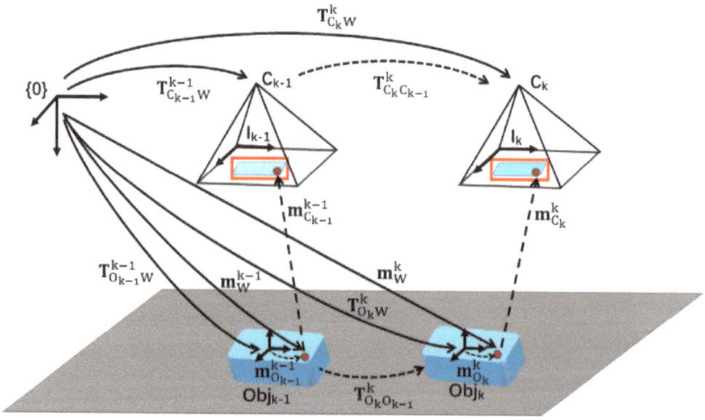

Figure 4. The dynamic object tracking process between airborne images.

Firstly, we need to achieve a spatiotemporal correlation of the same objects, namely object association. The image is divided into possible dynamic regions and static regions, using the semantic information obtained from the previous object detection step. In the static regions, a set of ORB features is extracted and tracked through the feature matching method for camera pose estimation and 3D mapping. Dynamic objects usually occupy a small proportion of UAV images, which makes it difficult to track them for aong time through ORB feature points. We sample every two points within an object region and track them.

The association of dynamic objects across consecutive frames is performed by employing a combined approach. First, we use the intersection over union (IoU) of the detected object boxes [56] to perform the object-wise matching. At the same time, point-wise matching within bounding boxes is conducted by the optical flow between consecutive frames. The combination of object matching and point matching for dynamic object association can be adapted to objects of different sizes and is more robust to occlusion.

For object-wise matching, the Kalman filter is initially employed to predict theocation of tracklets in the new frame. The IoU between the detection boxes and the predicted boxes is then computed as a measure of similarity to associate high-scoring detection boxes with the tracklets. To minimize missed detections and enhance trajectory consistency, we associateow-score detection boxes with unmatched tracklets.

For the point-wise matching, let ${}^{\mathrm{I}_k}\boldsymbol{\phi}^i \in \mathbb{R}^2$ be the optical flow produced by the movement of the camera and objects. It represents the displacement vector of pixel $\mathbf{P}^i_{\mathrm{I}_{k-1}}$ from frame I_{k-1} to I_k, and as follows:

$$ {}^{\mathrm{I}_k}\boldsymbol{\phi}^i = \tilde{\mathbf{P}}^i_{\mathrm{I}_k} - \mathbf{P}^i_{\mathrm{I}_{k-1}} \tag{2} $$

where $\tilde{\mathbf{P}}^i_{\mathrm{I}_k}$ is the correspondence of $\mathbf{P}^i_{\mathrm{I}_{k-1}}$ in I_k. We estimate scene flow based on optical flow, which can be used for dynamic object identification. Firstly, the scene flow $\mathbf{f}^i_k$ of a 3D point $\mathbf{m}^i_{\mathrm{W}}$ can be calculated through the camera pose $\mathbf{T}^k_{\mathrm{C}_k\mathrm{W}}$ as in [57]:

$$ \mathbf{f}^i_k = \mathbf{m}^{k-1,i}_{\mathrm{W}} - \mathbf{m}^{k,i}_{\mathrm{W}} = \mathbf{m}^{k-1,i}_{\mathrm{W}} - {\mathbf{T}^k_{\mathrm{C}_k\mathrm{W}}}^{-1} \cdot \mathbf{m}^{k,i}_{\mathrm{C}_k} \tag{3} $$

Unlike optical flow, scene flow can directly decide whether some structure is moving or not. In theory, the magnitude of the scene flow vector should be zero for all static 3D points. By calculating the scene flow of sampling points in an object to determine whether it is dynamic, if the value of the scene flow of a point is greater than the set threshold, the point is considered dynamic. If the proportion of dynamic points to all points in the object area is greater than the set threshold, the object is judged as a dynamic object.

Then, we predict the motion model of an object. Let $\mathbf{T}^k_{\mathrm{O}_k\mathrm{O}_{k-1}} \in \mathrm{SE}(3)$ describe the homogeneous transformation of the object between times $k-1$ and k, according to:

$$ \mathbf{T}^k_{\mathrm{O}_k\mathrm{O}_{k-1}} = \mathbf{T}^k_{\mathrm{O}_k\mathrm{W}} \cdot {\mathbf{T}^{k-1}_{\mathrm{O}_{k-1}\mathrm{W}}}^{-1} \tag{4} $$

In Figure 4, the above motion transformations are depicted as dashed curves. A point in its corresponding object coordinates is written as $\mathbf{m}^{k,i}_{\mathrm{O}_k} = \mathbf{T}^k_{\mathrm{O}_k\mathrm{W}} \cdot \mathbf{m}^{k,i}_{\mathrm{W}}$, substituting the object pose at time k from Equation (4), this becomes:

$$ \mathbf{m}^{k,i}_{\mathrm{W}} = {\mathbf{T}^k_{\mathrm{O}_k\mathrm{W}}}^{-1} \cdot \mathbf{m}^{k,i}_{\mathrm{O}_k} = {\mathbf{T}^{k-1}_{\mathrm{O}_{k-1}\mathrm{W}}}^{-1} \cdot {\mathbf{T}^k_{\mathrm{O}_k\mathrm{O}_{k-1}}}^{-1} \cdot \mathbf{m}^{k,i}_{\mathrm{O}_k} \tag{5} $$

Note that the relative positions of the points inside the rigid body remain unchanged:

$$ \mathbf{m}^{k,i}_{\mathrm{O}_k} = \mathbf{m}^{k-1,i}_{\mathrm{O}_{k-1}} = \mathbf{T}^{k-1}_{\mathrm{O}_{k-1}\mathrm{W}} \cdot \mathbf{m}^{k-1,i}_{\mathrm{W}} \tag{6} $$

Substituting Equation (6) into Equation (5):

$$ \mathbf{m}^{k,i}_{\mathrm{W}} = {\mathbf{T}^{k-1}_{\mathrm{O}_{k-1}\mathrm{W}}}^{-1} \cdot {\mathbf{T}^k_{\mathrm{O}_k\mathrm{O}_{k-1}}}^{-1} \cdot \mathbf{T}^{k-1}_{\mathrm{O}_{k-1}\mathrm{W}} \cdot \mathbf{m}^{k-1,i}_{\mathrm{W}} \tag{7} $$

Let ${}_{k-1}^{k}\mathbf{T}_{\mathrm{W}} = {\mathbf{T}_{\mathrm{O}_{k-1}\mathrm{W}}^{k-1}}^{-1} \cdot {\mathbf{T}_{\mathrm{O}_k\mathrm{O}_{k-1}}^{k}}^{-1} \cdot \mathbf{T}_{\mathrm{O}_{k-1}\mathrm{W}}^{k-1}$, which represents the motion of the 3D point on a rigid object. The point motion in the global reference frame is then expressed as:

$$\mathbf{m}_{\mathrm{W}}^{k,i} = {}_{k-1}^{k}\mathbf{T}_{\mathrm{W}} \cdot \mathbf{m}_{\mathrm{W}}^{k-1,i} \tag{8}$$

Based on the re-projection error, we solve the object motion ${}_{k-1}^{k}\mathbf{T}_{\mathrm{W}}$ by constructing a cost function. According to Equation (8), the error term is represented as:

$$\mathbf{e}_{repr}^{k,i} = \overset{\sim i}{\mathbf{P}}_{\mathrm{I}_k} - \mathbf{K} \cdot \mathbf{T}_{\mathrm{C}_k\mathrm{W}}^{k} \cdot {}_{k-1}^{k}\mathbf{T}_{\mathrm{W}} \cdot \mathbf{m}_{\mathrm{W}}^{k-1,i} = \overset{\sim i}{\mathbf{P}}_{\mathrm{I}_k} - \mathbf{K} \cdot \mathbf{G}^{k,i} \cdot \mathbf{m}_{\mathrm{W}}^{k-1,i} \tag{9}$$

where $\mathbf{G}^{k,i} = \mathbf{T}_{\mathrm{C}_k\mathrm{W}}^{k} \cdot {}_{k-1}^{k}\mathbf{T}_{\mathrm{W}} \in \mathrm{SE}(3)$. We parameterize the $\mathbf{G}^{k,i}$ by elements of theie-algebra $\mathbf{g}^{k,i} \in \mathrm{se}(3)$:

$$\mathbf{G}^{k,i} = \exp\left(\mathbf{g}^{k,i}\right) \tag{10}$$

The optimal solution is found via minimizing:

$$\mathbf{g}^{k,i*\mathrm{V}} = \underset{\mathbf{g}^{k,i\mathrm{V}}}{\operatorname{argmin}} \sum_{i}^{n_d} \rho_h\left(\mathbf{e}_i^{\mathrm{T}}\left(\mathbf{g}^{k,i}\right) \sum_p^{-1} \mathbf{e}_i\left(\mathbf{g}^{k,i}\right)\right) \tag{11}$$

where n_d represents the number of 3D–2D dynamic point correspondences. Here, ρ_h is the Huber function [58], and $\sum_p$ is the covariance matrix related to the re-projection error. The object motion, ${}_{k-1}^{k}\mathbf{T}_{\mathrm{W}} = {\mathbf{T}_{\mathrm{C}_k\mathrm{W}}^{k}}^{-1} \cdot \mathbf{G}^{k,i}$, can be recovered afterwards. This formulation enables us to jointly optimize the poses of the cameras and the dynamic objects, as well as the 3D map points.

4. Experimental Results

4.1. Experiment Setup

We collected a new dataset of visual monocular data using a UAV, as there are currently no publicly available UAV datasets specifically designed for outdoor scenarios that include dynamic objects. Our dataset aims to fill this gap in the research community by providing a valuable resource for studying and developing methods that address the challenges of dynamic object detection, tracking, and mapping in UAV-based visual systems. The data collection was conducted using the built-in monocular camera of the DJI Mini3 UAV, while the GNSS system provided navigation information. The 6D pose ground truth of the data was obtained through the aero triangulation method based on the photogrammetric software [59]. During data collection, the drone's camera was oriented toward the ground, and the flight altitude ranged between 30 and 50 m. The collected data encompassed dynamic vehicles, pedestrians, as well as static elements such as roads, buildings, and trees. The dataset is available at https://github.com/lemonhi/UAV_dataset/tree/main (accessed on 1 March 2024).

Our method is evaluated in terms of UAVocalization and object tracking performance. The evaluation is performed on our UAV dataset and KITTI tracking dataset [60]. We use UAV data for qualitative analysis of the method and the KITTI dataset for quantitative analysis of the method. It is worth noting that even if some tests are conducted on terrain images of the KITTI dataset, they can give an insight into the general performance of our method. Due to the non-deterministic nature of running the proposed method, such as RANSAC processing, we run the SLAM algorithm five times on each sequence and take median values as the demonstrating results.

As a suggestion from reference [36], we use the translational error E_t (meter) and the rotational error E_r (degree) as evaluation metrics for camera pose and object motion.

4.2. Test on Our UAV Dataset

Figure 5 illustrates the output of the proposed method on our UAV dataset, showcasing a spatiotemporal map that encompasses tracks for each detected dynamic object and camera as well as static map points. The first row presents the satellite maps of the test area (the yellow curves are the UAV's trajectories), while the bottom images show the reconstructed 3D map points (black), the trajectories of the cameras (green), and the traces of the detected objects (blue).

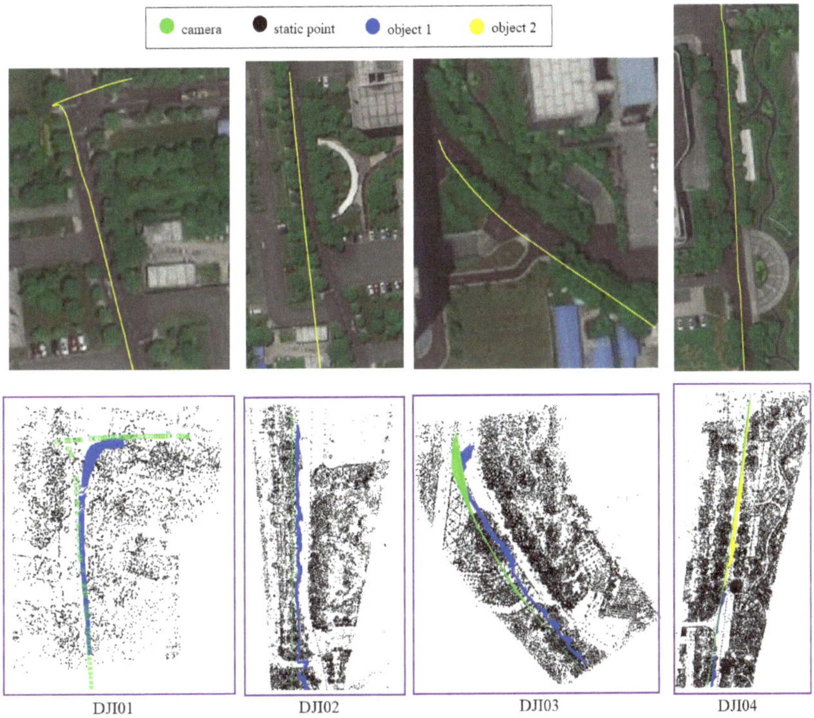

Figure 5. Illustration of the results on the UAV dataset. The first row presents the satellite maps of the test area, while the bottom images show the corresponding maps (2.8 points/m^2) and the trajectories of cameras and objects. (Image numbers and resolution: DJI01, 272 keyframes, 1920 × 1080; DJI02, 340 keyframes, 2720 × 1530; DJI03, 272 keyframes, 2720 × 1530; DJI04, 370 keyframes, 2720 × 1530).

Throughout the UAV flights, our method demonstrates effective tracking capabilities for both the UAV-carried camera and dynamic objects in the surrounding environment. The break of the object trajectories in the DJI01 sequence is due to a missed detection caused by tree shading. When the object is detected again, the method can recover the track and map it correctly.

Figure 6 displays the estimated camera trajectories of two sequences generated by our method, alongside their corresponding ground truth trajectories. Figure 7 displays the error plots for x, y, z separately for both trajectories. The algorithm can provide a suitable estimate for the pose of the UAV. By integrating depth map estimation and optical flow estimation into our tracking framework, it becomes more resilient to occlusion andoss, providing enhanced tracking performance in challenging scenarios. Due to the use of drone barometers as reference heights, there may be a certain gap between the measured height and the actual height, which can cause errors in scale estimation.

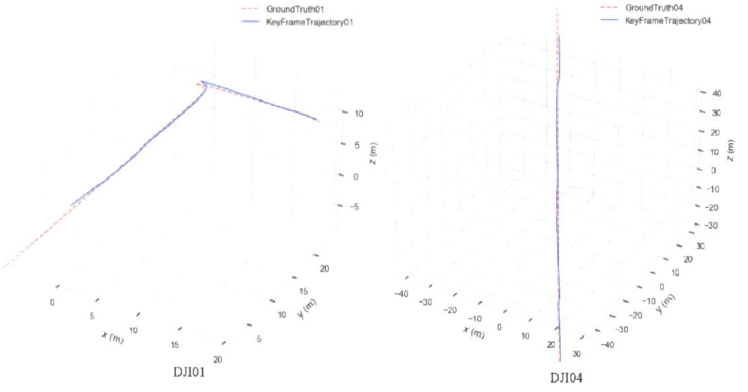

Figure 6. Trajectories on the sequences DJI01 and DJI04.

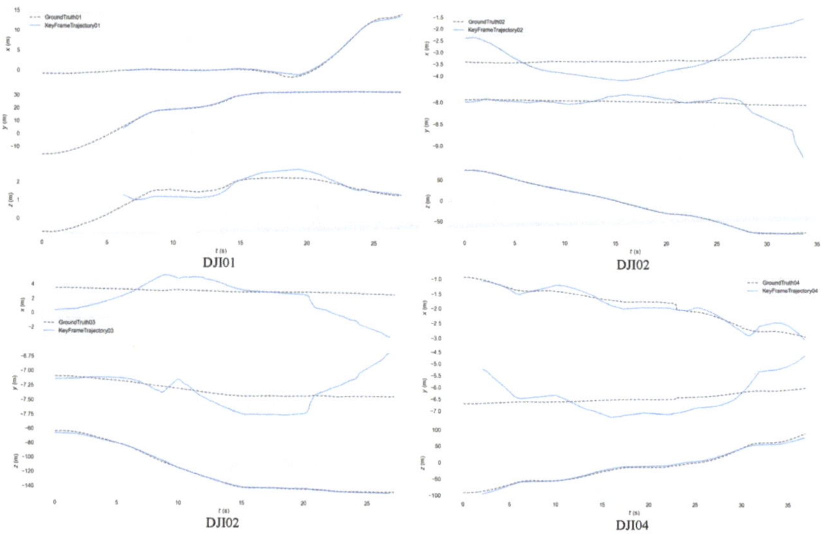

Figure 7. The error plots for x, y, z separately for both trajectories.

4.3. Evaluation on the KITTI Dataset

The KITTI tracking dataset is designed for autonomous driving scenarios, but it can provide a quantitative analysis basis for the validation of drone tracking algorithms. The KITTI tracking dataset contains 21 sequences in total with ground truth for camera poses and object traces. Among these sequences, some are not included in the evaluation of our method, as they contain no obvious dynamic objects. Finally, we chose sequence Seq.00, Seq.01, Seq.02, Seq.03, Seq.04, Seq.05, Seq.06, Seq.18, and Seq.20 as our evaluation data.

(1) Evaluation of the camera poses and object motion. Table 1 shows our results of both camera pose and object trace estimation compared to VDO-SLAM [36] and CubeSLAM [61] on nine image sequences. We directly used the experimental results in the paper for comparison, as we all tested using the same KITTI datasets. The CubeSLAM uses monocular images as the input to the method, while the data tested in the VDO-SLAM system include both monocular and stereo images. As our system is for monocular images, we chose the results of aearning-based monocular version of VDO-SLAM for comparison.

Table 1. Comparison of camera pose and object trace estimation with VDO-SLAM [36] and CubeSLAM [61] on 9 sequences from the KITTI dataset. The bold numbers indicate the best result.

	CubeSLAM [61]				VDO-SLAM [36]				Ours			
	Camera Pose		Object Trace		Camera Pose		Object Trace		Camera Pose		Object Trace	
Seq	E_r(deg)	E_t(m)	E_r(deg)	E_t(m)	E_r(deg)	E_t(m)	E_r(deg)	E_t(m)	E_r(deg)	E_t(m)	E_r(deg)	E_t(m)
00	-	-	-	-	0.1830	0.1847	2.0021	**0.3827**	**0.08240**	**0.08851**	**1.7187**	0.5425
01	-	-	-	-	0.1772	0.4982	**1.1833**	**0.3589**	**0.07378**	**0.1941**	1.4167	0.8396
02	-	-	-	-	0.0496	0.0963	**1.6833**	**0.4121**	**0.03120**	**0.06210**	1.4527	0.6069
03	**0.0498**	**0.0929**	3.6085	4.5947	0.1065	0.1505	**0.4570**	**0.2032**	0.08360	0.1559	1.4565	0.5896
04	0.0708	**0.1159**	5.5803	32.5379	0.1741	0.4951	3.1156	**0.5310**	**0.06888**	0.1755	**2.2280**	0.8898
05	**0.0342**	0.0696	3.2610	6.4851	0.0506	0.1368	**0.6464**	**0.2669**	0.1371	**0.0367**	1.0198	1.0022
06	-	-	-	-	0.0671	0.0451	**2.0977**	**0.2394**	**0.04546**	**0.02454**	2.4642	0.9311
18	0.0433	**0.0510**	3.1876	3.7948	0.1236	0.3551	**0.5559**	**0.2774**	**0.03618**	0.09566	2.1584	0.9624
20	0.1348	**0.1888**	3.4206	5.6986	0.3029	1.3821	**1.1081**	**0.3693**	**0.08530**	0.5838	1.1869	1.2102

The translation error E_t (meter) is computed as the L_2 norm of the translation component of relative pose error. The rotational error E_r (degree) is calculated as the angle of rotation in an axis-angle representation of the rotational component of relative pose error. In comparison with VDO-SLAM, our proposed method demonstrates competitive and high accuracy in estimating camera poses. However, when it comes to object pose estimation, our method exhibits slightly higher errors than VDO-SLAM. We attribute this weaker performance in object pose estimation to the inaccuracy resulting from object detection outcomes. The detection box encompasses a small portion of the static environment, and despite our utilization of the optical flow method for filtering, certain static points are still misclassified as dynamic object points. VDO-SLAM may face challenges when dealing with extensive object occlusion, while our system has a better performance by taking advantage of the optical follow estimation.

Our method has an errorevel similar to CUbeSLAM in camera pose estimation, which may be because we are both based on the ORB-SLAM2 framework. Additionally, our method achieves slightlyower errors in object motion estimation compared to CubeSLAM, perhaps due to theoss of information caused by CubeSLAM in the process of extracting geometric models, thereby introducing uncertainty.

Figure 8 illustrates the output of our method for three of the KITTI sequences. Meanwhile, Figure 9 presents both the output map and the corresponding input image of the method running up to a specific frame within the sequence highlighted in Figure 8. This visual representation provides a clearer depiction of the system's ability to detect and map dynamic objects. From the figures, it can be seen that our method performs relatively robustly inong-distance tracking of dynamic objects.

(2) Evaluation of the object tracking results. The performance of tracking dynamic objects is also demonstrated in our study. Figure 10 displays the results of object trackingength, which shows the selection of objects withonger trajectories. In the majority of sequences, our method achieves object trackingengths of 80% or higher. Notably, objects with trajectoryengths surpassing 200 frames, such as object 32 in Seq.05 and objects 3 and 4 in Seq.18, are successfully tracked by the system for over 80% of their duration. In this paper, optical flow estimation enables the detection and tracking of object motion by tracking pixel-level movement patterns between consecutive frames. This technique can help maintain the continuity of object tracking even in the presence of occlusions or temporaryoss of objects. The limited tracking performance observed in a small number of objects can be attributed to extensive occlusion or a significant distance from the camera.

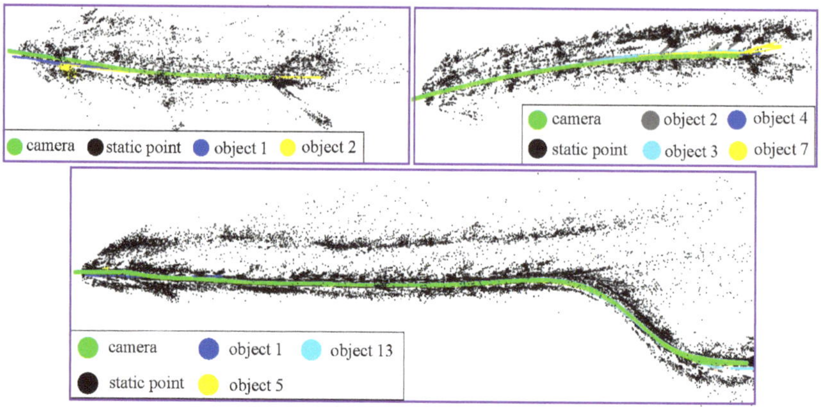

Figure 8. Illustration of the results on the KITTI dataset. (Topeft: Seq.03, top right: Seq.18, and bottom: Seq.20).

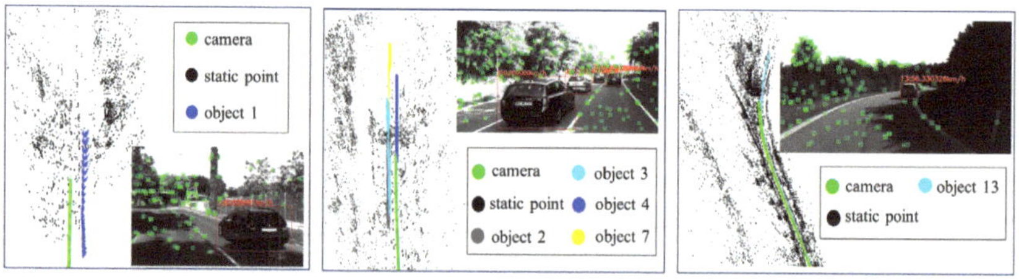

Figure 9. Illustration of system map for a certain frame and corresponding image. The bounding box and the speed of the objects are inferred in the image. The left figure represents Seq.03, the middle figure represents Seq.18, and the right figure represents Seq.20.

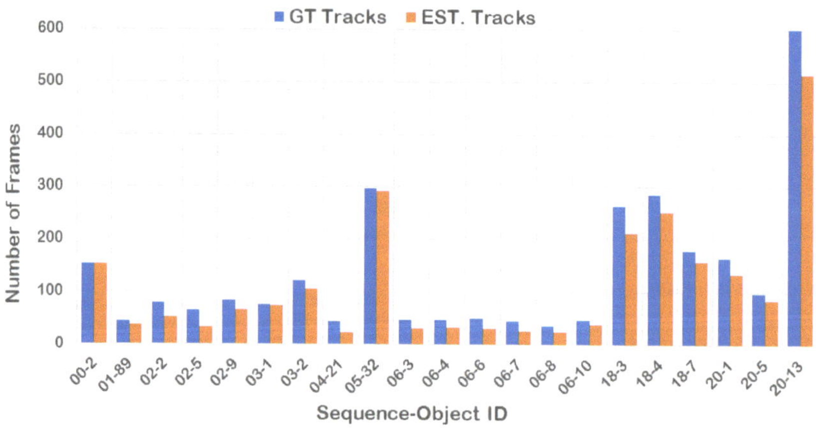

Figure 10. Tracking performance. Results of object trackingength for some selected objects (tracked for over 20 frames) due toimited space. The color bars represent the number of objects appearing in the image. "GT" refers to ground truth and "EST." refers to estimated values."Sequence" represents the sequence number of the KITTI dataset used in the experiment, and "Object id" represents the dynamic object id that appears in the sequence.

4.4. Timing Analysis

All the experiments were conducted on a desktop computer with an Intel Core i5 2.6 GHz CPU and 16 GB RAM. In this paper, the depth estimation and optical flow results are produced offline as input to the system. The timing of our method is highly dependent on the area size and number of detected objects in the scene. In KITTI sequencesike Seq.06, there are only two objects at a time as maximum. and it can thus run at 8 fps. However, Seq.18 can have up to 15 objects at a time, and its performance is seen as slightly compromised, running at 4 fps.

Due to the scale characteristics of UAV images, dynamic objects occupy fewer pixels in the UAV dataset compared to the KITTI dataset. Thus, the proposed method is able to run at the frame rate of 7–10 fps in our UAV datasets, which have resolutions of 1920 × 1080 or 2720 × 1530. The keyframe interval in our pipeline is around 15 frames. We do not include within these numbers the time of the monocular depth estimation and dense optical flow computation since it depends on the GPU power and the model complexity.

Like most frameworks that combine SLAM and dynamic object tracking, our system may encounter scalability issues when the number of dynamic objects in the scene increases significantly. Tracking aarge number of objects simultaneously can be computationally demanding and may impact the real-time performance of the system. As the complexity of the scene increases, the computational requirements mayimit the scalability of our system.

5. Conclusions

In this paper, we present a novel dynamic monocular SLAM method for UAV flight. The proposed approach exploits image-based semantic information to seamlessly integrate object tracking within the SLAM framework, eliminating the need for prior knowledge of object pose or geometry. Depth map estimation and optical flow estimation are designed to enhance target tracking capability, particularly in scenarios involving object occlusion andoss. To evaluate the proposed algorithm, extensive experiments are performed with various UAV-borne image sequences as well as the widely used KITTI dataset. Experimental results show that our method consistently delivers robust and accurate outcomes, particularly excelling in object motion estimation. The estimated motion information of the object can be further used for subsequent tasks, such as path planning and obstacle avoidance. Therefore, our framework has been proven to be suitable for unmanned aerial vehicle visual navigation applications.

Author Contributions: Conceptualization, M.L. and J.L.; methodology, M.L. and J.L.; validation, M.L., J.L., and Y.C.; resources, G.C.; writing, review, and editing, M.L. and J.L. All authors have read and agreed to the published version of the manuscript.

Funding: This research was funded by the National Natural Science Foundation of China under grant number 42271343.

Data Availability Statement: The original data presented in the study are openly available at https://github.com/lemonhi/UAV_dataset/tree/main (accessed on 1 March 2024).

Conflicts of Interest: The authors declare no conflicts of interest.

References

1. Balamurugan, G.; Valarmathi, J.; Naidu, V. Survey on UAV navigation in GPS denied environments. In Proceedings of the 2016 International Conference on Signal Processing, Communication, Power and Embedded System (SCOPES), Paralakhemundi, India, 3–5 October 2016; pp. 198–204.
2. Engel, J.; Schöps, T.; Cremers, D. SD-SLAM:arge-scale direct monocular SLAM. In Proceedings of the 13th European Conference, Zurich, Switzerland, 6–12 September 2014; pp. 834–849.
3. Forster, C.; Pizzoli, M.; Scaramuzza, D. SVO: Fast semi-direct monocular visual odometry. In Proceedings of the 2014 IEEE International Conference on Robotics and Automation (ICRA), Hong Kong, China, 31 May–7 June 2014; pp. 15–22.
4. Mur-Artal, R.; Montiel, J.M.M.; Tardos, J.D. ORB-SLAM: A versatile and accurate monocular SLAM system. *IEEE Trans. Robot.* **2015**, *31*, 1147–1163. [CrossRef]
5. Forster, C.; Zhang, Z.; Gassner, M.; Werlberger, M.; Scaramuzza, D. SVO: Semidirect visual odometry for monocular and multicamera systems. *IEEE Trans. Robot.* **2016**, *33*, 249–265. [CrossRef]
6. Mur-Artal, R.; Tardós, J.D. ORB-SLAM2: An open-source SLAM system for monocular, stereo, and RGB-D cameras. *IEEE Trans. Robot.* **2017**, *33*, 1255–1262. [CrossRef]

7. Saputra, M.R.U.; Markham, A.; Trigoni, N. Visual SLAM and structure from motion in dynamic environments: A survey. *ACM Comput. Surv. (CSUR)* **2018**, *51*, 37. [CrossRef]
8. Li, S.; Lee, D. RGB-D SLAM in dynamic environments using static point weighting. *IEEE Robot. Autom. Lett.* **2017**, *2*, 2263–2270. [CrossRef]
9. Sun, Y.; Liu, M.; Meng, M.Q.H. Improving RGB-D SLAM in dynamic environments: A motion removal approach. *Robot. Auton. Syst.* **2017**, *89*, 110–122. [CrossRef]
10. Bescos, B.; Fácil, J.M.; Civera, J.; Neira, J. DynaSLAM: Tracking, mapping, and inpainting in dynamic scenes. *IEEE Robot. Autom. Lett.* **2018**, *3*, 4076–4083. [CrossRef]
11. Xiao, L.; Wang, J.; Qiu, X.; Rong, Z.; Zou, X. Dynamic-SLAM: Semantic monocular visualocalization and mapping based on deepearning in dynamic environment. *Robot. Auton. Syst.* **2019**, *117*, 1–16. [CrossRef]
12. Bescos, B.; Neira, J.; Siegwart, R.; Cadena, C. Empty cities: Image inpainting for a dynamic-object-invariant space. In Proceedings of the 2019 International Conference on Robotics and Automation (ICRA), Montreal, QC, Canada, 20–24 May 2019; pp. 5460–5466.
13. Bešić, B.; Valada, A. Dynamic object removal and spatio-temporal RGB-D inpainting via geometry-aware adversarialearning. *IEEE Trans. Intell. Veh.* **2022**, *7*, 170–185. [CrossRef]
14. Beghdadi, A.; Mallem, M. A comprehensive overview of dynamic visual SLAM and deepearning: Concepts, methods and challenges. *Mach. Vis. Appl.* **2022**, *33*, 54. [CrossRef]
15. Fischler, M.A.; Bolles, R.C. Random sample consensus: A paradigm for model fitting with applications to image analysis and automated cartography. *Commun. ACM* **1981**, *24*, 381–395. [CrossRef]
16. Davison, A.J.; Reid, I.D.; Molton, N.D.; Stasse, O. MonoSLAM: Real-time single camera SLAM. *IEEE Trans. Pattern Anal. Mach. Intell.* **2007**, *29*, 1052–1067. [CrossRef]
17. Klein, G.; Murray, D. Parallel tracking and mapping for small AR workspaces. In Proceedings of the 2007 6th IEEE and ACM International Symposium on Mixed and Augmented Reality, Nara, Japan, 13–16 November 2007; pp. 225–234.
18. Rublee, E.; Rabaud, V.; Konolige, K.; Bradski, G. ORB: An efficient alternative to SIFT or SURF. In Proceedings of the 2011 International Conference on Computer Vision, Barcelona, Spain, 6–13 November 2011; pp. 2564–2571.
19. Zhong, F.; Wang, S.; Zhang, Z.; Wang, Y. Detect-SLAM: Making object detection and SLAM mutually beneficial. In Proceedings of the 2018 IEEE Winter Conference on Applications of Computer Vision (WACV), Lake Tahoe, NV, USA, 12–15 March 2018; pp. 1001–1010.
20. Bescos, B.; Campos, C.; Tardós, J.D.; Neira, J. DynaSLAM II: Tightly-coupled multi-object tracking and SLAM. *IEEE Robot. Autom. Lett.* **2021**, *6*, 5191–5198. [CrossRef]
21. Li, A.; Wang, J.; Xu, M.; Chen, Z. DP-SLAM: A visual SLAM with moving probability towards dynamic environments. *Inf. Sci.* **2021**, *556*, 128–142. [CrossRef]
22. Morelli, L.; Ioli, F.; Beber, R.; Menna, F.; Remondino, F.; Vitti, A. COLMAP-SLAM: A framework for visual odometry. *Int. Arch. Photogramm. Remote Sens. Spat. Inf. Sci.* **2023**, *48*, 317–324.
23. Azimi, A.; Ahmadabadian, A.H.; Remondino, F. PKS: A photogrammetric key-frame selection method for visual-inertial systems built on ORB-SLAM3. *ISPRS J. Photogramm. Remote Sens.* **2022**, *191*, 18–32. [CrossRef]
24. Jian, R.; Su, W.; Li, R.; Zhang, S.; Wei, J.; Li, B.; Huang, R. A semantic segmentation basedidar SLAM system towards dynamic environments. In Proceedings of the Intelligent Robotics and Applications: 12th International Conference (ICIRA 2019), Shenyang, China, 8–11 August 2019; pp. 582–590.
25. Zhou, B.; He, Y.; Qian, K.; Ma, X.; Li, X. S4-SLAM: A real-time 3DIDAR SLAM system for ground/watersurface multi-scene outdoor applications. *Auton. Robot.* **2021**, *45*, 77–98. [CrossRef]
26. Liu, W.; Anguelov, D.; Erhan, D.; Szegedy, C.; Reed, S.; Fu, C.Y.; Berg, A.C. SSD: Single shot multibox detector. In Proceedings of the 14th European Conference, Amsterdam, The Netherlands, 11–14 October 2016; pp. 21–37.
27. He, K.; Gkioxari, G.; Dollár, P.; Girshick, R. Mask R-CNN. In Proceedings of the 2017 IEEE International Conference on Computer Vision (ICCV), Venice, Italy, 22–29 October 2017; pp. 2961–2969.
28. Wang, C.C.; Thorpe, C.; Thrun, S. Online simultaneousocalization and mapping with detection and tracking of moving objects: Theory and results from a ground vehicle in crowded urban areas. In Proceedings of the 2003 IEEE International Conference on Robotics and Automation (Cat. No.03CH37422), Taipei, Taiwan, 14–19 September 2003; pp. 842–849.
29. Wangsiripitak, S.; Murray, D.W. Avoiding moving outliers in visual SLAM by tracking moving objects. In Proceedings of the 2009 IEEE International Conference on Robotics and Automation, Kobe, Japan, 12–17 May 2009; pp. 375–380.
30. Kundu, A.; Krishna, K.M.; Jawahar, C. Realtime multibody visual SLAM with a smoothly moving monocular camera. In Proceedings of the 2011 International Conference on Computer Vision, Barcelona, Spain, 6–13 November 2011; pp. 2080–2087.
31. Reddy, N.D.; Singhal, P.; Chari, V.; Krishna, K.M. Dynamic body VSLAM with semantic constraints. In Proceedings of the 2015 IEEE/RSJ International Conference on Intelligent Robots and Systems (IROS), Hamburg, Germany, 28 September–2 October 2015; pp. 1897–1904.
32. Bârsan, I.A.; Liu, P.; Pollefeys, M.; Geiger, A. Robust dense mapping forarge-scale dynamic environments. In Proceedings of the 2018 IEEE International Conference on Robotics and Automation (ICRA), Brisbane, QLD, Australia, 21–25 May 2018; pp. 7510–7517.

33. Huang, M.; Wu, J.; Zhiyong, P.; Zhao, X. High-precision calibration of wide-angle fisheyeens with radial distortion projection ellipse constraint (RDPEC). *Mach. Vis. Appl.* **2022**, *33*, 44. [CrossRef]
34. Huang, J.; Yang, S.; Zhao, Z.; Lai, Y.K.; Hu, S.M. ClusterSLAM: A SLAM backend for simultaneous rigid body clustering and motion estimation. In Proceedings of the 2019 IEEE/CVF International Conference on Computer Vision (ICCV), Seoul, Republic of Korea, 27 October–2 November 2019; pp. 5875–5884.
35. Henein, M.; Zhang, J.; Mahony, R.; Ila, V. Dynamic SLAM: The need for speed. In Proceedings of the 2020 IEEE International Conference on Robotics and Automation (ICRA), Paris, France, 31 May–31 August 2020; pp. 2123–2129.
36. Zhang, J.; Henein, M.; Mahony, R.; Ila, V. VDO-SLAM: A visual dynamic object-aware SLAM system. *arXiv* **2020**, arXiv:2005.11052. [CrossRef]
37. Shan, M.; Wang, F.; Lin, F.; Gao, Z.; Tang, Y.Z.; Chen, B.M. Google map aided visual navigation for UAVs in GPS-denied environment. In Proceedings of the 2015 IEEE International Conference on Robotics and Biomimetics (ROBIO), Zhuhai, China, 6–9 December 2015; pp. 114–119.
38. Zhuo, X.; Koch, T.; Kurz, F.; Fraundorfer, F.; Reinartz, P. Automatic UAV image geo-registration by matching UAV images to georeferenced image data. *Remote Sens.* **2017**, *9*, 376. [CrossRef]
39. Volkova, A.; Gibbens, P.W. More robust features for adaptive visual navigation of UAVs in mixed environments: A novelocalisation framework. *J. Intell. Robot. Syst.* **2018**, *90*, 171–187. [CrossRef]
40. Kim, Y. Aerial map-based navigation using semantic segmentation and pattern matching. *arXiv* **2021**, arXiv:2107.00689. [CrossRef]
41. Couturier, A.; Akhloufi, M.A. A review on absolute visualocalization for UAV. *Robot. Auton. Syst.* **2021**, *135*, 103666. [CrossRef]
42. Qin, T.; Shen, S. Robust initialization of monocular visual-inertial estimation on aerial robots. In Proceedings of the 2017 IEEE/RSJ International Conference on Intelligent Robots and Systems (IROS), Vancouver, BC, Canada, 24–28 September 2017; pp. 4225–4232.
43. Qin, T.; Li, P.; Shen, S. VINS-mono: A robust and versatile monocular visual-inertial state estimator. *IEEE Trans. Robot.* **2018**, *34*, 1004–1020. [CrossRef]
44. Fu, Q.; Wang, J.; Yu, H.; Ali, I.; Guo, F.; He, Y.; Zhang, H. PL-VINS: Real-time monocular visual-inertial SLAM with point andine features. *arXiv* **2020**, arXiv:2009.07462. [CrossRef]
45. Yuan, W.; Gu, X.; Dai, Z.; Zhu, S.; Tan, P. New CRFs: Neural window fully-connected CRFs for monocular depth estimation. *arXiv* **2022**, arXiv:2203.01502. [CrossRef]
46. Kalman, R.E. A new approach toinear filtering and prediction problems. *J. Basic Eng.* **1960**, *82D*, 35–45. [CrossRef]
47. Redmon, J.; Divvala, S.; Girshick, R.; Farhadi, A. You onlyook once: Unified, real-time object detection. In Proceedings of the 2016 IEEE Conference on Computer Vision and Pattern Recognition (CVPR), Las Vegas, NV, USA, 27–30 June 2016; pp. 779–788.
48. Zaidi, S.S.A.; Ansari, M.S.; Aslam, A.; Kanwal, N.; Asghar, M.; Lee, B. A survey of modern deepearning based object detection models. *Digit. Signal Process.* **2022**, *126*, 103514. [CrossRef]
49. Lin, T.Y.; Maire, M.; Belongie, S.; Hays, J.; Perona, P.; Ramanan, D.; Dollár, P.; Zitnick, C.L. Microsoft COCO: Common objects in context. In Proceedings of the 13th European Conference, Zurich, Switzerland, 6–12 September 2014; pp. 740–755.
50. Zhu, P.; Wen, L.; Du, D.; Bian, X.; Fan, H.; Hu, Q.; Ling, H. Detection and tracking meet drones challenge. *IEEE Trans. Pattern Anal. Mach. Intell.* **2021**, *44*, 7380–7399. [CrossRef]
51. Eigen, D.; Puhrsch, C.; Fergus, R. Depth map prediction from a single image using a multi-scale deep network. *arXiv* **2014**, arXiv:1406.2283. [CrossRef]
52. Sun, D.; Yang, X.; Liu, M.Y.; Kautz, J. PWC-net: CNNs for optical flow using pyramid, warping, and cost volume. In Proceedings of the 2018 IEEE/CVF Conference on Computer Vision and Pattern Recognition, Saltake City, UT, USA, 18–23 June 2018; pp. 8934–8943.
53. Mayer, N.; Ilg, E.; Hausser, P.; Fischer, P.; Cremers, D.; Dosovitskiy, A.; Brox, T. Aarge dataset to train convolutional networks for disparity, optical flow, and scene flow estimation. In Proceedings of the 2016 IEEE Conference on Computer Vision and Pattern Recognition (CVPR), Las Vegas, NV, USA, 27–30 June 2016; pp. 4040–4048.
54. Butler, D.J.; Wulff, J.; Stanley, G.B.; Black, M.J. A naturalistic open source movie for optical flow evaluation. In Proceedings of the 12th European Conference on Computer Vision, Florence, Italy, 7–13 October 2012; pp. 611–625.
55. Geiger, A.; Lenz, P.; Urtasun, R. Are we ready for autonomous driving? The KITTI vision benchmark suite. In Proceedings of the 2012 IEEE Conference on Computer Vision and Pattern Recognition, Providence, RI, USA, 16–21 June 2012; pp. 3354–3361.
56. Zhang, Y.; Sun, P.; Jiang, Y.; Yu, D.; Weng, F.; Yuan, Z.; Luo, P.; Liu, W.; Wang, X. ByteTrack: Multi-object tracking by associating every detection box. In Proceedings of the 17th European Conference, Tel Aviv, Israel, 23–27 October 2022; pp. 1–21.
57. Lv, Z.; Kim, K.; Troccoli, A.; Sun, D.; Rehg, J.M.; Kautz, J. earning rigidity in dynamic scenes with a moving camera for 3D motion field estimation. In Proceedings of the 15th European Conference on Computer Vision, Munich, Germany, 8–14 September 2018; pp. 468–484.
58. Huber, P.J. Robust estimation of aocation parameter. In *Breakthroughs in Statistics: Methodology and Distribution*; Springer: Berlin/Heidelberg, Germany, 1992; pp. 492–518.
59. Agistoft, LLC. Agisoft Metashape. 2023. Available online: https://www.agisoft.com/zh-cn/downloads/installer (accessed on 1 May 2023).

60. Geiger, A.; Lenz, P.; Stiller, C.; Urtasun, R. Vision meets robotics: The KITTI dataset. *Int. J. Robot. Res.* **2013**, *32*, 1231–1237. [CrossRef]
61. Yang, S.; Scherer, S. CubeSLAM: Monocular 3D object SLAM. *IEEE Trans. Robot.* **2019**, *35*, 925–938. [CrossRef]

drones

MDPI

Article

Event-Assisted Object Tracking on High-Speed Drones in Harsh Illumination Environment

Yuqi Han [1], Xiaohang Yu [2], Heng Luan [3] and Jinli Suo [1,*]

1 Department of Automation, Tsinghua University, Beijing 100084, China; yqhan@mail.tsinghua.edu.cn
2 Tsinghua-UC Berkeley Shenzhen Institute, Shenzhen 518071, China; yuxh21@mails.tsinghua.edu.cn
3 Research and Development Center, TravelSky Technology Ltd., Beijing 101318, China; luanheng@travelsky.com.cn
* Correspondence: jlsuo@tsinghua.edu.cn

Abstract: Drones have been used in a variety of scenarios, such as atmospheric monitoring, fire rescue, agricultural irrigation, etc., in which accurate environmental perception is of crucial importance for both decision making and control. Among drone sensors, the RGB camera is indispensable for capturing rich visual information for vehicle navigation but encounters a grand challenge in high-dynamic-range scenes, which frequently occur in real applications. Specifically, the recorded frames suffer from underexposure and overexposure simultaneously and degenerate the successive vision tasks. To solve the problem, we take object tracking as an example and leverage the superior response of event cameras over a large intensity range to propose an event-assisted object tracking algorithm that can achieve reliable tracking under large intensity variations. Specifically, we propose to pursue feature matching from dense event signals and, based on this, to (i) design a U-Net-based image enhancement algorithm to balance RGB intensity with the help of neighboring frames in the time domain and then (ii) construct a dual-input tracking model to track the moving objects from intensity-balanced RGB video and event sequences. The proposed approach is comprehensively validated in both simulation and real experiments.

Keywords: drones; harsh illumination; image enhancement; event-assisted object tracking; multi-sensor fusion

Citation: Han, Y.; Yu, X.; Luan, H.; Suo, J. Event-Assisted Object Tracking on High-Speed Drones in Harsh Illumination Environment. *Drones* **2024**, *8*, 22. https://doi.org/10.3390/drones8010022

Academic Editor: Dongdong Li, Gongjian Wen, Yangliu Kuai and Runmin Cong

Received: 30 November 2023
Revised: 7 January 2024
Accepted: 15 January 2024
Published: 16 January 2024

1. Introduction

As lightweight, flexible, and cost-effective [1–3] platforms, drones have often been used in a variety of remote tasks, such as surveillance [4,5], detection [6], and delivery [7]. In such applications, drones need to accurately perceive the surrounding environments to support subsequent decisions and actions. In general, common sensors used on UAVs include visible-wavelength optical cameras [8], LiDAR [9], NIR/MIR cameras [10], etc. Each type of sensor has its own advantages and disadvantages, so multi-mode sensing has been the typical solution in this field. Among the various sensors, the visible-wavelength camera is an indispensable sensing unit due to its high resolution, capability of collecting rich information, and low cost of construction.

As one of the most important tasks of a drone, object tracking [11–14] has been widely studied. Broadly speaking, object-tracking algorithms take either the RGB frame as input or its combination with other sensing modes. RGB-only methods [15–18] prevail in frame-based object tracking but are limited in harsh illumination scenarios. Some researchers proposed to incorporate information from event-based cameras, which show superior performance in both low-light and high-dynamic-range scenes. To fuse the information from RGB frames and event sequences, Mitrokin et al. [19] proposed a time-image representation to combine temporal information of the event stream, and Chen et al. [20] improved event representation by proposing a synchronous Time-Surface with Linear Time Decay representation. These approaches exhibit promising performance in object tracking with high time consistency.

However, the above methods are difficult to apply on 24/7 UAVs due to the limited sensing capability of RGB sensors in cases with complex illumination. Because overexposure and underexposure both lead to the image quality degrading greatly and hamper accurate tracking, the reliable drone-based sensing of harshly lit scenes is quite challenging [21,22]. Taking the video in Figure 1 as an example, when capturing a car traveling through a tunnel, there exist a large intensity range in each frame and abrupt variation among different frames; the car is even undetectable in some frames by both tracking algorithms and human vision systems due to underexposure. Fortunately, drones are subjected to continuously varying illumination while in flight, causing recordings with different quality for a target region. Considering that the feature points of neighboring video frames are mostly consistent [23,24], we are inspired to compensate low-quality images with guidance from high-quality counterparts, achieving continuously high-quality videos as well as robust downstream tasks. One of the most crucial problems is to recognize the matching features in adjacent frames that undergo abrupt intensity changes.

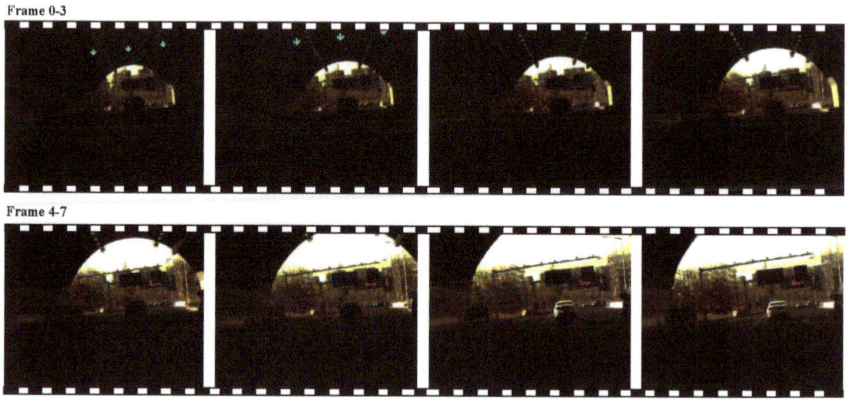

Figure 1. A typical high-dynamic-range RGB video of a car driving through a tunnel, in which the car is almost invisible in the last 5∼6 frames due to underexposure.

To increase the image quality under harsh illumination, researchers have made a lot of explorations in recent years. One most common way is to reconstruct HDR images by merging the set of multi-exposure LDR images [25]. For dynamic scenes, image alignment is required to address the inconsistency among frames with different exposures. Kang et al. [26] initially aligned neighboring frames with the reference frame and merged these aligned images to craft an HDR image. Later works [27,28] modified it by adding a motion estimation block and a refinement stage. Differently, Kalantari et al. [29] proposed a patch-based optimization technique, synthesizing absent exposures within each image before reconstructing the ultimate HDR image. Gryaditskaya et al. [30] enhanced this method by introducing an adaptive metering algorithm capable of adjusting exposures, thereby mitigating artifacts induced by motion. Instead of capturing frames with different exposure times, some methods use deep neural networks to reconstruct the HDR image from a single input image. However, due to relying on a fixed reference exposure, the reconstruction is strongly ill-posed and cannot achieve high between-frame consistency. Additionally, many existing HDR video reconstruction methods focus on developing some special hardware, such as scanline exposure/ISO [31–33], per-pixel exposure [34], modulo camera [35], etc., but these new cameras are still being research and not ready for commercial use in a near future. Some other recent approaches work under the deep-optics scheme and focus on jointly optimizing both the optical encoder and CNN-based decoder for HDR imaging challenges. The above methods usually make assumptions about the lighting conditions, which might not hold in real scenes. Additionally, most of these algorithms need ground-truth high-dynamic-range images for supervised network training and exhibit limited performance in scenes different from the training data. Hence, these methods are enlightening but difficult to be directly applied on practical UAV platforms working in open environments.

The event camera, also known as neuromorphic vision sensor, is an emerging technique that records intensity changes exceeding the threshold asynchronously [36,37]. In recent years, event signals have been used in a variety of high-speed tasks due to their high sensitivity and fast response, such as high-speed tracking [38–41], frame interpolation [42,43], optical flow estimation [44–46], motion detection [47], etc. Unlike conventional optical camera sensors, event cameras output the "events" indicating that there occurs sufficiently large intensity variation at certain positions and instants and also indicate the polarity of the change. Considering that an event camera can record the motion over a large intensity range and is insensitive to abrupt intensity changes, we propose to use event signals to explicitly align the RGB frames and thus compensate for the quality degradation harming the successive object tracking. In other words, with the consistent description of event signals, we enhance low-quality images under guidance from their high-quality counterparts and achieve continuous high-quality scene perception. Specifically, we match the key points occurring at different instants [48] and utilize the matching to balance the intensity change in sequential RGB frames. Afterward, we construct a fusion network to aggregate the enhanced RGB frames and event signals for robust object tracking.

The contributions of this paper are as follows:

- We propose an event-assisted robust object-tracking algorithm working in high-dynamic-range scenes, which successfully integrates the information from an event camera and an RGB camera to overcome the negative impact of harsh illumination on tracking performance. As far as we know, this is the first work of object tracking under harsh illumination using dual-mode cameras.
- We construct an end-to-end deep neural network to enhance the high-dynamic-range RGB frames and conduct object tracking sequentially, and the model is built in an unsupervised manner. According to the quantitative experiment, the proposed solution improves tracking accuracy by up to 39.3%.
- We design an approach to match the feature points occurring at different time instants from the dense event sequence, which guides the intensity compensation in high-dynamic-range RGB frames. The proposed feature alignment can register the key points in high-dynamic-range frames occurring within a 1 s window.
- The approach demonstrates superb performance in a variety of harshly lit environments, which validates the effectiveness of the proposed approach and largely broadens the practical applications of drones.

In the following, we first introduce the framework and algorithm design for the proposed event-assisted object tracking in Section 2, including event-based cross-frame alignment, RGB image enhancement, and dual-mode object tracking. In Section 3, we present the experimental settings, including the datasets and training details. The qualitative results and quantitative results are discussed in Section 4. Further, we present the results for the real-world data as well as the ablation study. Finally, in Section 5, we summarize the paper, discuss the limitation of the proposed solution, and highlight future work to be conducted on efficient collaborative sensing around drones.

2. Framework and Algorithm Design

This section presents the details of the proposed event-assisted robust object-tracking approach working under harsh illumination. Here, we first briefly introduce the framework and then describe the design of three key modules, including the retrieval of feature registration across frames, the enhancement of high-dynamic-range frames, and the successive dual-mode object tracking.

The basic idea of the proposed approach is to utilize the reliable motion cue perception capability of event cameras to prevent the quality degradation of RGB frames and then combine the event signals and the enhanced RGB video to boost the successive tracking performance suffering from overexposure and underexposure. The whole framework of the proposed event-assisted object-tracking approach is shown in Figure 2; it consists of mainly three key modules:

(i) Retrieving the motion trajectories of key feature points from the dense event sequence. We divide the event sequence into groups occurring in overlapping, short time windows, and the key points from Harris corner detection in each event group can construct some motion trajectories. Further, we integrate these short local trajectories to figure out the motion over a longer period across the RGB frames, even under harsh illumination.

(ii) Enhancing the high-dynamic-range RGB frame according to inter-frame matching and information propagation. Based on the matching among feature points across frames, we build a deep neural network to compensate for the overexposed or underexposed regions using neighboring frames with higher-visibility reference frames to guide low-visibility objective frames. In implementation, we build a U-Net-based neural network for image enhancement.

(iii) Tracking the target objects by fusing information from both RGB and event inputs. We design a tracking model taking dual-mode inputs to aggregate the information from the enhanced RGB frames and event sequences to locate the motion trajectories. Specifically, we construct 3D CNNs for feature extraction, fuse the features from two arms using the self-attention mechanism, and then employ an MLP to infer the final object motion.

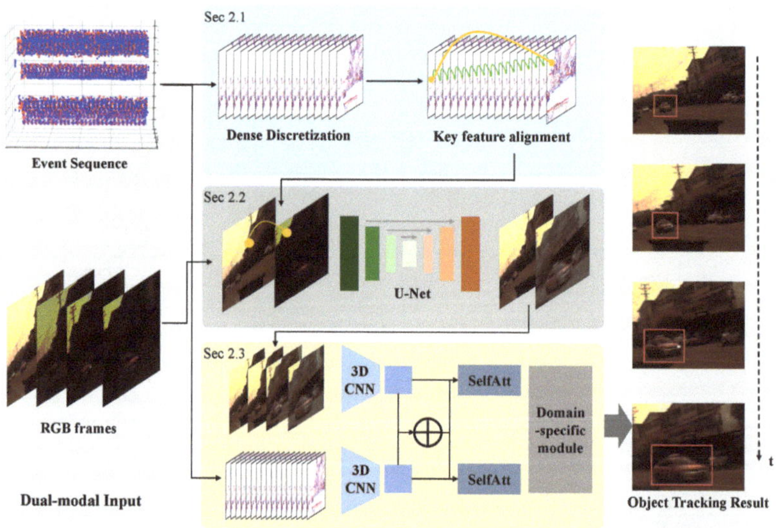

Figure 2. The framework and working flow of the event-assisted robust tracking algorithm under harsh illumination. The whole pipeline is fully automatic and consists of three key steps, with the first one including conventional optimization and the latter two being implemented by deep neural networks.

2.1. *Event-Based Cross-Frame Alignment*

Event-based key feature extraction and matching are conducted here to utilize the stable event signals under harsh illumination for cross-frame alignment of the degraded RGB video, facilitating frame compensation using corresponding positions with decent quality in neighboring frames. We locate the key features of moving objects from the event sequence, as illustrated in Figure 3.

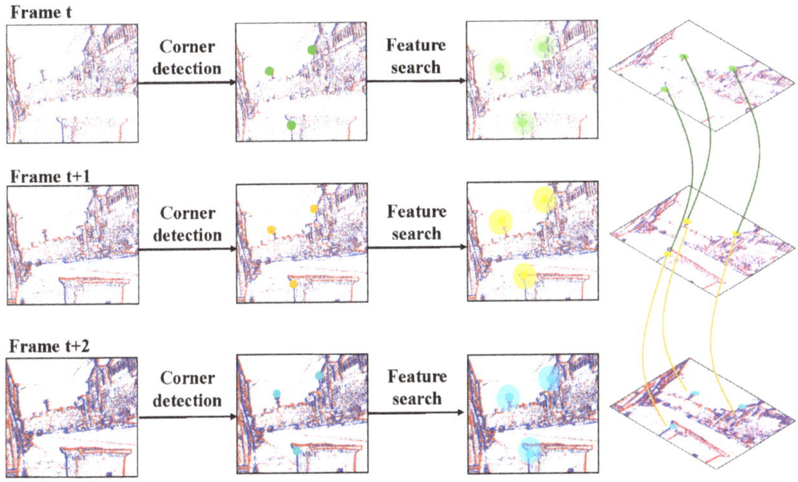

Figure 3. The illustration of event-based key point alignment. We locate the key feature points through Harris detection and search the matching counterparts locally (circular candidate regions are highlighted with different colors) to constitute the motion trajectories, as shown in the right column.

Given a time duration T, we assume that there are N RGB frames and S event signals. We define the captured RGB frames as $\{I_0, I_1, \cdots, I_N\}$, and the corresponding time stamps are defined as $\{T_0, T_1, \cdots, T_N\}$. Event signal s is defined as a quadruple x_s, y_s, t_s, P_s, where x_s, y_s denote the coordinates of s; t_s presents the response time instant; and P_s indicates the polarity of intensity change. Firstly, we divide the S event signals into $K \times N$ groups along the time dimension and project each group into $K \times N$ 2D images, named event frame. We adopt the Harris corner detection algorithm for the above event frames to extract individual key feature points. Further, we align the key feature points at different time instants. Assuming that the shape of the moving objects is fixed within a very short time slot, i.e., the key features in adjacent frames are similar, we construct a small circular search region with radius r around each key feature. In other words, the key feature at the eth frame matches the features inside the searching circle of the $e + 1$th frame.

For the nth RGB frame, we first align the event frames between $n \times S$ and $(n+1) \times S$. From the displacement between the features of multiple event frames, one can construct the moving trajectory of the key event feature points, which reflects the displacement of the corresponding key features in the RGB frame. Naturally, we can eventually infer the position of the corresponding key feature from nth to $n + 1$th RGB frames.

2.2. RGB Image Enhancement

After matching the feature points in different RGB frames, we enhance the underexposed and overexposed regions utilizing the high-visibility counterparts to adjust the intensity and supplement the details. For description simplicity, we define the low-visibility frames as the objective and the high-visibility frames as the reference. To achieve enhancement, there are two core issues to be addressed: (i) how to determine the objective frame that needs to be enhanced; (ii) how to design the learning model to improve the visibility to match the reference frame while preserving the original structure of the objective frame.

We first estimate the visibility of the frames to determine which frames are highly degraded. Intuitively, since harsh illumination leads to local overexposure or underexposure, which is usually of lacking texture, we use information richness to characterize the degeneration degree. In implementation, we define the visibility (V_i) of input RGB image R_i as the difference from its low-pass-filtered version ($\hat{R}_i$), i.e.,

$$V_i = Var(R_i - \hat{R}_i), \tag{1}$$

where $Var(\cdot)$ denotes the variance calculation.

In general, we divide the frames into groups and conduct compensation within each group. We iteratively find the objective frame with the lowest visibility score and the reference frame with the highest visibility and then conduct enhancement. The iteration ends when the number of iterations exceeds a predetermined number P or the difference between the visibility of the target and the reference frame smaller than η. In our experiments, we set $P = 10$ and $\eta = 0.1$.

For enhancement, we designed a U-Net-shaped network structure inferring the enhanced frame from the objective and reference frames, as shown in Figure 4. (The corresponding optimization process is detailed in [49].) The network consists of a three-layer encoder for feature extraction and a three-layer decoder. Skip connections are used to facilitate the preservation of spatial information. The network is trained in an unsupervised manner. We define the loss function based on aligned feature points. Considering that the enhanced frame is expected to be similar to the reference image around the key feature points and close to the original frame at other locations, we define a combinational loss function. To guarantee the former similarity, we minimize the MSE difference, and we use the LPIPS loss for the latter. Denoting the reference image by I_{ref}, the original objective image as I_{obj}, and the output as I_{out}, we define the loss function as

$$L = MSE(I_{\text{ref}}(k) - I_{\text{out}}(k)) + \alpha LPIPS(I_{\text{obj}}(\neg k) - I_{\text{out}}(\neg k)) \tag{2}$$

where k denotes the positions of key features and $\neg k$ denotes the remaining pixels; α is the hyper-parameter, which is set to 0.05 during training.

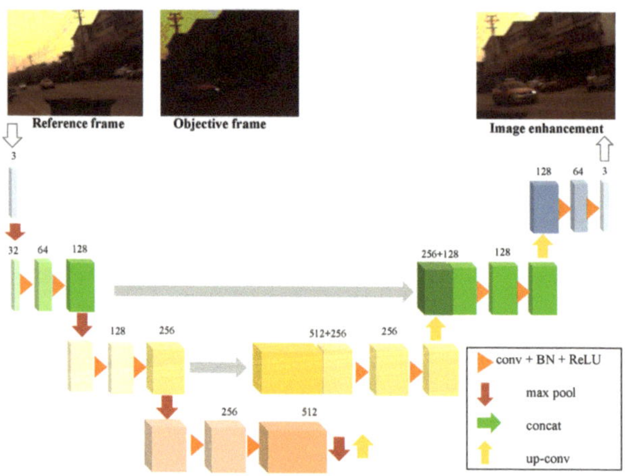

Figure 4. The structure of the RGB image enhancement module. We input the captured RGB frame into the U-Net network, which comprises a three-layer encoder for feature extraction and a three-layer decoder for image enhancement. The network includes skip connections to connect encoder and decoder layers, facilitating the preservation of spatial information. This diagram showcases convolutional, pooling, upsampling, and downsampling layers, with the following key operations: conv denotes convolution; BN denotes batch normalization; ReLU refers to the ReLU activation function; max pool denotes the max pooling operation; concat and Up-conv denote the concatenation and transposed convolution, respectively.

2.3. Dual-Mode Object Tracking

To leverage the motion cues in both the event sequence and the enhanced RGB frames, we construct a dual-mode tracking module for reliable object tracking. The proposed dual-mode tracking module is based on RT-MDNet [50]. The module consists of a shared feature mapping network aiming at constructing the shared representation to distinguish the object from the background and a domain-specific network focusing on domain-independent information extraction. Different from RT-MDNet [50], the proposed dual-mode design focuses on feature fusion from two types of inputs and constructs two self-attention modules to highlight the combinational representation from two individual inputs.

The architecture of the network is shown in Figure 5. We first construct two individual 3D CNNs to extract features from the inputs and output feature vectors of the same size. Subsequently, we concatenate the two feature vectors and use convolution to obtain a combinational representation of the fused features. Subsequently, we construct the self-attention network to retrieve the information underlying independent feature inputs. (Please refer to [51] for the steps of the optimization process.) A two-layer fully connected MLP is used to output the common feature. We refer to RT-MDNet [50] to construct the domain-specific layer afterward, outputting the final tracking results.

During model training, for each detection bounding box, a cross-entropy loss function is constructed to ensure that the target and background are separated as much as possible, and the same also applies to multiple domains. In the latter, fine-tuning stage, we apply different strategies for the first frame and the subsequent ones of a given sequence. For the first frame, we choose multiple bounding boxes following a Gaussian distribution to conduct domain-specific adaption, while for the subsequent frames, we build random samples based on the results from the previous frame and search for the proper bounding box through regression.

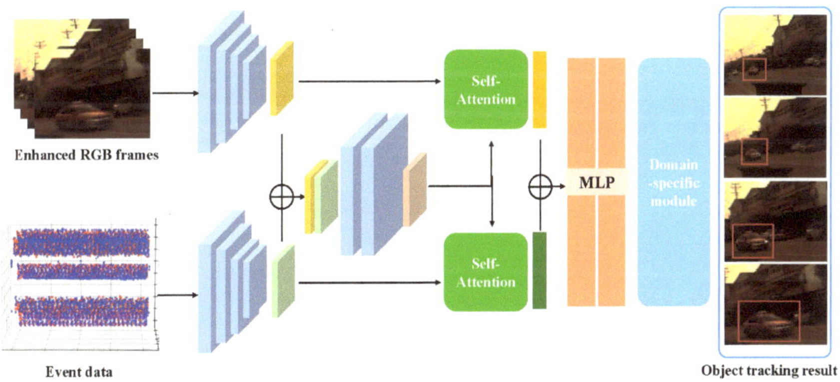

Figure 5. The structure of the object-tracking module. The RGB frames and event sequence are individually fed into two 3D CNN modules for feature extraction, and the extracted features are concatenated and sent to another CNN module for fusion. Then, the individual and fused features are separately sent to the self-attention network. Finally, two MLPs are applied to derive the object detection and tracking results.

3. Experimental Settings

Datasets. We verify the proposed method on both simulated and real datasets. We use VisEvent [52] as the simulated data and mimic harsh illumination by modifying the brightness and contrast of the RGB frames. Specifically, we modify the luminance and contrast as follows: We let the luminance vary linearly, quadratically, or exponentially across the frames, and the image contrast undergoes a linear change with different slopes. We first randomly select 1/3 of the data for luminance modification and then apply contrast modification to 1/3 randomly selected videos. Two examples from the simulated dataset are shown in Figure 6. The first scene mimics the brightness changes in the underexposed scenes, and the second scene simulates overexposure, through modification of image brightness and contrast. One can see that we can generate videos under complex illumination from the original counterpart with roughly uniform illuminance. In the generated high-dynamic-range RGB frames, the textures of some regions are invisible in some frames due to either underexposure or overexposure. In contrast, the contours across the whole field of view are recorded decently.

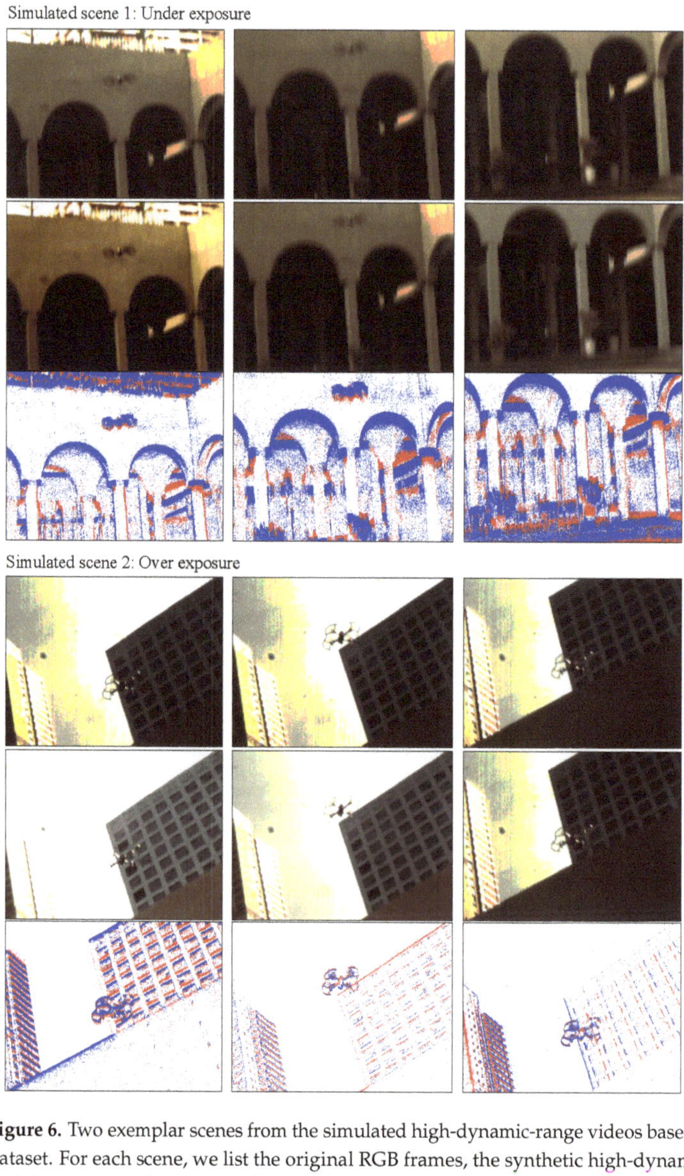

Figure 6. Two exemplar scenes from the simulated high-dynamic-range videos based on the VisEvent dataset. For each scene, we list the original RGB frames, the synthetic high-dynamic-range frames, and the corresponding events from top to bottom. The first scene has a linear increase in intensity and a linear decrease in contrast to mimic underexposure in the 1st frame. The second sequence undergoes linearly decreasing intensity to mimic overexposure in the first frame.

For the real-world data, we captured some typical nighttime traffic scenes with a pair of registered cameras (one RGB and the other events). The scenes consist of complex illumination (e.g., traffic lights, neon signs, etc.) and large intensity variations. From the two exemplar scenes in Figure 7, it can be seen that these scenarios exhibit large illuminance variations and the traffic participants are almost invisible in some frames, due to either underexposure or overexposure. This challenging dataset can be directly used to test the effectiveness of the proposed object-tracking algorithm in real scenarios, as shown in Figure 7.

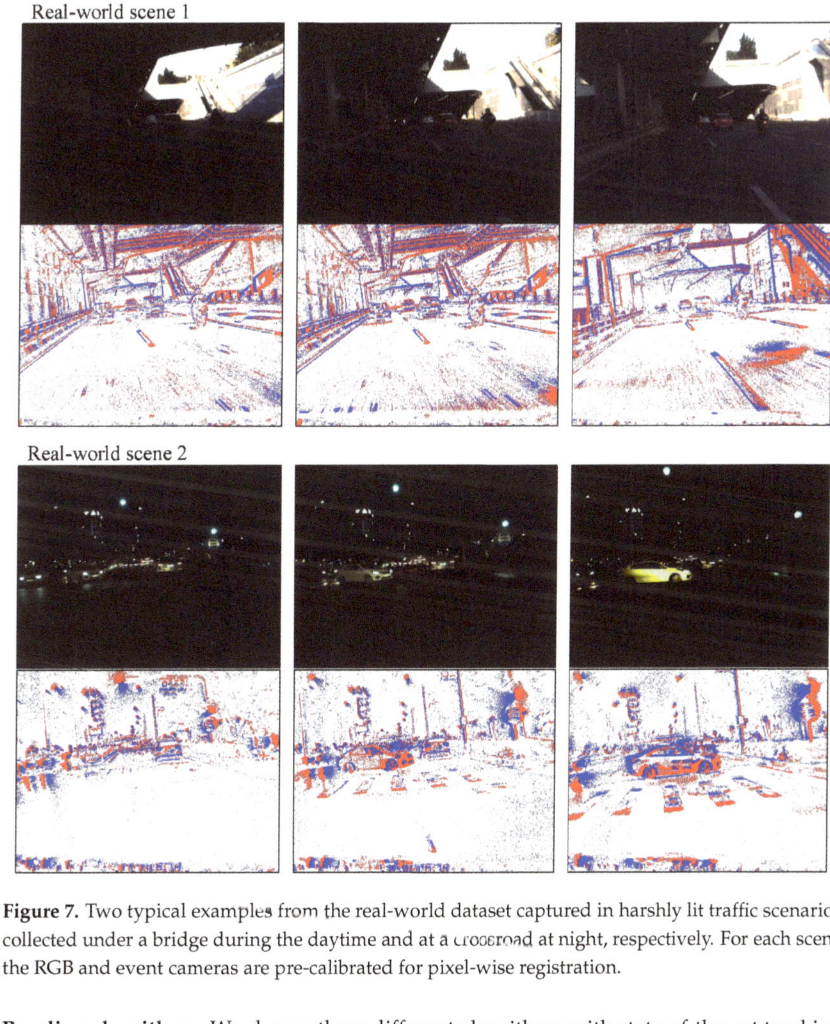

Figure 7. Two typical examples from the real-world dataset captured in harshly lit traffic scenarios, collected under a bridge during the daytime and at a crossroad at night, respectively. For each scene, the RGB and event cameras are pre-calibrated for pixel-wise registration.

Baseline algorithms. We choose three different algorithms with state-of-the-art tracking performance as baselines for the proposed solution, i.e., RT-MDNet [50], Siamrpn++ [53], and VisEvent [52]. RT-MDNet [50] and Siamrpn++ [53] are two RGB-input trackers performing well under normal illumination. So far, there are few objective algorithms specially developed for harsh illumination scenarios; we chose the above two robust and widely used tracking solutions as baselines. VisEvent [52] constructs a two-modality neural network fusing RGB and event signals, and we compare the proposed solution with VisEvent [52] to verify the effectiveness of the image enhancement module under harsh illumination. This benchmark has input similar to our method's and exhibits state-of-the-art performance, serving as a good option to validate the proposed image enhancement module under harsh illumination.

Training. Training is implemented on the NVIDIA 3090 for about 4.7 h. We set the input image size as well as the spatial resolution of the event sequence to 640 $\times$ 480 pixels and seven continuous RGB frames ($\sim$350 ms) for intensity balancing. We use the Adam optimizer, with the learning rate being 5×10^{-4}, the momentum being 0.9, and the weight decay being 5×10^{-4}.

4. Results

In this section, we construct a series of experiments to verify the effectiveness of the proposed method on two tasks in high-dynamic-range scenes: image enhancement and object tracking. We first show the visual and quantitative performance against some baseline algorithms. Also, we give the qualitative results based on the real data to further show the visual difference between the proposed solution and baselines. Finally, we conduct ablation experiments to quantify the contribution of the key module of the algorithms.

4.1. *Results Based on Simulated Data*

In this subsection, we validate our approach in terms of image enhancement and object-tracking accuracy, based on simulated data. Here, we give both qualitative and quantitative experimental results to comprehensively analyze the effectiveness of the proposed solution. For the qualitative results, we show the result of image enhancement first and compare the object-tracking performance with that of the baseline algorithms afterwards. For the quantitative results, we compare the precision plot (PP) and success plot (SP) to assess the tracking performance.

4.1.1. Qualitative Results

Figure 8 shows the qualitative results for an exemplar video from the simulated dataset. The top row shows the raw RGB sequence, with large intensity changes both within and across frames. In this scene, a person runs from a location with strong illumination toward a destination with a large shadow. Due to the extremely dark intensity, it is challenging to recognize their silhouette in the last frame. We enhance the RGB frames according to the temporal matching extracted from the event signals, and the results are shown in the middle row. The enhanced version is of much more balanced intensity and can highlight the human profile even under weak illumination.

We further show the object-tracking result in the bottom row. The bounding boxes of our approach and the other three competitors are overlaid, with different colors. When sufficiently illuminated, all the algorithms can track the object with high accuracy. RT-MDNet, VisEvent, and the proposed algorithm are comparable, while there exists some deviation in the bounding box output by Siamrpn++ tracking. When the light becomes weak, the proposed algorithm can still identify the person's location, while RT-MDNet's and VisEvent's bounding boxes deviate. When the light is extremely weak, only the proposed method, RT-MDNet, and VisEvent can track the object, because of high sensitivity and robustness to abrupt intensity changes in the event signals. In comparison, the RGB image in RT-MDNet and VisEvent is not enhanced and thus reduces the final tracking accuracy, while our approach demonstrates reliable tracking consistently.

4.1.2. Quantitative Results

We introduce the typical matrix PP and SP here to evaluate accuracy in object tracking. Specifically, the PP indicates the frame percentage where the deviation between the estimated object center location and ground truth is less than the determined threshold. The SP denotes the frame percentage where the IoU between the estimated bounding box and the ground-truth bounding boxes is higher than the determined threshold. Table 1 shows the PPs and SPs of our approach and three state-of-the-art object-tracking algorithms.

Since there is no ground truth for the real data, we only conduct quantitative analysis on simulated data.

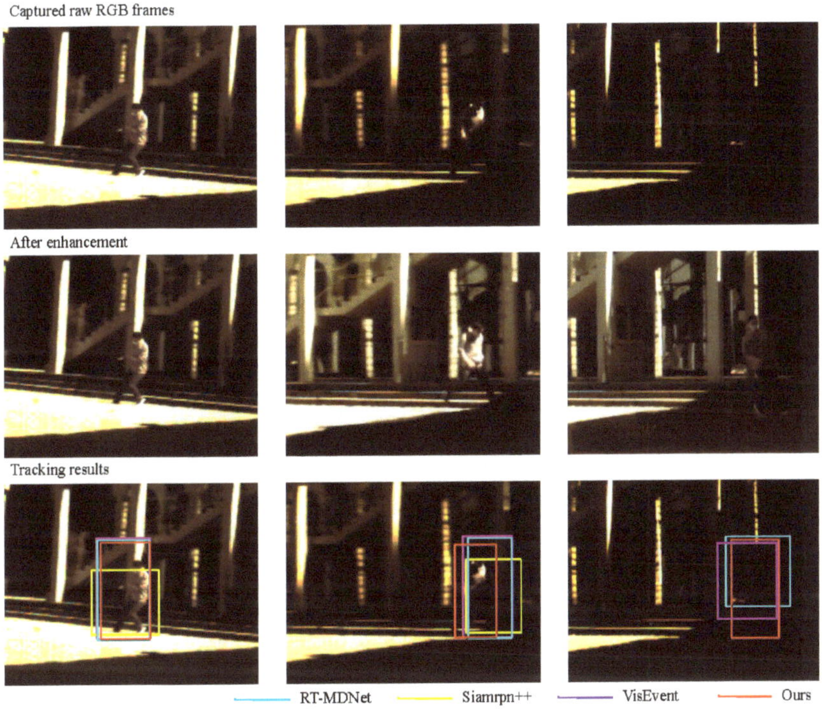

Figure 8. Performance of RGB enhancement in object tracking in a typical exemplar scene in the simulated dataset. (**Top**) The captured RGB video frames. (**Middle**) The corresponding enhanced images obtained with the proposed method. (**Bottom**) The tracking results of different object-tracking algorithms.

Table 1. The quantitative performance of different object-tracking algorithms on the simulated dataset, in terms of PPs and SPs.

	Our Algorithm	VisEvent	Siamrpn++	RT-MDNet
PP	**0.783**	0.712	0.390	0.405
SP	**0.554**	0.465	0.232	0.321

According to Table 1, the proposed algorithm demonstrates the tracking results with the highest accuracy. Even under harsh illumination, we can track the target object continuously, while Siamrpn++ and RT-MDNet show poor tracking results under the same conditions. Moreover, though VisEvent takes the event signal as the input, it ignores the influence of the low-quality RGB frames and produces inferior tracking accuracy. From the ranking, we can draw two conclusions: first, the event signals can help address performance degeneration in high-dynamic-range scenes; secondly, enhancing the degraded RGB frames can further raise accuracy in object tracking.

4.2. Results Based on Real-World Data

To investigate the performance of our approach in real high-dynamic-range scenes, we test our algorithm on some videos under challenging illumination, with one typical example being shown in Figure 9. The video is captured at a tunnel entrance, and the frames in the top row show a car traveling through the tunnel. When the car enters the tunnel, it is difficult to capture images with high visual quality due to insufficient light, and the car turns indistinguishable in the last frame. The middle row shows the result of image enhancement, demonstrating that the visual quality of the RGB frames is largely increased compared with the raw input.

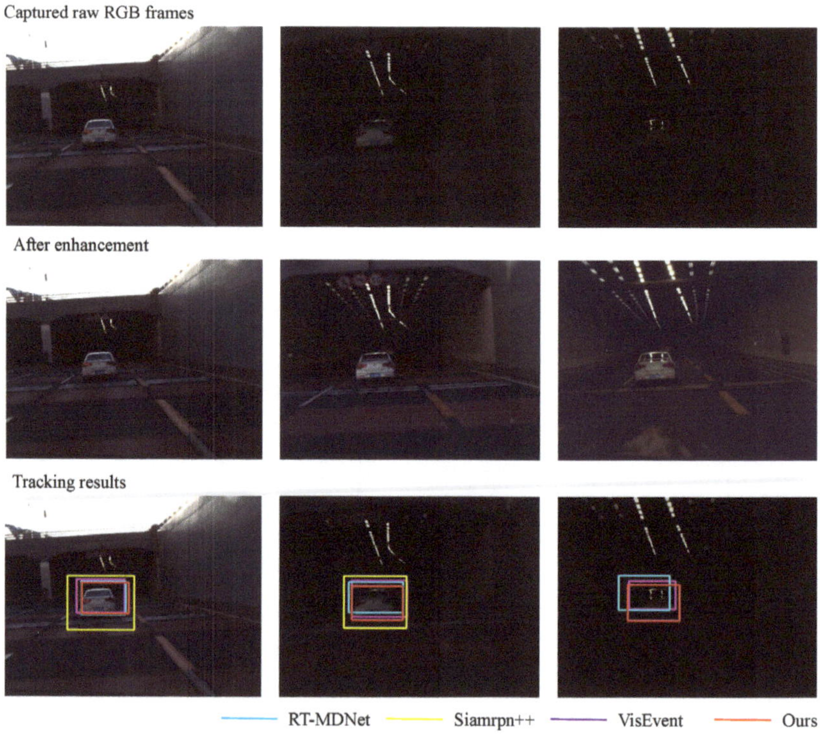

Figure 9. Demonstration of our image enhancement of the tracking result and performance comparison with existing object-tracking algorithms on a real-world high-dynamic-range scene—a white car driving through a tunnel. (**Top**) The captured RGB frames. (**Middle**) Our enhanced RGB images. (**Bottom**) The tracking results of different algorithms.

The tracking results are shown in the bottom row of Figure 9. All four algorithms can track the car at high brightness. When the light becomes weaker, the performance of the two RGB-based tracking algorithms decreases: Siamrpn++ cannot track the car, and RT-MDNet produces a bounding box with a large offset; on the other hand, VisEvent can achieve relatively higher robustness, but the bounding box is not accurate. On the contrary, we can achieve reliable tracking over the whole sequence. Based on the above experiments, we can further verify that (i) the illumination condition affects accuracy in object tracking and (ii) the event signal can assist object tracking under harsh illumination.

4.3. Ablation Studies

The ablation experiment focuses on validating the contribution of event-based temporal alignment to RGB image enhancement and object tracking. In the proposed approach, we use Harris corner detection to retrieve key feature points from the dense event sequence, and here, we compare its performance against two methods: using random event signals as key features and using the detected Harris corner points from the RGB images rather than event signals.

From the upper row in Figure 10, one can see that there exist large intensity variations within each frame and abrupt changes among frames, which is quite challenging for object-tracking algorithms and even human vision systems, especially in the third frame. Here, we adopt the person in the third frame as the tracking target, and the results with different key feature guidance are shown in the bottom row. One can see that the proposed alignment strategy performs best in terms of both the quality of the enhanced image and object-tracking accuracy. In comparison, the result produced through registration from random event signals slightly enhances image quality and results in a looser bounding box, while registration from RGB frames provides little help, which again validates the strategy of introducing event cameras for such harshly lit scenes. The inferior performance of the two benchmarking implementations is mainly attributed to the fact that they cannot identify the temporal matching properly due to the lack of descriptive features.

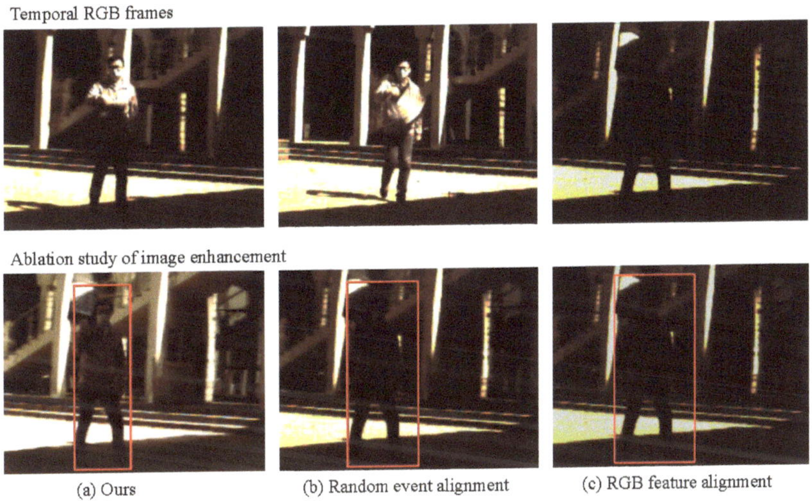

Figure 10. An example showing the results of ablation studies. The upper row displays the RGB frames of a high-dynamic-range scene. The lower row shows the image enhancement and object-tracking results based on three different types of temporal registration guidance, with the person in the third frame (the darkest and most challenging one) as the target object. From left to right: key feature alignment using the proposed event-based Harris corner points, random event signals, and Harris corner points in RGB frames.

5. Summary and Discussions

Visible-wavelength optical cameras provide rich scene information for the environmental sensing of drones. However, harsh illumination causes high dynamic ranges (e.g., at nighttime, at entrances or exits, etc.) and hampers reliable environmental perception. In order to extend the applicability of visible-wavelength cameras in real scenes, we propose a dual-sensing architecture that leverages the advantages of event cameras to increase the imaging quality of the RGB sensor as well as the successive object-tracking performance.

The proposed event-assisted robust object tracker exploits two main features of event signals, i.e., robust imaging under complex illumination and fast response. These advantageous and unique features support extracting the continuous trajectories of corner points to guide the temporal registration of high-dynamic-range RGB frames. Registration plays a central role in compensating the intensity changes. Experimentally, the proposed event-assisted robust object tracking can work quite well in a high-dynamic-range environment that goes beyond the capability of RGB cameras.

The performance of the proposed algorithm is superior to both the counterpart taking only the RGB frames as input and that directly taking two inputs, and the advantages hold in a wide range of applications. From the comparison, we can obtain the following two conclusions: (i) Under harsh illumination, the quality of RGB images greatly affects performance in downstream tasks. In order to ensure the robustness of performance in tasks such as object tracking, the RGB frames need to be enhanced first. (ii) Event signals, as lightweight and efficient sensors, can be used to capture critical information in high-speed-moving scenes. In addition, event signals are insensitive to lighting conditions and can be used for scene sensing under extreme illumination.

Limitations. The proposed algorithm mainly has two limitations. First, because of the involved complex calculations, it is difficult to deploy the algorithm into a UAV due to the limited arithmetic power. To achieve UAV deployment, it is necessary to further optimize the network structure for lightweight computation. Second, since the event camera can only capture the intensity changes in the scene, it is difficult to sense the targets being relatively stationary with respect to the event camera. Therefore, other complementary sensors need to be equipped for highly robust object tracking.

Potential extensions. In the future, we will dig deeper into the characteristics of event signals and construct neural networks that are more compatible with event signals to realize lightweight network design and efficient learning. In addition, we will integrate sensing units such as LIDAR and IMUs to achieve depth-aware 3D representation of scenes.

Author Contributions: Y.H. and J.S. conceived this project. Y.H. designed the framework and the network architecture. Y.H. and X.Y. implemented the event-based key feature alignment as well as temporal RGB image enhancement. H.L. collected the dataset and conducted the comparison experiments. X.Y. designed and conducted the ablation studies and analyzed the experimental results. J.S. dominated the discussion of this work. J.S. supervised this research and finally approved the version to be submitted. All authors have read and agreed to the published version of the manuscript.

Funding: This work was supported by Ministry of Science and Technology of China (Grant No. 2020AAA0108202), National Natural Science Foundation of China (grant number 61931012, 62171258).

Data Availability Statement: The data are available from the corresponding author upon reasonable request.

Conflicts of Interest: Author Heng Luan was employed by the company Research and Development Center, TravelSky Technology Ltd. The remaining authors declare that the research was conducted in the absence of any commercial or financial relationships that could be construed as a potential conflict of interest.

References

1. Zhang, J.; Hu, J.; Lian, J.; Fan, Z.; Ouyang, X.; Ye, W. Seeing the forest from drones: Testing the potential of lightweight drones as a tool for long-term forest monitoring. *Biol. Conserv.* **2016**, *198*, 60–69. [CrossRef]
2. Duffy, J.P.; Cunliffe, A.M.; DeBell, L.; Sandbrook, C.; Wich, S.A.; Shutler, J.D.; Myers-Smith, I.H.; Varela, M.R.; Anderson, K. Location, location, location: Considerations when using lightweight drones in challenging environments. *Remote Sens. Ecol. Conserv.* **2018**, *4*, 7–19. [CrossRef]
3. Zhang, Y.; He, D.; Li, L.; Chen, B. A lightweight authentication and key agreement scheme for Internet of Drones. *Comput. Commun.* **2020**, *154*, 455–464. [CrossRef]
4. McNeal, G.S. Drones and the future of aerial surveillance. *Georg. Wash. Law Rev.* **2016**, *84*, 354.
5. Akram, M.W.; Bashir, A.K.; Shamshad, S.; Saleem, M.A.; AlZubi, A.A.; Chaudhry, S.A.; Alzahrani, B.A.; Zikria, Y.B. A secure and lightweight drones-access protocol for smart city surveillance. *IEEE Trans. Intell. Transp. Syst.* **2021**, *23*, 19634–19643. [CrossRef]
6. Guvenc, I.; Koohifar, F.; Singh, S.; Sichitiu, M.L.; Matolak, D. Detection, tracking, and interdiction for amateur drones. *IEEE Commun. Mag.* **2018**, *56*, 75–81. [CrossRef]
7. Bamburry, D. Drones: Designed for product delivery. *Des. Manag. Rev.* **2015**, *26*, 40–48. [CrossRef]
8. Panda, S.S.; Rao, M.N.; Thenkabail, P.S.; Fitzerald, J.E. Remote Sensing Systems—Platforms and Sensors: Aerial, Satellite, UAV, Optical, Radar, and LiDAR. In *Remotely Sensed Data Characterization, Classification, and Accuracies*; CRC Press: Boca Raton, FL, USA , 2015; pp. 37–92.
9. Jeong, N.; Hwang, H.; Matson, E.T. Evaluation of low-cost lidar sensor for application in indoor UAV navigation. In Proceedings of the IEEE Sensors Applications Symposium, Seoul, Republic of Korea, 12–14 March 2018; pp. 1–5.
10. Bellon-Maurel, V.; McBratney, A. Near-infrared (NIR) and mid-infrared (MIR) spectroscopic techniques for assessing the amount of carbon stock in soils—Critical review and research perspectives. *Soil Biol. Biochem.* **2011**, *43*, 1398–1410. [CrossRef]

11. Chen, P.; Dang, Y.; Liang, R.; Zhu, W.; He, X. Real-time object tracking on a drone with multi-inertial sensing data. *IEEE Trans. Intell. Transp. Syst.* **2017**, *19*, 131–139. [CrossRef]
12. Wen, L.; Zhu, P.; Du, D.; Bian, X.; Ling, H.; Hu, Q.; Liu, C.; Cheng, H.; Liu, X.; Ma, W.; et al. Visdrone-SOT2018: The vision meets drone single-object tracking challenge results. In Proceedings of the European Conference on Computer Vision Workshops, Munich, Germany, 8–14 September 2018; pp. 469–495.
13. Bartak, R.; Vykovský, A. Any object tracking and following by a flying drone. In Proceedings of the Mexican International Conference on Artificial Intelligence, Cuernavaca, Mexico, 25–31 October 2015; pp. 35–41.
14. Zhang, H.; Wang, G.; Lei, Z.; Hwang, J.N. Eye in the sky: Drone-based object tracking and 3D localization. In Proceedings of the ACM International Conference on Multimedia, Nice, France, 21–25 October 2019; pp. 899–907.
15. Bertinetto, L.; Valmadre, J.; Henriques, J.F.; Vedaldi, A.; Torr, P.H. Fully-convolutional siamese networks for object tracking. In Proceedings of the European Conference on Computer Vision Workshops, Amsterdam, The Netherlands, 8–10 and 15–16 October 2016; pp. 850–865.
16. Danelljan, M.; Bhat, G.; Shahbaz Khan, F.; Felsberg, M. ECO: Efficient convolution operators for tracking. In Proceedings of the IEEE/CVF Conference on Computer Vision and Pattern Recognition, Honolulu, HI, USA, 21–26 July 2017; pp. 6638–6646.
17. Dai, K.; Wang, D.; Lu, H.; Sun, C.; Li, J. Visual tracking via adaptive spatially-regularized correlation filters. In Proceedings of the IEEE/CVF Conference on Computer Vision and Pattern Recognition, Long Beach, CA, USA, 15–20 June 2019; pp. 4670–4679.
18. Li, P.; Chen, B.; Ouyang, W.; Wang, D.; Yang, X.; Lu, H. GradNet: Gradient-guided network for visual object tracking. In Proceedings of the IEEE/CVF Conference on Computer Vision and Pattern Recognition, Long Beach, CA, USA, 15–20 June 2019; pp. 6162–6171.
19. Mitrokhin, A.; Fermüller, C.; Parameshwara, C.; Aloimonos, Y. Event-based moving object detection and tracking. In Proceedings of the IEEE/RSJ International Conference on Intelligent Robots and Systems, Madrid, Spain, 1–5 October 2018; pp. 1–9.
20. Chen, H.; Suter, D.; Wu, Q.; Wang, H. End-to-end learning of object motion estimation from retinal events for event-based object tracking. In Proceedings of the AAAI Conference on Artificial Intelligence, New York, NY, USA, 7–12 February 2020; Volume 34, pp. 10534–10541.
21. Burdziakowski, P.; Bobkowska, K. Lighting Conditions—Accuracy Considerations. *Sensors* **2021**, *21*, 3531. [CrossRef]
22. Wisniewski, M.; Rana, Z.A.; Petrunin, I. Drone Model Classification Using Convolutional Neural Network Trained on Synthetic Data. *J. Imaging* **2022**, *8*, 218. [CrossRef] [PubMed]
23. Onzon, E.; Mannan, F.; Heide, F. Neural auto-exposure for high-dynamic range object detection. In Proceedings of the IEEE/CVF CVPR Conference on Computer Vision and Pattern Recognition, Nashville, TN, USA, 20–25 June 2021; pp. 7710–7720.
24. Mahlknecht, F.; Gehrig, D.; Nash, J.; Rockenbauer, F.M.; Morrell, B.; Delaune, J.; Scaramuzza, D. Exploring Event Camera-Based Odometry for Planetary Robots. *IEEE Robot. Autom. Lett.* **2022**, *7*, 8651–8658.
25. Debevec, P.E.; Malik, J. Recovering high dynamic range radiance maps from photographs. In *Seminal Graphics Papers: Pushing the Boundaries*; Association for Computing Machinery: New York, NY, USA, 2023; Volume 2, pp. 643–652.
26. Kang, S.B.; Uyttendaele, M.; Winder, S.; Szeliski, R. High dynamic range video. *ACM Trans. Graph.* **2003**, *22*, 319–325. [CrossRef]
27. Mangiat, S.; Gibson, J. High dynamic range video with ghost removal. In Proceedings of the Applications of Digital Image Processing, San Diego, CA, USA, 1–5 August 2010; National Council of Teachers of Mathematics: Reston, VA, USA, 2010; Volume 7798, pp. 307–314.
28. Mangiat, S.; Gibson, J. Spatially adaptive filtering for registration artifact removal in HDR video. In Proceedings of the IEEE International Conference on Image Processing, Brussels, Belgium, 11–14 September 2011; pp. 1317–1320.
29. Kalantari, N.K.; Ramamoorthi, R. Deep high dynamic range imaging of dynamic scenes. *ACM Trans. Graph.* **2017**, *36*, 1–12. [CrossRef]
30. Gryaditskaya, Y. High Dynamic Range Imaging: Problems of Video Exposure Bracketing, Luminance Calibration and Gloss Editing. Ph.D. Thesis, Saarland University, Saarbrücken, Germany , 2016.
31. Hajisharif, S.; Kronander, J.; Unger, J. Adaptive dualISO HDR reconstruction. *EURASIP J. Image Video Process.* **2015**, *2015*, 41. [CrossRef]
32. Heide, F.; Steinberger, M.; Tsai, Y.T.; Rouf, M.; Pająk, D.; Reddy, D.; Gallo, O.; Liu, J.; Heidrich, W.; Egiazarian, K.; et al. Flexisp: A flexible camera image processing framework. *ACM Trans. Graph.* **2014**, *33*, 1–13. [CrossRef]
33. Cai, J.; Gu, S.; Zhang, L. Learning a deep single image contrast enhancer from multi-exposure images. *IEEE Trans. Image Process.* **2018**, *27*, 2049–2062. [CrossRef]
34. Nayar, S.K.; Mitsunaga, T. High dynamic range imaging: Spatially varying pixel exposures. In Proceedings of the IEEE/CVF Conference on Computer Vision and Pattern Recognition, Hilton Head, SC, USA, 15 June 2000; Volume 1, pp. 472–479.
35. Zhao, H.; Shi, B.; Fernandez-Cull, C.; Yeung, S.K.; Raskar, R. Unbounded high dynamic range photography using a modulo camera. In Proceedings of the IEEE International Conference on Computational Photography, Houston, TX, USA, 24–26 April 2015; pp. 1–10.
36. Gallego, G.; Delbrück, T.; Orchard, G.; Bartolozzi, C.; Taba, B.; Censi, A.; Leutenegger, S.; Davison, A.J.; Conradt, J.; Daniilidis, K.; et al. Event-based vision: A survey. *IEEE Trans. Pattern Anal. Mach. Intell.* **2020**, *44*, 154–180. [CrossRef]
37. Muglikar, M.; Gehrig, M.; Gehrig, D.; Scaramuzza, D. How to calibrate your event camera. In Proceedings of the IEEE/CVF Conference on Computer Vision and Pattern Recognition, Nashville, TN, USA, 20–25 June 2021; pp. 1403–1409.

38. Lagorce, X.; Meyer, C.; Ieng, S.H.; Filliat, D.; Benosman, R. Asynchronous event-based multikernel algorithm for high-speed visual features tracking. *IEEE Trans. Neural Netw. Learn. Syst.* **2014**, *26*, 1710–1720. [CrossRef]
39. Rebecq, H.; Ranftl, R.; Koltun, V.; Scaramuzza, D. High speed and high dynamic range video with an event camera. *IEEE Trans. Pattern Anal. Mach. Intell.* **2019**, *43*, 1964–1980. [CrossRef]
40. Brandli, C.; Muller, L.; Delbruck, T. Real-time, high-speed video decompression using a frame-and event-based DAVIS sensor. In Proceedings of the IEEE International Symposium on Circuits and Systems, Melbourne, Australia, 1–5 June 2014; pp. 686–689.
41. Ni, Z.; Pacoret, C.; Benosman, R.; Ieng, S.; RÉGNIER*, S. Asynchronous event-based high speed vision for microparticle tracking. *J. Microsc.* **2012**, *245*, 236–244. [CrossRef]
42. Tulyakov, S.; Gehrig, D.; Georgoulis, S.; Erbach, J.; Gehrig, M.; Li, Y.; Scaramuzza, D. Time lens: Event-based video frame interpolation. In Proceedings of the IEEE/CVF Conference on Computer Vision and Pattern Recognition, Nashville, TN, USA, 20–25 June 2021; pp. 16155–16164.
43. Tulyakov, S.; Bochicchio, A.; Gehrig, D.; Georgoulis, S.; Li, Y.; Scaramuzza, D. Time lens++: Event-based frame interpolation with parametric non-linear flow and multi-scale fusion. In Proceedings of the IEEE/CVF Conference on Computer Vision and Pattern Recognition, New Orleans, LA, USA, 18–24 June 2022; pp. 17755–17764.
44. Pan, L.; Liu, M.; Hartley, R. Single image optical flow estimation with an event camera. In Proceedings of the IEEE/CVF Conference on Computer Vision and Pattern Recognition, Seattle, WA, USA, 13–19 June 2020; pp. 1669–1678.
45. Bardow, P.; Davison, A.J.; Leutenegger, S. Simultaneous optical flow and intensity estimation from an event camera. In Proceedings of the IEEE/CVF Conference on Computer Vision and Pattern Recognition, Las Vegas, NV, USA, 27–30 June 2016; pp. 884–892.
46. Wan, Z.; Dai, Y.; Mao, Y. Learning dense and continuous optical flow from an event camera. *IEEE Trans. Image Process.* **2022**, *31*, 7237–7251. [CrossRef] [PubMed]
47. Akolkar, H.; Ieng, S.H.; Benosman, R. Real-time high speed motion prediction using fast aperture-robust event-driven visual flow. *IEEE Trans. Pattern Anal. Mach. Intell.* **2020**, *44*, 361–372. [CrossRef] [PubMed]
48. Ramesh, B.; Zhang, S.; Lee, Z.W.; Gao, Z.; Orchard, G.; Xiang, C. Long-term object tracking with a moving event camera. In Proceedings of the British Machine Vision Conference, Newcastle, UK, 3–6 September 2018; p. 241.
49. Ronneberger, O.ß.; Fischer, P.; Brox, T. U-Net: Convolutional networks for biomedical image segmentation. In Proceedings of the Medical Image Computing and Computer-Assisted Intervention—MICCAI 2015: 18th International Conference, Munich, Germany, 5–9 October 2015; Proceedings, Part III 18; Springer International Publishing: Cham, Switzerland, 2015; pp. 234–241.
50. Jung, I.; Son, J.; Baek, M.; Han, B. Real-time MDNet. In Proceedings of the European Conference on Computer Vision, Munich, Germany, 8–14 September 2018; pp. 83–98.
51. Vaswani, A.; Shazeer, N.; Parmar, N.; Uszkoreit, J.; Jones, L.; Gomez, A.N.; Kaiser, L.; Polosukhin, I. Attention is all you need. In Proceedings of the 31st Conference on Neural Information Processing Systems (NIPS 2017), Long Beach, CA, USA, 4–9 December 2017; Volume 30.
52. Wang, X.; Li, J.; Zhu, L.; Zhang, Z.; Chen, Z.; Li, X.; Wang, Y.; Tian, Y.; Wu, F. VisEvent: Reliable Object Tracking via Collaboration of Frame and Event Flows. *arXiv* **2023**, arXiv:2108.05015.
53. Li, B.; Wu, W.; Wang, Q.; Zhang, F.; Xing, J.; Yan, J. SiamRPN++: Evolution of Siamese Visual Tracking With Very Deep Networks. In Proceedings of the IEEE/CVF Conference on Computer Vision and Pattern Recognition, Long Beach, CA, USA, 15–20 June 2019; pp. 4277–4286.

drones

MDPI

Article

Typical Fault Detection on Drone Images of Transmission Lines Based on Lightweight Structure and Feature-Balanced Network

Gujing Han [1,2,*], Ruijie Wang [1,2], Qiwei Yuan [1,2], Liu Zhao [1,2], Saidian Li [1,2], Ming Zhang [1,2], Min He [3] and Liang Qin [3]

1 Department of Electronic and Electrical Engineering, Wuhan Textile University, Wuhan 430200, China; 2115053005@mail.wtu.edu.cn (R.W.); 2115053017@mail.wtu.edu.cn (Q.Y.); 2115363112@mail.wtu.edu.cn (S.L.); 2115053033@mail.wtu.edu.cn (L.Z.); zhangming@wtu.edu.cn (M.Z.)
2 State Key Laboratory of New Textile Materials and Advanced Processing Technologies, Wuhan Textile University, Wuhan 430200, China
3 School of Electrical and Automation, Wuhan University, Wuhan 430072, China; whuhemin@whu.edu.cn (M.H.); qinliang@whu.edu.cn (L.Q.)
* Correspondence: gjhan@wtu.edu.cn

Abstract: In the context of difficulty in detection problems and the limited computing resources of various fault scales in aerial images of transmission line UAV inspections, this paper proposes a TD-YOLO algorithm (YOLO for transmission detection). Firstly, the Ghost module is used to lighten the model's feature extraction network and prediction network, significantly reducing the number of parameters and the computational effort of the model. Secondly, the spatial and channel attention mechanism scSE (concurrent spatial and channel squeeze and channel excitation) is embedded into the feature fusion network, with PA-Net (path aggregation network) to construct a feature-balanced network, using channel weights and spatial weights as guides to achieving the balancing of multi-level and multi-scale features in the network, significantly improving the detection capability under the coexistence of multiple targets of different categories. Thirdly, a loss function, NWD (normalized Wasserstein distance), is introduced to enhance the detection of small targets, and the fusion ratio of NWD and CIoU is optimized to further compensate for the loss of accuracy caused by the lightweightedness of the model. Finally, a typical fault dataset of transmission lines is built using UAV inspection images for training and testing. The experimental results show that the TD-YOLO algorithm proposed in this article compresses 74.79% of the number of parameters and 66.92% of the calculation amount compared to YOLOv7-Tiny and increases the mAP (mean average precision) by 0.71%. The TD-YOLO was deployed into Jetson Xavier NX to simulate the UAV inspection process and was run at 23.5 FPS with good results. This study offers a reference for power line inspection and provides a possible way to deploy edge computing devices on unmanned aerial vehicles.

Keywords: TD-YOLO; Ghost module; feature-balanced network; NWD loss

Citation: Han, G.; Wang, R.; Yuan, Q.; Zhao, L.; Li, S.; Zhang, M.; He, M.; Qin, L. Typical Fault Detection on Drone Images of Transmission Lines Based on Lightweight Structure and Feature-Balanced Network. *Drones* **2023**, *7*, 638. https://doi.org/10.3390/drones7100638

Academic Editor: Diego González-Aguilera

Received: 5 September 2023
Revised: 13 October 2023
Accepted: 14 October 2023
Published: 17 October 2023

1. Introduction

1.1. Research Background

Due to the complex and diverse environments in which transmission lines are erected, they are exposed to the wind, sun, rain, snow, and ice all year round, which can easily cause different degrees of failure and damage to power equipment [1,2]. In recent years, UAV inspection has been an important mode of inspection of transmission lines at home and abroad. This inspection mode can effectively overcome the disadvantages of manual inspection, such as "expensive, slow, difficult, and dangerous", and has the advantages of safety, high efficiency, flexible control, fewer restricted conditions, and low cost. However, UAV inspections are bound to generate a large number of inspection images [3,4]. For the inspection of electrical equipment in a large number of UAV aerial images, the method of manually checking the fault results is mainly used, which consumes a lot of labor costs and is likely to cause missed inspections or false inspections. Therefore, it is of great significance to carry out research on artificial intelligence-based inspection methods under the background of UAV inspection big data. At present, target detection based on deep learning is an important research direction in the field of computer vision. While the drone is inspecting the transmission line, the deep learning algorithm carried out by the drone is used to detect faults in the aerial images, which saves time. The human work conducted after the drone inspection also ensures the accuracy of the inspection [5,6].

1.2. Methods Based on Deep Learning and Its Limitations

Typical fault detection algorithms for transmission lines in UAV inspection, based on deep learning, are divided into two categories [7]: one is the two-stage detection algorithm, and representative algorithms include R-CNN [8], Fast R-CNN [9], Faster R-CNN [10], and Cascade R-CNN [11]. Compared with the traditional algorithm, the two-stage detection algorithm has significantly improved accuracy. However, because the detection process needs to be completed in two steps, the speed could be faster, and the application range could be narrower. The other is a one-stage detection algorithm, which directly predicts the category and location of the target through the target detection network. Representative algorithms include SSD (single-shot multibox detector) [12] and the YOLO series (You Only Look Once) [13–19]. The SSD algorithm has contributed to the idea of a one-stage detection algorithm. Still, because it does not have an FPN (feature pyramid network), the accuracy is not enough. At present, the most researched one-stage algorithm is mainly the YOLO series.

However, the current typical fault detection of transmission lines based on deep learning still has three limitations. The first limitation is the lack of detection accuracy due to aerial scale shifts during drone inspections, resulting in seriously missed inspections. To address this problem, literature [20] proposed three improved strategies based on Faster R-CNN for transmission line multi-target detection, including the adaptive image preprocessing algorithm, area-based non-maximum suppression algorithm, and cut detection scheme, to achieve accurate localization and recognition of multiple targets in complex backgrounds. Literature [21] introduced a Gaussian function to improve the non-maximum value suppression method and reduce the missed detection of partially occluded fault targets. Literature [22] introduced YOLOv5 to detect 12 types of fault samples in transmission lines and adopted CBAM (convolutional block attention module) and bi-FPN (bi-directional feature pyramid network) improvement strategies to integrate target multi-scale features effectively. This method can accurately detect multi-scale fault targets in transmission lines in complex environments. Based on YOLOv5, literature [23] proposed a transmission line small-target fault detection network that integrates prior knowledge and an attention model. Compared with the literature [21], a more advanced target detection model is used to enhance the precise detection of small targets. The parameters of the improved models in the above literature are large, which is inconvenient for deployment and application on UAVs.

The second limitation is the large number of parameters derived while improving the model's accuracy, making it difficult to deploy on UAVs. In response to this problem, the literature [24] proposed a lightweight model embedded in the double attention mechanism combined with MobelieNetV2 to detect multiple foreign objects on the transmission line. This method has high accuracy and detection speed, and its lightweight model idea lays the groundwork for model deployment. Literature [25] replaced the backbone network of YOLOv4 with a lightweight network, MobileNetV3, which is used to detect insulators and their damage in transmission lines. Literature [26] selects the pruned YOLOv4-Tiny model and combines the attention mechanism to realize the insulator research and defect detection under the hardware end. The lightweight improvement strategies for the model in the above literature are mainly divided into replacing the lightweight backbone, using lightweight convolution, and model pruning. However, the selected basic algorithm is relatively backward, with room for improvement.

The third limitation is that the single detection object leads to low inspection efficiency. Literature [27] improved Faster R-CNN (FPN). It proposed Pin-FPN, which uses various data-enhancement methods to detect pin defect faults in transmission lines and can achieve the accurate detection of small targets. Literature [28] improved YOLOv5 to detect bird nests in transmission lines and improved the detection effect of bird nests in complex backgrounds through the attention mechanism. Literature [29] combines the feature pyramid structure based on R-CNN to position insulators in complex backgrounds accurately. Literature [30] improves YOLOv5 to detect insulators and their damage in transmission lines and uses a lightweight network to reduce the model's size and increase the speed. Literature [31] adds CAT-BiFPN and ACmix attention mechanisms based on YOLOv7 to detect various defects of insulation, and the detection effect is better for targets of different scales. Judging from the current research results, the detection objects are only faults of insulators, bird's nests [32], and fittings, and there are few kinds of research on multiple types of fault inspections. The efficiency is low if applied to actual transmission line UAV inspections. Therefore, there is an urgent need for a typical fault detection algorithm for transmission lines with the advantages of convenient deployment, fast inference speed, high precision, and high inspection efficiency.

1.3. This Work

Based on the above problem analyses, this paper proposes a TD-YOLO algorithm (a lightweight object detection network that can detect multi-scale faults in real-time). The network adopts a structure combining the context lightweight structure and the feature-balanced network, which effectively solves the problems that different faults are difficult to detect simultaneously, occupy too many computing resources, and the detection speed is too slow in the detection process. Specifically, the innovations and contributions of this paper are as follows:

(1) To solve the problem that the calculation resources of the algorithm carried by the UAV are limited and the fault cannot be accurately detected, this paper proposes a new context lightweight structure (C2fGhost) from the perspective of the model lightweight, which will be calculated. While the volume is compressed by 43%, the mAP is increased by 0.14%. In addition, we combine the advantages of the Ghost module, SPPCSPC structure, and convolution, and propose two lightweight structures, GhostSPPCSPC and GhostConv. Compared with the original model, the calculation amount of the improved model is reduced by 69%, and the number of parameters is reduced by 75.7%.

(2) To solve the problem that it is difficult to detect different fault scales during the UAV inspection process, a feature-balanced network is proposed. Based on the attention mechanism and PA-Net, the network can better integrate deep information and shallow information and effectively improve the problem that it is difficult to detect targets of different scales at the same time.

(3) To solve the problem that it is difficult to detect small targets in aerial images, NWD was initially used to replace the positioning loss function in the model, and it was found that the calculation amount of the model increased suddenly, and the training time was greatly increased. Then, a loss function was proposed for the fusion of NWD and CIoU in proportion, and the best fusion ratio (70%NWD + 30%CIoU) was found. While reducing the number of parameters and training time, the accuracy is higher than that of all NWD loss functions. By using the missed detection rate to measure the detection effect of small targets, the test results show that the missed detection rate of the defects decreased by 6.76%, and the missed detection rate of anti-vibration hammer corrosion decreased by 14.61%.

(4) Deploy the algorithm in this paper to the embedded device Jeston Xavier NX to simulate the UAV inspection process and put forward the deployment condition limit index. The accuracy of the algorithm in the embedded device reached 93.5%, and the detection speed reached (23.5 $\pm$ 2.2) FPS. Meet the accuracy and real-time performance of drone inspections.

2. Materials and Methods

2.1. Datasets

The dataset used in this paper is provided by the State Grid Corporation of China. The dataset records fault images of transmission lines taken by M300-RTK. There are 3824 pictures in total. Each picture contains one or more targets. The target labels include four types of typical faults of transmission lines: Corrosion of insulators, insulator defects, bird's nests, and anti-vibration hammers, corresponding to 'Insulator', 'Defect', 'Nest', and 'Fzc_xs' in the first row of Table 1. At the same time, the number of labels corresponding to each category is shown in the second row of Table 1. LabelImg software is used to label the image, and the dataset is divided by a ratio of 8:1:1 (training set: validation set: test set). The number of categories in each group is higher than that of the standard VOC2017 dataset in the production of the VOC format dataset; therefore, this dataset has the same training ability as the standard dataset in the sample size. Some faults are shown in Figure 1.

Table 1. Fault abbreviation and quantity.

Fault Abbreviation	Insulator	Defect	Nest	Fzc_xs
Numbers	4556	1333	1525	7287

(a) (b)

Figure 1. (**a**,**b**) The typical fault sample diagram was selected in this paper.

2.2. Overview of YOLOv7 Methods

The YOLOv7 algorithm is a new YOLO series algorithm proposed after the YOLOv4 and YOLOv5 algorithms. The detection speed and accuracy of YOLOv7, in the range of 5FPS to 160FPS, are ahead of the current mainstream target detection algorithms. YOLOv7-Tiny is a lightweight version of YOLOv7. The overall structure is shown in Figure 2. The model structure consists of three parts: feature extraction network (backbone), feature fusion network (neck), and prediction network (head).

For the feature extraction network, YOLOv7-Tiny adopts the ELAN (efficient layer aggregation networks) structure, which is an efficient layer aggregation network. ELAN is mainly composed of VOV-Net and CSP-Net. Its function is to avoid using too many transition layers and reduce those that are unnecessary. The necessary parameters shorten the feature extraction path and increase the extraction efficiency.

The feature fusion network still uses the PA-Net structure in YOLOv5. The top-down and bottom-up paths can extract multi-scale features from feature maps at different levels, capturing rich semantic and spatial information.

The prediction network consists of three convolution modules that output target classification information, localization information, and confidential information, and three prediction heads with different detection scales (80 × 80, 40 × 40, 20 × 20). Through three pieces of information, the model's loss function can make better predictions on the classification and location of the target. The model loss calculation formula is as follows:

$$\begin{aligned} L_{cls} &= \sum_{t=0}^{S\times S}\sum_{j=0}^{B} I_{ij}^{obj} \sum_{c\in classes} [p_i'(c)\log(p_i(c))] \\ &+ \sum_{t=0}^{S\times S}\sum_{j=0}^{B} I_{ij}^{obj} \sum_{c\in classes} [(1-p_i'(c))\log(1-p_i(c))] \end{aligned} \quad (1)$$

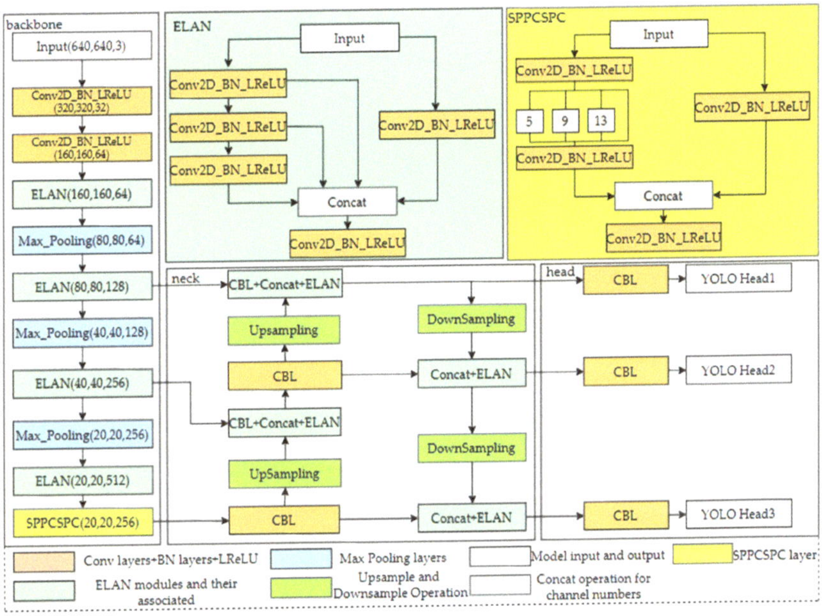

Figure 2. YOLOv7-Tiny structure diagram.

Equation (1) is the classification loss function of the model, denoted as L_{cls}. Where $S \times S$ is the image input size 640 × 640, i represents the i-th square of the feature map, j represents the j-th prediction box predicted by the square, $c \in classes$ represents the correct category, p_i (c) and p_i' (c) represent the predicted confidence score and the actual confidence score, respectively.

$$\begin{aligned} S_{IoU} &= \frac{A\cap B}{A\cup B} \\ v &= \frac{4}{\pi^2}\left(\arctan\frac{w^{gt}}{h^{gt}} - \arctan\frac{w}{h}\right)^2 \\ \alpha &= \frac{v}{(1-S_{IoU})+v} \\ L_{box} &= 1 - S_{IoU} + \frac{\rho^2(A,B)}{c^2} + \alpha v \end{aligned} \quad (2)$$

Equation (2) is the locus loss function of the target box, also known as the regression loss, notated as L_{box}, which is mainly used as the $CIoU$ loss function [33]. In Figure 3, box A is the real box, box B is the prediction box, and S_{IoU} is the intersection ratio between the real box and the prediction box; box M is the smallest external rectangle containing box A and box B. Where $\rho^2(A, B)$ is the Euclidean distance between the centroids of the real box and the predicted box, i.e., the length of d in the diagram; c in Equation (2) is the diagonal length of the smallest outer matrix M that encloses box AB; w^{gt} and h^{gt} are the width and height of box A of the real box, and w and h are the width and height of box B of the predicted box. Compared with the traditional IoU, the $CIoU$ introduces a penalty term

v, which can better handle targets with different aspect ratios; it can measure the distance between the predicted box and the real box more accurately and improve the accuracy of target detection for the situation that boxes of different sizes have different overlap when the IoU values are the same, i.e., the problem of scale sensitivity.

$$\begin{aligned} L_{conf} &= \sum_{i=0}^{S\times S}\sum_{j=0}^{B} I_{ij}^{obj}[C'_i \log(C_i) + (1 - C'_i)\log(1 - C_i)] \\ &- \sum_{i=0}^{S\times S}\sum_{j=0}^{B} I_{ij}^{nobj}[C'_i \log(C_i) + (1 - C'_i)\log(1 - C_i)] \end{aligned} \tag{3}$$

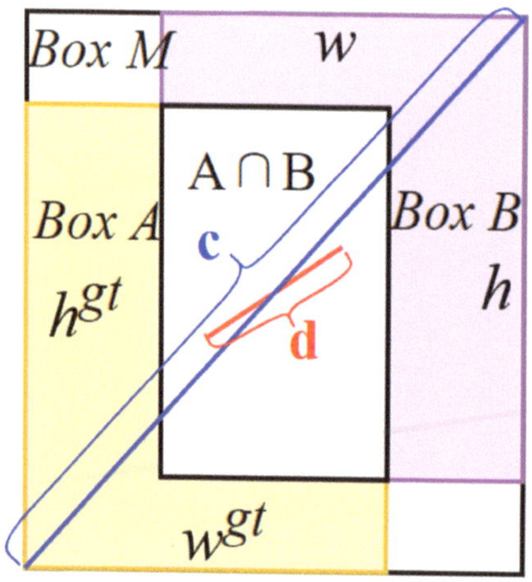

Figure 3. Calculation diagram of *CIoU*.

Equation (3) is the confidence loss function of the target, denoted as L_{conf}. Among them, *obj* and *nobj* represent the presence or absence of the target in the grid, and C_i and C_i' represent the categories of the real box and the predicted box. Then, the total loss function of YOLOv7-Tiny is composed of the addition of the three according to a certain ratio, such as Equation (4).

$$L_{total} = 0.5 \times L_{cls} + 0.05 \times L_{box} + L_{conf} \tag{4}$$

Finally, during prediction, a large number of redundant prediction frames are eliminated after non-maximum value suppression and other processing operations, and finally, the prediction category with the highest confidence score is output, and the coordinate information of the target is returned by positioning the target.

2.3. The Overall Architecture of TD-YOLO

During the test, it was found that YOLOv7-Tiny runs at a slow speed on the embedded device. The detection of complex and variable-scale faults and tiny target faults in the transmission line inspection process has missed detection and false detection, and the accuracy is low. Therefore, this paper proposes a TD-YOLO algorithm. The structure is shown in Figure 4.

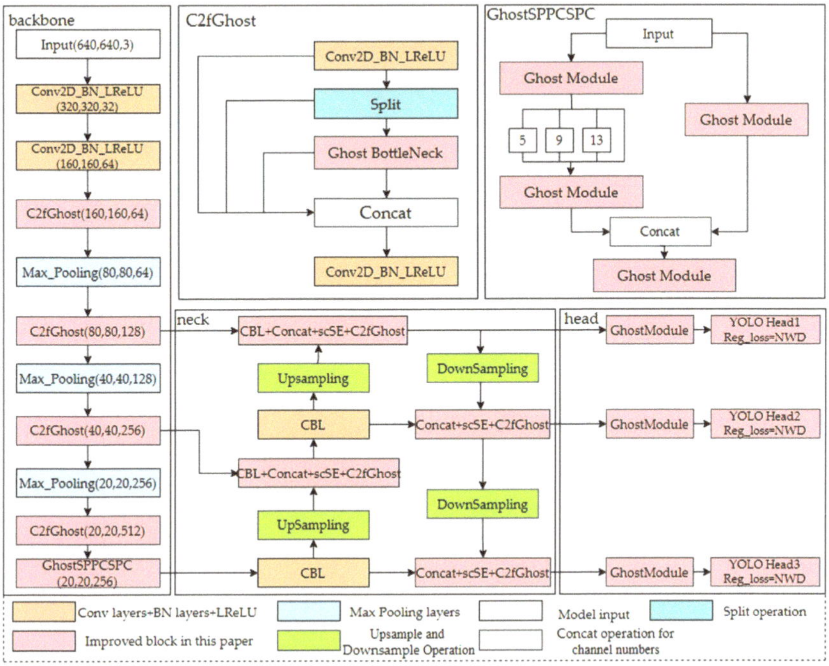

Figure 4. TD-YOLO structure diagram.

2.3.1. Various Improvements of Model Lightweight Based on the Ghost Module

Due to the limited computational resources required for UAV-carried embedded devices, the deployment of a model with many parameters to the UAV for detection is slow. It cannot meet the real-time detection requirements of this paper. Therefore, the approach of this paper is to consider the characteristics of each part of the YOLOv7 model, combined with the Ghost lightweight module (the Ghost structure is shown in Figure 5), and design a light optimization strategy that is best suited to fit with each part of the network. Based on the above analysis, this paper proposes the C2fGhost structure in the feature extraction network, the GhostSPPCSPC structure in the feature fusion network, and the Ghost (head) part combined with the Ghost module in the prediction part.

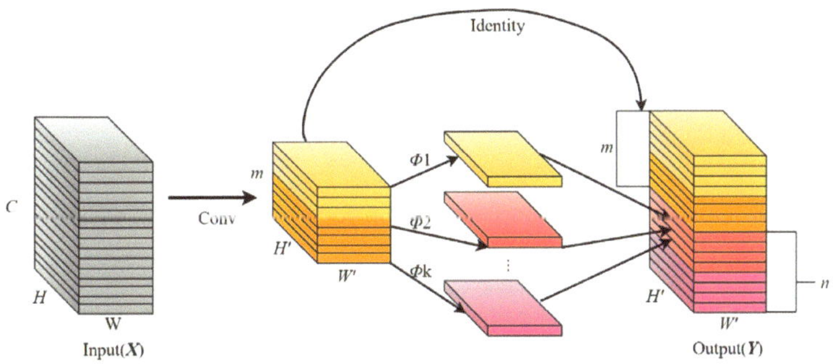

Figure 5. Ghost module structure.

Compared with the unnecessary, redundant feature maps generated in the normal convolution process, the Ghost module uses simple and easy-to-operate linear operations to enhance features and increase channels' mining information from original features with a small computational cost, which is a lightweight and efficient convolution module. The principle of the Ghost module is shown in Equation (5) [34]:

$$Y = X \times f, X \in R^{C \times H \times W}, Y \in R^{C' \times H' \times m}$$
$$y_{ij} = \phi_{ij}(y_i),\ \forall i = 1, \cdots, m, j = 1, \cdots, s \tag{5}$$

As can be seen from Equation (5), the Ghost module operates by first generating m original feature maps using fewer convolution kernels in the common convolution way (*) and later generating the remaining n feature maps by performing a simple linear transformation Φ on the already developed, $m \leq n$.

Firstly, to address the problem of information redundancy caused by the multi-layer intersection of ELAN modules, this paper designs a C2fGhost structure based on the idea of residuals combined with a lightweight module. The original C2f structure (shown in Figure 6b) continues the advantages of the ELAN structure of multi-gradient triage while adding the residual branch of BottleNeck to enable the model to learn a richer feature representation. Based on the Ghost module for C2f, this paper is further improved by replacing BottleNeck with Ghost BottleNeck (shown in Figure 7).

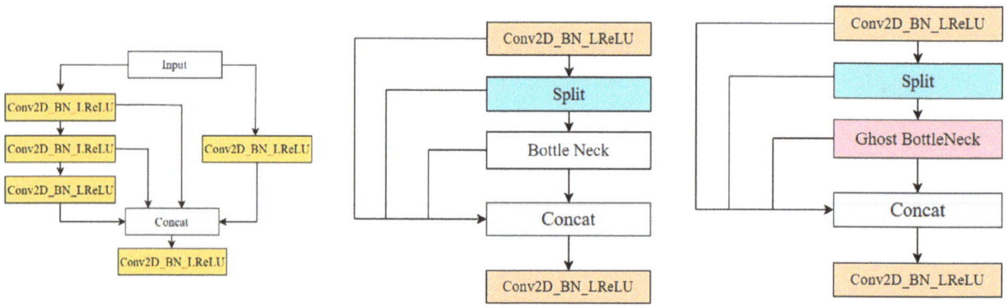

(a) (b) (c)

Figure 6. (**a**) ELAN module structure diagram; (**b**) C2f module structure diagram; (**c**) C2fGhost structure diagram.

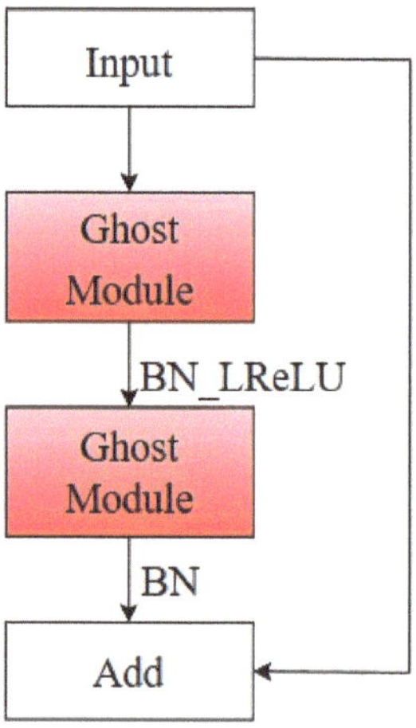

Figure 7. Ghost bottleneck structure.

The C2fGhost structure connects features at different levels to achieve multi-scale perception and strengthen the model's ability to detect targets with medium-scale changes in transmission lines. At the same time, through the residual branch of Ghost BottleNeck, the model can learn richer feature representations and still, the advantages of low complexity and a small amount of calculation of the Ghost module are retained. Then, while retaining the original structure of SPP, the ghost replacement is performed on some convolutions to achieve the purpose of lightweighting the model, which is denoted as GhostSPPCSPC. Finally, the convolution module that is in front of the three different scale detection heads in the head part is replaced by the Ghost module, and the model is further simplified, which is recorded as GhostConv(Head), and the calculation amount and model parameters are significantly reduced.

2.3.2. Improvement of Multi-Scale Feature Fusion Based on Feature-Balanced Network

In the inspection of transmission lines, the scale of fault targets spans large scales, and it is challenging to detect multi-type faults and multi-scale features. Different detection targets can be effectively identified if a higher weight ratio is assigned to the detection targets, improving detection accuracy. The attention mechanism refers to the behavior of human beings to selectively pay attention to the important parts of the received information. It can assign different proportions of weights according to different detection objects and solve the problem that multi-scale features are challenging to identify. However, a single spatial or channel attention mechanism has limitations, and it is stretched in target detection tasks with frequent scale changes. Therefore, this paper chooses the currently widely used attention mechanism, scSE [35], that combines spatial and channels. Compared with the attention mechanism CBAM [36], which also belongs to the combination of spatial and channel mechanisms, it is primarily used in the medical field of high-precision segmentation. It has the advantage of accurate recognition of fault multi-scale information. Its structure is shown in Figure 8.

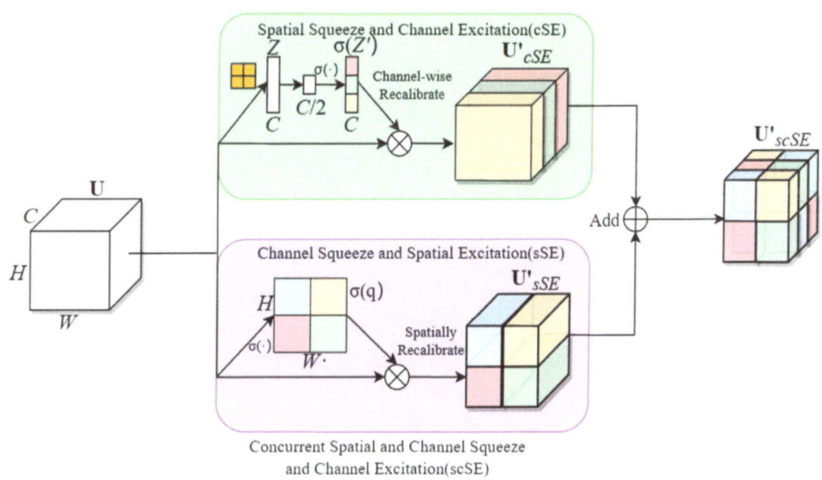

Figure 8. The scSE structure diagram [35].

The scSE process principle is shown in Equation (6). The calculation of the scSE attention mechanism consists of two steps, cSE and sSE. In cSE, the input feature map U is transformed into a feature map of $1 \times 1 \times C$ after global pooling Z. It is then normalized using a sigmoid function, noted as activations $\sigma\ (Z_i)$, and these activations are adaptively adjusted to ignore the less important channels and emphasize the important ones, and finally, the calibrated feature map (U'_{cSE}) is obtained by channel-wise multiplication. In the sSE part, U undergoes a $1 \times 1 \times 1$ convolution into a $1 \times H \times W$ feature map, with each value $\sigma(q_i, j)$ corresponding to the relative importance of the spatial information (i, j) for a given feature map. This recalibration provides the more important relevant spatial locations and ignores the irrelevant ones. The final output of the two is summed to obtain scSE [35].

$$
\begin{aligned}
&\mathrm{U} = [\mathrm{u1}, \mathrm{u2}, \cdots, \mathrm{uC}],\ ui \in R^{H\times W} \\
&Z_k = AvgPool2D(\mathrm{U}) = \frac{1}{H\times W}\sum_{i}^{H}\sum_{j}^{W} u_k(i,j),\ Z \in R^{1\times 1\times C} \\
&\mathrm{U}\prime_{cSE} = F_{cSE}(\mathrm{U}) = [\sigma(\mathrm{Z}_1)\mathrm{u1}, \sigma(\mathrm{Z}_2)\mathrm{u2}, \cdots, \sigma(\mathrm{Z}_C)\mathrm{uC}] \\
&q = \mathrm{W}_{sq}\cdot \mathrm{U},\ \mathrm{W}_{sq} \in R^{1\times 1\times C\times 1}, q \in R^{H\times W} \\
&\mathrm{U}'_{sSE} = F_{sSE}(\mathrm{U}) = \left[\sigma(q_{1,1})\mathrm{u}^{1,1}, \cdots, \sigma(q_{i,j})\mathrm{u}^{i,j}, \cdots, \sigma(q_{H,W})\mathrm{u}^{i,j}\right], u^{i,j} \in R^{1\times 1\times C} \\
&\mathrm{U}'_{scSE} = \mathrm{U}'_{cSE} + \mathrm{U}'_{sSE}
\end{aligned} \tag{6}
$$

However, there is still the problem of the complex fusion of features at different scales in the model. Hence, this paper addresses the problem by proposing a feature-balanced network (FBN) that combines PA-Net with the scSE attention mechanism. The feature-balanced network forms the neck part of the improved algorithm, and the structure is shown in Figure 9.

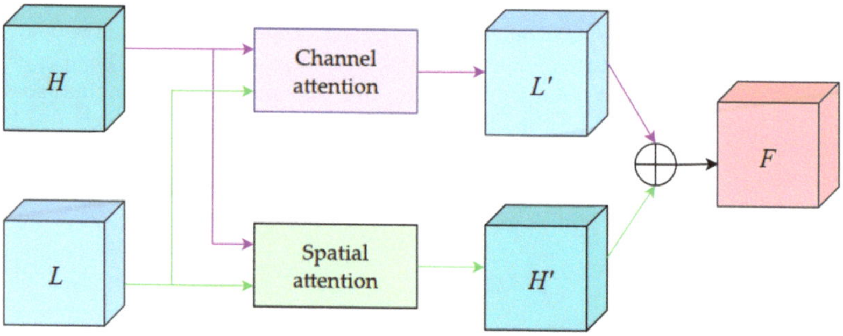

Figure 9. FBN structure diagram.

The entire network takes the high-level feature map H and the low-level feature map L as output and fuses the output features of the two branches. In the channel attention branch, high-level feature maps guide low-level features with channel attention masks. The channel attention cSE enhances the network's feature extraction in transmission lines, leading to a low-level feature map L' with rich semantic information. In the spatial attention branch, a spatial attention mask guides the high-level feature map using the low-level feature map. The spatial attention module sSE strengthens the capture of spatial information, resulting in a high-level feature map H' with spatial information. Finally, after the two are fused, a feature quantity containing spatial and channel information is output, and then the deep and shallow features are fused through PA-Net to balance the multi-scale features.

2.3.3. Small Target Detection Optimization Based on NWD Loss Function

When the object-to-image ratio is less than 0.1, it can be called a small object, a relative definition of small objects [34]. The anti-vibration hammer corrosion and insulator damage in the detection objects of this paper can be divided into small target ranges, as shown in Figure 9. Also, in Table 2 of 4.5, the results show that the detection accuracy of the anti-vibration hammer is the lowest. Hence, the detection optimization for small targets is the focus and difficulty of this paper. *To solve this problem, TD-YOLO first introduces the NWD loss function for small object detection to replace part of the CIoU of the localization loss in the YOLOv7-Tiny loss function. Secondly, it explores the fusion ratio of NWD and CIoU so that the algorithm can improve the detection accuracy of small objects while retaining the advantage of the fast training speed of CIoU, effectively reducing the amount of calculation of the model.*

Table 2. Comparison of lightweight ablation experiments based on Ghost modules.

	mAP (%)	FLOPs (G)	Params (MB)
YOLOv7-Tiny	92.79	13	12.3
YOLOv7-Tiny-C2fGhost	**92.93**	7.5	7.3
YOLOv7-Tiny-GhostSPPCSPC	92.84	10.3	9.5
YOLOv7-GhostConv(Head)	92.81	10.3	9.3
YOLOv7-Tiny-C2fGhost -GhostSPPCSPC	92.55	7	6.15
YOLOv7-Tiny-C2fGhost -GhostConv(Head)	92.74	4.7	4.3
YOLOv7-Tiny-C2fGhost-GhostSPPCSPC-GhostConv(Head)	91.98	**4.1**	**3**

CIoU is very sensitive to the position deviation of small targets that occupy fewer pixels [37]. If there is a slight position deviation in the position of the tiny target, the intersection of union (IoU) will drop significantly, greatly affecting the model accuracy. Taking Figure 10a as an example, damaged insulators belong to small objects, while insulators belong to ordinary objects, and the bounding boxes generated by them are shown in Figure 11. Box A represents the ground-truth bounding box, and boxes B and C represent the predicted bounding boxes with 1-pixel and 4-pixel diagonal deviation, respectively; thus, the corresponding intersection ratios can be calculated.

(**a**) (**b**)

Figure 10. (**a**) Example of a broken insulator in a small target; (**b**) example of vibration hammer rust in small targets.

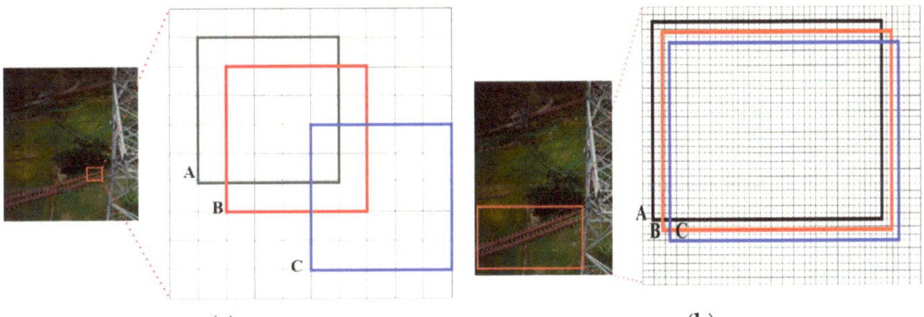

(**a**) (**b**)

Figure 11. (**a**) IoU transformation of small targets; (**b**) IoU transformation of normal targets.

For the small target in Figure 11a, the IoU changes as follows:

$$IoU = \frac{|A \cap B|}{|A \cup B|} = 0.53 \Rightarrow IoU = \frac{|A \cap C|}{|A \cup C|} = 0.06 \quad (7)$$

For the normal target in Figure 11b, the IoU changes as follows:

$$IoU = \frac{|A \cap B|}{|A \cup B|} = 0.9 \Rightarrow IoU = \frac{|A \cap C|}{|A \cup C|} = 0.65 \tag{8}$$

It can be seen from Equations (7) and (8) that for small targets, a minor position deviation leads to a significant IoU drop (from 0.53 to 0.06). The IoU drop (from 0.9 to 0.65) is not evident for ordinary objects under the same position deviation. This means that the CIoU is very sensitive to the position deviation of small targets that occupy fewer pixels. If there is a slight position deviation in the position of the tiny target, the IoU will drop significantly, which will greatly affect the model's accuracy.

Therefore, TD-YOLO chooses the NWD loss function that is insensitive to objects of different scales. NWD uses a two-dimensional Gaussian distribution to model the peripheral bounding box of the object, which can better describe the weight of different pixels, where the importance of pixels decreases from the center to the boundary. Bounding box A and bounding box B can be converted into the distribution distance between two Gaussian distributions. This new measurement method can evaluate the similarity between the model boundary and the Gaussian distribution and can more accurately judge the position information between the two boxes. To continuously improve the performance of the detector, the principle of NWD is shown in Equation (9) [38].

$$\begin{aligned} &\mu = \begin{bmatrix} c_x \\ c_y \end{bmatrix}, \Sigma = \begin{bmatrix} \frac{w^2}{4} & 0 \\ 0 & \frac{h^2}{4} \end{bmatrix} \\ &W_2^2(N_a, N_b) = \left\| \left(\left[cx_a, cy_a, \frac{w_a}{2}, \frac{h_a}{2} \right]^T, \left[cx_b, cy_b, \frac{w_b}{2}, \frac{h_b}{2} \right] \right)^T \right\|_2^2 \\ &NWD(N_a, N_b) = \exp\left(-\frac{\sqrt{W_2^2(N_a, N_b)}}{C} \right) \end{aligned} \tag{9}$$

In Equation (9), cx_a, cy_a, w_a, h_a, cx_b, cy_b, w_b, and h_b are the center coordinates, height, and width of bounding boxes A and B, and according to box A = (cx_a, cy_a, w_a, h_a), box B = (cx_b, cy_b, w_b, h_b) can construct the inscribed ellipse of frame A and frame B; then, model the two-dimensional Gaussian distribution N (μ, Σ) according to the Gaussian density, and the Gaussian distribution of frame A and frame B is N_a, N_b; C is the constraint quantity of the dataset, and the calculation of NWD is realized through this process. NWD is a better way to measure the similarity between two frames, and its insensitivity to differently scaled targets makes it more suitable for detecting small targets, which improves the accuracy of detecting anti-vibration hammer corrosion and insulator breakage significantly in this paper.

3. Experimental Results

3.1. Experimental Environment

This paper adopts the deep learning framework based on the PyTorch 1.7.1 environment; the environment is Ubuntu 20.04, python 3.7.11, CUDA = 11.4, and the training graphics card is configured as an NVIDIA RTXA6000/48 G graphics card. The processor is an Intel Xeon Platinum 8171 M CPU@2.60 GHz. The RAM is 96 G. The graphics card used by the local test computer is an NVIDIA RTX 3060 Ti, the processor is an AMD Ryzen5 5600 X, and the RAM is 32 G.

3.2. Training Process and Parameter Settings

In this paper, the backbone network is significantly modified in the improvement process; therefore, pre-training weights are not applicable. To reduce the likelihood of the model falling into a local optimum, a stochastic gradient descent (SGD) optimizer is used. The training batch was set to 8, and 300 rounds were trained. A cosine annealing learning rate was used, and a decaying learning rate was applied to the bias layer to improve the convergence speed of the model to enhance the diversity of the data with the robustness of the model itself. Figure 12a–c show the three loss curves before and after the model's improvement. It can be seen that the improved model has improved compared to the original model, especially in Figure 12b. For the dataset containing more small targets in this paper, the improvement of the localization loss effect after replacing the NWD is particularly obvious. From Figure 12d, it can be seen that the improved model has a significant improvement in mAP, which verifies the feasibility of the improved algorithm in this paper.

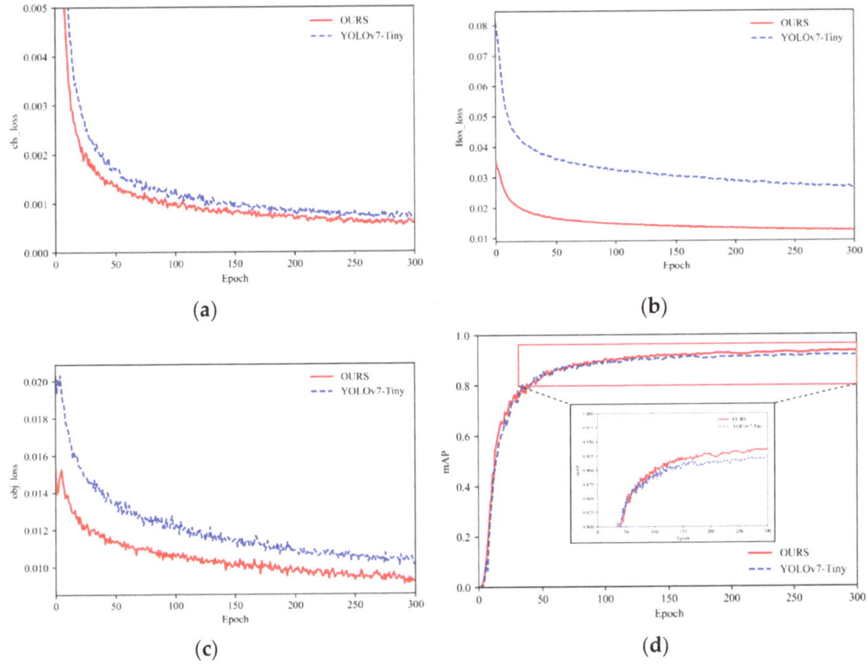

Figure 12. (**a**) Comparison chart of classification loss curves; (**b**) comparison of positioning loss curves; (**c**) comparison of loss-of-confidence curves; (**d**) mAP curve comparison chart.

3.3. Performance Evaluation Indicators

To better evaluate the missed detection of small targets caused by the difference in scale transformation, this paper introduces the missed detection rate (miss rate) [39] and the indicators for the conventional evaluation of the advantages of target detection algorithms: mean average precision (mAP), inference delay (speed), model size (params), and number of floating point operations (FLOPs).

$$Precision = \frac{TP}{TP + FP} \tag{10}$$

$$Recall = \frac{TP}{TP + FN} \tag{11}$$

$$missRate = \frac{FN}{TP + FN} \tag{12}$$

$$mAP = \frac{\sum_i^N AP_i}{N} \tag{13}$$

In Equations (10)–(13): *TP*, *FP*, and *FN* represent the number of correct detections, false detections, and missed detections; *AP* is the integral of the P–R curve; and N is the detection category. Figure 13 is the mAP curve drawn by the improved algorithm in this paper.

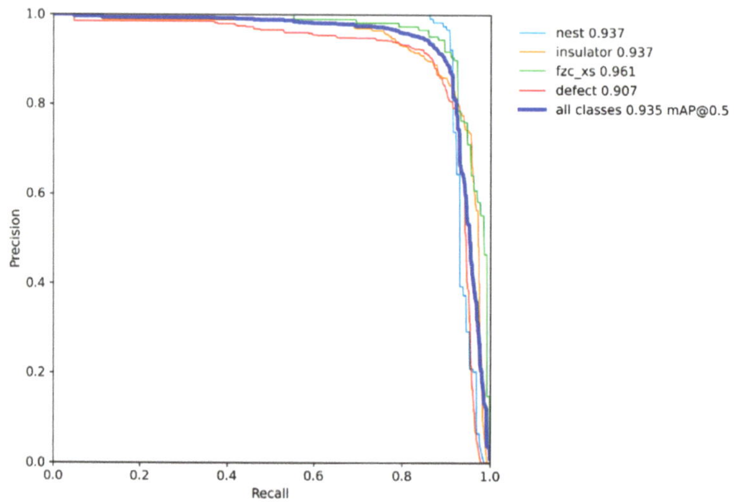

Figure 13. This paper improved the algorithm mAP curve.

4. Experimental Discussion

4.1. Validation of Model Lightweight Effects

To evaluate the impact of different improvement strategies on the detection performance of YOLOv7-Tiny, comparative experiments are carried out on the typical fault dataset of transmission lines. First, the model is improved based on Ghost Module lightweight, and the test results are shown in Table 2.

From Table 2, it can be seen that the C2fGhost improvement, due to its structural excellence, still improves mAP by 0.14% compared to YOLOv7-Tiny, with a reduced number of parameters and computation, and the GhostSPPCSPC and GhostConv(Head) improvements only replace part of the ordinary convolution, with a reduced number of parameters and computation and a slight accuracy. The three Ghost-based lightweight improvements were then subjected to ablation experiments, and after ablation for the latter two, while retaining C2fGhost, it was found that the replaced convolution in YOLOv7-C2fGhost-GhostConv(Head) involved a change in the number of channels of the three scale detection heads, the computational power decreased by 63.9%, and the number of parameters decreased by 65.1%. In terms of accuracy (mAP), since the convolution in the prediction part mainly generates a series of feature mappings that contain information on the position, category, and size of the object, and the ones in the Ghost module can obtain this information through another residual branch, then, based on this, the decrease in accuracy is not significant with fewer convolution layers, and the mAP decreases by 0.05%. The final three-improvement ablation experiment, therefore, results in a 67.7% decrease in model computation, a 76.7% decrease in the number of parameters, and a 0.81% decrease in accuracy.

4.2. Validation of Feature-Balanced Network Validity and Comparison of Similar Attention Mechanisms

The impact of feature-balancing networks on model size, computational effort, and accuracy, as well as a comparison of the attention mechanism scSE used in the FBN with

CBAM, which is also a combination of spatial and channel attention, previously used, is shown in Table 3 [39].

Table 3. Experimental results of feature-balanced networks embedding different attention mechanisms.

Models	Map (%)	FLOPs (G)	Params (MB)
YOLOv7-Tiny-Ghost	91.98	4.1	3
YOLOv7-Tiny-Ghost-FBN(CBAM) [40]	92.18	4.4	3.1
YOLOv7-Tiny-Ghost-FBN(scSE)	92.31	4.2	3.1

It can be seen in Table 3 that based on YOLOv7-Tiny-Ghost, CBAM and scSE are, respectively, added to form a feature-balanced network with different attention mechanisms. The mAP of the former increased by 0.2%, and the latter increased by 0.33%; the amount of calculation and the amount of parameters increased by 0.3 G, 0.1 G, and 0.1 MB, respectively. While the accuracy improved, the amount of calculation and the number of parameters did not increase significantly; however, the reason why scSE is ahead of CBAM is its better channel-attention mechanism structure and its parallel connection method. The former increases the accuracy, and the latter reduces the amount of calculation, which is why scSE is chosen in this paper.

To further verify its effectiveness, this paper visualizes the Grad-CAM heat map for the following typical situations, and the test results are shown in Figure 14. It can be seen in Figure 14 that in Figure 14a,b, the thermal region of the improved model is enlarged, which means that the model assigns more weights to the targets to be detected, and the darker the color, the more weights are allocated. Figure 14c shows that the model before the improvement assigns incorrect weights to areas with no detection target. Although the improved model has fewer thermal areas than before, it accurately identifies the thermal area.

YOLOv7-Tiny-Ghost	YOLOv7-Tiny-Ghost-scSE

(**a**) Improved model with larger thermal region

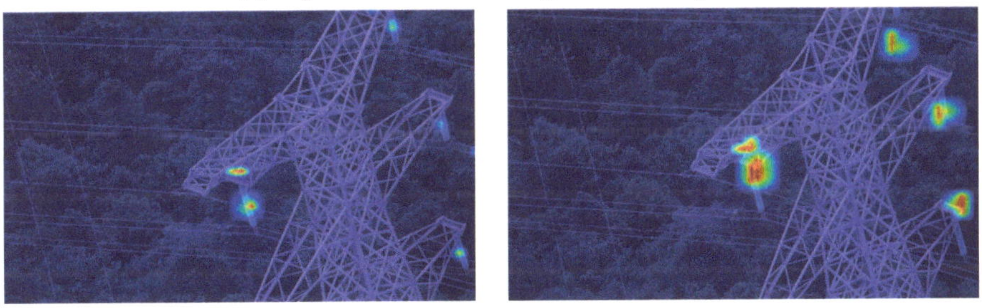

(**b**) Improved model with more thermal region

Figure 14. *Cont.*

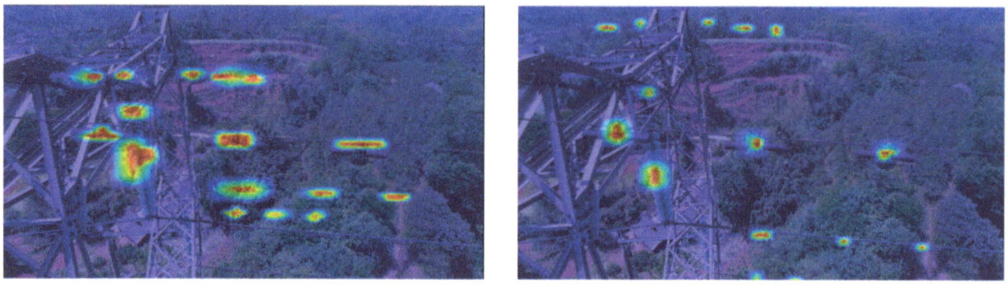

(c) Improved model with more accurate thermal region

Figure 14. Comparison of the results of Grad-CAM after adding scSE.

4.3. Validation of the Effect of NWD Loss Function and the Effect of NWD on the Model with Different Fusion Ratios

In this paper, CIou is replaced with an NWD loss function with better detection accuracy for small targets, and the training time is found to increase substantially after training. Then, an improvement strategy of mixing different proportions of NWD with CIoU is proposed to retain the accuracy of NWD while speeding up the training time. Finally, the models with loss functions fused in different proportions are retrained and tested on a typical fault dataset of transmission lines. The proportion of NWD loss functions in the experiments was set to 100%, 90%, 80%, 70%, and 60%, respectively, and the model performance for different fusion proportions is shown in Table 4. The 90% NWD + 10% CIoU in the table is the localization loss function consisting of 90% of the NWD loss function and 10% of the CIoU loss function together, and the others are similar.

Table 4. Experimental results after fusion of NWD with CIoU at different ratios.

Models	Training Time /(h)	mAP /(%)	Miss Rate (Fzc_xs)/(%)	Miss Rate (Defect)/(%)
YOLOv7-Tiny-Ghost	11.2	91.98	16.96	23.07
−(100%NWD)	24.5	92.92	11.03	10.24
−(90%NWD + 10%CIoU)	23	92.53	14.23	13.84
−(80%NWD + 20%CIoU)	21.5	92.83	14.35	11.31
−(70%NWD + 30%CIoU)	20	93.18	10.20	8.46
−(60%NWD + 40%CIoU)	18.5	92.5	13.04	12.3
−(50%NWD + 50%CIoU)	17	91.8	13.99	14.6

Figure 15 shows the test results of models with different fusion ratios on the dataset. It can be seen in Table 4 and Figure 16 that as the proportion of NWD decreases, the training time also gradually increases, and mAP presents a process of rising first and then falling, and 70% is the critical value. The mAP is 1.2% higher than the initial model; the training time decreases as the proportion of NWD decreases. This study adopts a fusion ratio model of (70%NWD + 30%CIoU) to balance the training time and model accuracy. The detection effect of small targets is improved, the missed detection rate of anti-vibration hammer corrosion is reduced by 6.76%, and the missed detection rate of insulator damage is reduced by 14.61%, proving the method's effectiveness and feasibility in this paper.

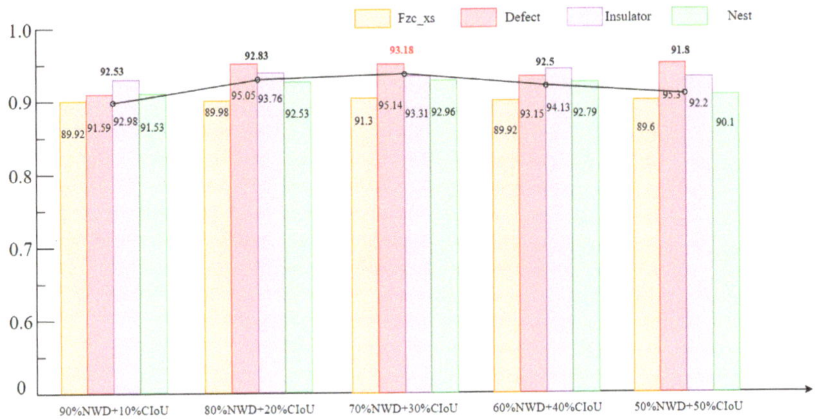

Figure 15. Test results of models with different fusion proportions on datasets.

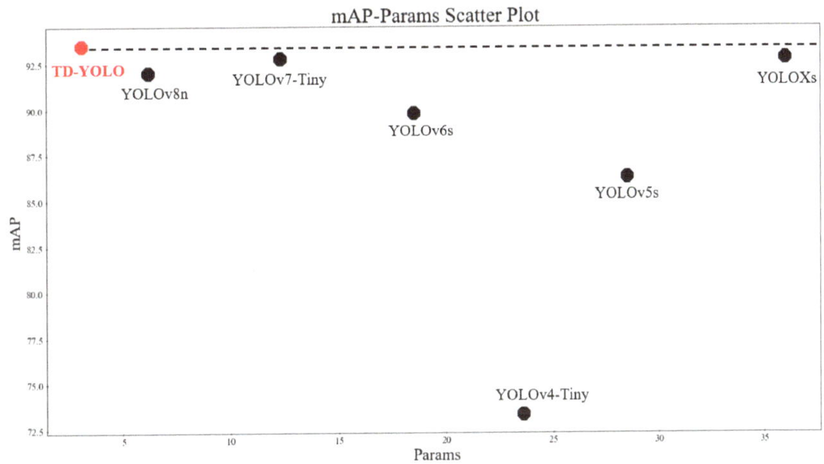

Figure 16. mAP-Params scatter plots of different models.

4.4. Comparison of Ablation Experiments

Table 5 is based on YOLOv7-Tiny and the comparison of the experimental results before and after adding the improvement strategy proposed in this paper. Among them, YOLOv7-Tiny is recorded as Algorithm 1.

Table 5. Ablation experiment results.

Models	Ghost	FBN	NWD	Fzc_xs (AP%)	Defect (AP%)	Insulator (AP%)	Nest (AP%)	mAP (%)	Parmas (MB)	FLOPs (G)
Algorithm 1				90.81	94.67	92.85	92.84	92.79	12.3	13
Algorithm 2	√			89.35	92.87	93.15	92.55	91.98	3	4.2
Algorithm 3	√	√		89.38	93.4	93.9	92.71	92.31	3.1	4.2
Algorithm 4	√		√	89.7	95.94	93.18	91.07	92.47	3	4.2
Algorithm 5	√	√	√	**90.7**	**96.1**	**93.7**	**93.7**	**93.5**	3.1	4.3

It can be seen in Table 5 that Algorithm 1 is the initial YOLOv7-Tiny, and Algorithm 2 optimizes the lightweight structure of the Ghost module based on Algorithm 1, the amount of calculation is reduced by 67.7%, the amount of parameters is reduced by 75.6%, and mAP is only reduced by 0.81%. For Algorithm 3 and Algorithm 4, based on Algorithm 2, the scSE attention mechanism is added to form a feature-balanced network and the NWD loss function is added to enhance the detection effect of small targets. Compared with Algorithm 2, Algorithm 3 has improved AP values for all detected objects. The problem of low accuracy, caused by scale transformation in the detection process, has been greatly improved; compared with Algorithm 2, Algorithm 4 has greatly improved the accuracy of small-target anti-vibration hammer corrosion and insulator damage, which also verifies the effectiveness of NWD for small target detection. Algorithm 5 is TD-YOLO, which combines three improvement strategies. The accuracy of each type of detection object is improved. Compared with Algorithm 2, the number of parameters remains unchanged, and the amount of calculation only increases by 0.1 G.

4.5. Horizontal Comparison of Experimental Results

To verify the model's performance and detection effect of the algorithm (TD-YOLO) in this paper, the original model and the other eight models were selected for comparison, as shown in Table 6.

Table 6. Comparison of various indicators of different models on the test set.

Models	Fzc_xs (AP%)	Defect (AP%)	Insulator (AP%)	Nest (AP%)	mAP (%)	Inference (ms)	Params (MB)
Faster R-CNN	55.72	85.76	89.34	80.18	77.75	78	114
YOLOv4	83.74	86.48	91.87	81.89	86	22.8	256
YOLOv4-Tiny	62.58	75.33	84.15	71.18	73.31	6.28	23.6
YOLOv5s	87.86	83.94	91.33	82.05	86.3	13	28.5
YOLOXs	90.84	95.42	**96.18**	88.63	92.77	15	36
YOLOv6s	89.6	88.1	92.6	88.8	89.8	9	18.5
YOLOv7-Tiny	90.81	94.67	92.85	92.84	92.79	5	12.3
YOLOv8n	90.6	93.8	92.8	90.9	92	4	6.2
TD-YOLO	**90.7**	**96.1**	93.7	**93.7**	**93.5**	**3.5**	**3.1**

It can be seen in Table 5 that the accuracy and speed of the second-stage algorithm Faster R-CNN have a significant gap compared with the first-stage algorithm YOLO series, especially for tiny target anti-vibration hammer corrosion, with only a 55.72% mAP. From the algorithm extension of YOLOv4 to YOLOv4-Tiny, the YOLO series algorithms are developing towards becoming lightweight. In the table, YOLOv5s, YOLOXs, YOLOv6s, YOLOv7-Tiny, and YOLOv8n are all their corresponding lightweight versions, and the accuracy is gradually increasing. For the model, the number of parameters gradually decreases; TD-YOLO compares with the original algorithm, mAP is improved by 0.71%, and the number of model parameters is reduced by 74.8%. Further, we analyzed the position of the improved algorithm in the current mainstream lightweight algorithm and drew the data as a parameter-precision floating-point diagram, as shown in Figure 16. It can be seen from the verification results on the transmission line fault detection data that the performance of TD-YOLO is in a leading position compared with the other YOLO series lightweight algorithms in various indicators.

To further verify the advantages of the proposed algorithm, three representative scenarios are selected to verify the model, namely, target faults under shadow occlusion, multi-scale target faults, and multiple small target faults [41,42]. In the experiment, it was compared with Faster R-CNN, the mainstream lightweight algorithm in Table 6, and our TD-YOLO algorithm. The detection results are shown in Figure 17.

(**a**) Faster R-CNN

(**b**) YOLOv4-Tiny

(**c**) YOLOv5s

(**d**) YOLOv6s

(**e**) YOLOXs

(**f**) YOLOv7-Tiny

Figure 17. *Cont.*

(g) TD-YOLO

Figure 17. Comparison of three representative scene detection effects in different model test sets.

5. Edge-Side Deployment

The edge deployment object uses Jetson Xavier NX, which has 384 CUDA cores, 48 Tensor cores, and two NVIDIA engines. It can run multiple modern neural networks in parallel, processing high-resolution data from multiple sensors simultaneously. It can be mounted onto a UAV to simulate the inspection conditions of UAVs. Real-time data collection is performed by calling the hardware camera, and the test results are shown in Table 7. It can be seen in Table 7 that the improved model reduces the inference delay by 12 ms compared with the original YOLOv7-Tiny, and the real-time detection speed increases by 4.8 FPS, reaching 23.5 ± 2.2 FPS. The simulation of the live drone inspection image is shown in Figure 18. The detection results meet the typical faults of transmission lines in the process of UAV inspection testing requirements. Finally, we explored whether the hardware parameters met the conditions for UAV deployment, and the test results are shown in Table 8 [43].

Table 7. Test results on the Jetson Xavier NX before and after the improved model.

Models	Inference (ms)	NMS (ms)	Speed (FPS)	mAP (%)
Algorithm 1	50 ± 4	4.5 ± 1.5	18.3 ± 1.8	92.79
Algorithm 2	33 ± 3	4.5 ± 1.5	26.7 ± 2.3	91.98
Algorithm 3	35.7 ± 2.8	4.5 ± 1.5	24.8 ± 2.4	92.31
Algorithm 4	34.9 ± 2.1	4.5 ± 1.5	25.3 ± 2.2	92.47
Algorithm 5	38 ± 3	4.5 ± 1.5	23.5 ± 2.2	**93.5**

Figure 18. Simulation of live drone inspection image.

Table 8. Comparison of indicators of Jeston Xavier NX and M300-RTK.

Indicators	Jeston Xavier NX	M300-RTK	Effective
Weight	260 g	Maximum load of 2.7 kg	√
Form Factor	70 mm × 45 mm	180 mm × 130 mm	√
Power Consumption	Maximum 15 W	Rated power 17 W	√
Frame Rate	23.5 ± 2.2 FPS	Maximum 30 FPS	√

The name of the algorithm in Table 7 is the same as that in Table 5. Algorithm 1 is the YOLOv7-Tiny model, and Algorithm 5 is TD-YOLO after the ablation experiment.

As can be seen from Table 8, the embedded devices tested in this paper are all suitable for deployment in the UAVs used for transmission line inspection, which further validates the feasibility of the algorithms in this paper.

6. Conclusions

1. This paper proposes a typical fault detection algorithm for transmission lines based on a lightweight module and a feature-balanced network. Through the Ghost module, YOLOv7-Tiny is reorganized in a lightweight way to reduce the parameters and computation of the model so that it can meet the deployment conditions. Through the introduction of the scSE attention mechanism and PA-Net to form a feature-balancing network, the information of the upper and lower layers is better integrated, which, to a certain extent, reduces the missed detection caused by the insufficient feature expression capability during the scale transformation process of faults. The NWD loss function is used to replace part of the CIoU to improve the detection of small target faults while ensuring the training speed of the model.

2. Based on the self-built dataset, the model designed in this paper has obvious advantages in terms of detection accuracy and detection speed compared with the lightweight models of the same stage, and the effectiveness of the model's improvement is verified by the mobile hardware.

3. The self-built dataset in this paper mainly includes transmission line equipment faults (typically broken insulators), transmission line foreign object faults (typically bird's nests), and transmission line metalwork faults (typically anti-vibration hammer corrosion), and the fault types are not limited to these typical faults. Further research will be carried out by adding fault-type detection to make the model more universal.

Author Contributions: Conceptualization, G.H. and R.W.; methodology, G.H.; software, R.W.; validation, R.W., Q.Y. and S.L.; formal analysis, G.H.; investigation, L.Z.; resources, M.H., S.L. and L.Q.; data curation, R.W., Q.Y. and L.Q.; writing—original draft preparation, R.W.; writing—review and editing, G.H., R.W. and L.Q.; visualization, M.H.; supervision, M.Z. and L.Q. All authors have read and agreed to the published version of the manuscript.

Funding: This work was supported by the National Key R&D Program of China (No.2020YFB0905900).

Conflicts of Interest: The authors declare no conflict of interest.

Code: https://github.com/wangruijie123/Drones-YOLOv7 (accessed on 10 October 2023).

Test Videos: https://www.youtube.com/@chrisD-zg9kc/featured (accessed on 10 October 2023).

References

1. He, M.; Qin, L.; Deng, X. Transmission Line Segmentation Solutions for UAV Aerial Photography Based on Improved UNet. *Drones* **2023**, *7*, 274. [CrossRef]
2. Sui, Y.; Ning, P.; Niu, P. Review on Mounted UAV for Transmission Line Inspection. *Power Syst. Technol.* **2021**, *9*, 3636–3648.
3. Lunze, J.; Richter, J. Reconfigurable Fault-tolerant Control: A Tutorial Introduction. *Eur. J. Control* **2008**, *14*, 359–386. [CrossRef]
4. Merrill, W.; DeLaat, J.; Bruton, W. Advanced detection, isolation, and accommodation of sensor failures–Real-time evaluation. *J. Guid. Control Dyn.* **1988**, *11*, 517–526. [CrossRef]
5. Liu, C.; Wu, Y. Research progress of vision detection methods based on deep learning for transmission lines. *Proc. CSEE* **2022**, *8*, 31.

6. Khodayar, M.; Liu, G.; Wang, J. Deep learning in power systems research: A review. *CSEE J. Power Energy Syst.* **2021**, *3*, 209–220.
7. Chen, C.; Zheng, Z.; Xu, T.; Guo, S.; Feng, S.; Yao, W.; Lan, Y. YOLO-Based UAV Technology: A Review of the Research and Its Applications. *Drones* **2023**, *7*, 190. [CrossRef]
8. Girshick, R.; Donahue, J.; Darrell, T. Rich feature hierarchies for accurate object detection and semantic segmentation. In Proceedings of the 2014 IEEE Conference on Computer Vision and Pattern Recognition (CVPR), Columbus, OH, USA, 23–28 June 2014; IEEE: Piscataway, NJ, USA, 2014; pp. 580–587.
9. Girshick, R. Fast R-CNN. In Proceedings of the 2015 IEEE International Conference on Computer Vision (ICCV), Santiago, Chile, 7–13 December 2015; IEEE: Piscataway, NJ, USA, 2015; pp. 1440–1448.
10. Ren, S.; He, K.; Girshick, R. Faster R-CNN: Towards Real-Time Object Detection with Region Proposal Networks. *IEEE Trans. Pattern Anal. Mach. Intell.* **2017**, *39*, 1137–1149. [CrossRef]
11. Cai, Z.; Vasconcelos, N. Cascade R-CNN: Delving Into High Quality Object Detection. In Proceedings of the IEEE Conference on Computer Vision and Pattern Recognition (CVPR), Salt Lake City, UT, USA, 18–23 June 2018; pp. 6154–6162.
12. Liu, W. SSD: Single Shot MultiBox Detector. In *Computer Vision-ECCV*; Lecture Notes in Computer Science, 9905; Springer: Cham, Switzerland, 2016.
13. Redmon, J.; Divvala, S.; Girshick, R. You Only Look Once: Unified, Real-Time Object Detection. In Proceedings of the IEEE Conference on Computer Vision and Pattern Recognition (CVPR), Las Vegas, NV, USA, 27–30 June 2016; pp. 779–788.
14. Redmon, J.; Farrhadi, A. YOLO9000: Better, Faster, Stronger. In Proceedings of the IEEE Conference on Computer Vision and Pattern Recognition (CVPR), Honolulu, HI, USA, 21–26 July 2017; pp. 6517–6525.
15. Redmon, J.; Farrhadi, A. YOLOv3: An Incremental Improvement. *arXiv* **2018**, arXiv:1804.02767.
16. Bochkovskiy, A.; Wang, C.; Liao, H. YOLOv4: Optimal Speed and Accuracy of Object Detection. *arXiv* **2020**, arXiv:2004.10934.
17. YOLOv5. Available online: https://github.com/ultralytics/yolov5 (accessed on 23 July 2023).
18. Li, C.; Li, L.; Jiang, H. YOLOv6: A Single-Stage Object Detection Framework for Industrial Applications. *arXiv* **2022**, arXiv:2209.02976.
19. Wang, C.; Bochkovskiy, A. YOLOv7: Trainable bag-of-freebies sets new state-of-the-art for real-time object detectors. In Proceedings of the IEEE/CVF Conference on Computer Vision and Pattern Recognition (CVPR), Vancouver, BC, Canada, 17–24 June 2023; pp. 7464–7475.
20. Bai, J.; Zhao, R.; Gu, F. Multi-target Detection and Fault Recognition Image Processing Method. *High Volt. Eng.* **2019**, *11*, 3504–3511.
21. Hao, S.; Ma, R.; Zhao, X. Fault Detection of YOLOv3 Transmission Line Based on Convolutional Block Attention Model. *Power Syst. Technol.* **2021**, *8*, 2979–2987.
22. Hao, S.; Yang, L.; Ma, X. YOLOv5 Transmission Line Fault Detection Based on Attention Mechanism and Cross-scale Feature Fusion. *Proc. CSEE* **2023**, *6*, 2319–2331.
23. Hao, S.; Zhang, X.; Ma, X. Small Target Fault Detection Method for Transmission Line Based on PKAMNet. *High Volt. Eng.* **2023**, *3*, 1–10.
24. Qiu, Z.; Zhu, X.; Liao, C. A Lightweight YOLOv4-EDAM Model for Accurate and Real-time Detection of Foreign Objects Suspended on Power Lines. *IEEE Trans. Power Deliv.* **2022**, *38*, 1329–1340. [CrossRef]
25. Deng, F.; Xie, Z.; Mao, W. Research on edge intelligent recognition method oriented to transmission line insulator fault detection. *Int. J. Electr. Power Energy Syst.* **2022**, *139*, 108054. [CrossRef]
26. Han, G.; He, M. Insulator detection and damage identification based on improved lightweight YOLOv4 network. *Energy Rep.* **2021**, *7*, 187–197. [CrossRef]
27. Li, X.; Liu, H.; Liu, G. Transmission Line Pin Defect Detection Based on Deep Learning. *Power Syst. Technol.* **2021**, *8*, 2988–2995.
28. Zhang, H.; Qi, Q.; Zhang, J. Bird nest detection method for transmission lines based on improved YOLOv5. *Power Syst. Prot. Control* **2023**, *2*, 151–159.
29. Zhao, W.; Xu, M.; Cheng, X. An Insulator in Transmission Lines Recognition and Fault Detection Model Based on Improved Faster RCNN. *IEEE Trans. Instrum. Meas.* **2021**, *70*, 1–8. [CrossRef]
30. Chen, K.; Liu, X.; Jia, L. Insulator Defect Detection Based on Lightweight Network and Enhanced Multi-scale Feature Fusion. *High Volt. Eng.* **2023**, *2*, 1–12.
31. Kang, J.; Wang, Q.; Liu, W. Detection Model of Aerial Photo Insulator Multi-defect by Integrating CAT-BiFPN and Attention Mechanism. *High Volt. Eng.* **2023**, *2*, 1–15.
32. Li, H.; Dong, Y.; Liu, Y.; Ai, J. Design and Implementation of UAVs for Bird's Nest Inspection on Transmission Lines Based on Deep Learning. *Drones* **2022**, *6*, 252. [CrossRef]
33. Zheng, Z.; Wang, P.; Ren, D. Enhancing Geometric Factors in Model Learning and Inference for Object Detection and Instance Segmentation. *arXiv* **2020**, arXiv:2005.03572v4. [CrossRef]
34. Han, K.; Wang, Y.; Tian, Q. GhostNet: More Features From Cheap Operations. In Proceedings of the IEEE/CVF Conference on Computer Vision and Pattern Recognition (CVPR), Seattle, WA, USA, 13–19 June 2020; pp. 1580–1589.
35. Roy, A.; Navab, N.; Wachinger, C. Recalibrating Fully Convolutional Networks with Spatial and Channel' Squeeze & Excitation' Blocks. *IEEE Trans. Med. Imaging* **2019**, *2*, 540–549.
36. Woo, S.; Park, J.; Lee, J. Convolutional Block Attention Module. In Proceedings of the European Conference on Computer Vision (ECCV), Munich, Germany, 8–14 September 2018; pp. 3–19.

37. Dong, G.; Xie, W.; Huang, X. Review of Small Object Detection Algorithms Based on Deep Learning. *Comput. Eng. Appl.* **2023**, *11*, 16–27.
38. Wang, J.; Xu, C.; Yang, W. A Normalized Gaussian Wasserstein Distance for Tiny Object Detection. *arXiv* **2021**, arXiv:2110.13389.
39. Blanke, M.; Kinnaert, M.; Lunze, J.; Staroswiecki, M. *Diagnosis and Fault-Tolerant Control*; Springer: Berlin/Heidelberg, Germany, 2003.
40. Han, G.; Wang, R.; Yuan, Q.; Li, S.; Zhao, L.; He, M.; Yang, S.; Qin, L. Detection of Bird Nests on Transmission Towers in Aerial Images Based on Improved YOLOv5s. *Machines* **2023**, *11*, 257. [CrossRef]
41. Ding, S. *Model-Based Fault Diagnosis Techniques*; Springer: London, UK, 2013. [CrossRef]
42. Frank, P.; Ding, S.; Marcu, T. Model-based fault diagnosis in technical processes. *Trans. Inst. Meas. Control* **2000**, *22*, 57–101. [CrossRef]
43. Li, Y.; Fan, Q.; Huang, H.; Han, Z.; Gu, Q. A Modified YOLOv8 Detection Network for UAV Aerial Image Recognition. *Drones* **2023**, *7*, 304. [CrossRef]

drones

Article

Relative Localization within a Quadcopter Unmanned Aerial Vehicle Swarm Based on Airborne Monocular Vision

Xiaokun Si [1], Guozhen Xu [1], Mingxing Ke [1], Haiyan Zhang [1], Kaixiang Tong [2] and Feng Qi [1,*]

[1] College of Electronic Engineering, National University of Defense Technology, Hefei 230031, China
[2] Beijing Space Information Relay and Transmission Technology Center, Beijing 102300, China
* Correspondence: qifeng17@nudt.edu.cn

Abstract: Swarming is one of the important trends in the development of small multi-rotor UAVs. The stable operation of UAV swarms and air-to-ground cooperative operations depend on precise relative position information within the swarm. Existing relative localization solutions mainly rely on passively received external information or expensive and complex sensors, which are not applicable to the application scenarios of small-rotor UAV swarms. Therefore, we develop a relative localization solution based on airborne monocular sensing data to directly realize real-time relative localization among UAVs. First, we apply the lightweight YOLOv8-pose target detection algorithm to realize the real-time detection of quadcopter UAVs and their rotor motors. Then, to improve the computational efficiency, we make full use of the geometric properties of UAVs to derive a more adaptable algorithm for solving the P3P problem. In order to solve the multi-solution problem when less than four motors are detected, we analytically propose a positive solution determination scheme based on reasonable attitude information. We also introduce the maximum weight of the motor-detection confidence into the calculation of relative localization position to further improve the accuracy. Finally, we conducted simulations and practical experiments on an experimental UAV. The experimental results verify the feasibility of the proposed scheme, in which the performance of the core algorithm is significantly improved over the classical algorithm. Our research provides viable solutions to free UAV swarms from external information dependence, apply them to complex environments, improve autonomous collaboration, and reduce costs.

Keywords: UAV swarm; relative localization; Perspective-n-Point; GNSS-denied environments; YOLO; keypoint detection

Citation: Si, X.; Xu, G.; Ke, M.; Zhang, H.; Tong, K.; Qi, F. Relative Localization within a Quadcopter Unmanned Aerial Vehicle Swarm Based on Airborne Monocular Vision. *Drones* **2023**, 7, 612. https://doi.org/10.3390/drones7100612

Academic Editor: Diego González-Aguilera

Received: 31 August 2023
Revised: 26 September 2023
Accepted: 26 September 2023
Published: 29 September 2023

1. Introduction

Small multi-rotor UAVs have the advantages of good maneuverability, rich expansion functions, and great intelligence potential, but the limited performance of a single aircraft and poor survivability have also been exposed in use [1]. Swarming can compensate for the weaknesses of a single UAV while further leveraging its strengths [2]. Currently, UAV swarms have shown great value and potential in missions such as aerial Internet of Things (IoT) [3,4], relay communication support [5,6], aerial light shows, regional security [7], and military operations [8], which have become one of the inevitable trends in the development of UAV applications. Accurate real-time position information is the basis for UAVs to accomplish a variety of air-to-ground missions. In addition to absolute position information, it also involves the relative position relationship between each UAV within a swarm. It is no exaggeration to say that relative location information is no less important than absolute location information from a swarm perspective. It enables UAVs to maintain planned formations, avoid collisions with each other, and accomplish coordinated maneuvers [9]. Therefore, precise relative localization is a must for swarm UAVs, which is of great significance in reducing the swarm's reliance on absolute position information and improving the swarm's ability to survive in hazardous environments.

In recent years, solutions based on various hardware and methods have emerged for relative localization problems. While they show good performance, the different characteristics and conditions of use make many of these solutions inappropriate for small multi-rotor UAV swarms. Currently, the acquisition of relative localization information between UAVs still relies heavily on the absolute position data of each UAV from the Global Navigation Satellite System (GNSS) [10]. In addition, similar problems exist with relative localization via motion capture systems, simultaneous localization and mapping (SLAM) [11,12], and ground-based ultra-wide band (UWB) localization systems [13]. They all need to first obtain their respective position coordinates in the same spatial coordinate system from external infrastructure or environmental information and then solve for the relative localization information based on this. These methods have obvious drawbacks. Firstly, once absolute localization has failed, relative localization will also not be possible, for example, when encountering a GNSS-denied environment, when the coverage of ground-based localization stations is exceeded or when the environmental features required for SLAM are not evident. Secondly, errors in absolute localization will be superimposed and magnified during the conversion to relative localization information [14]. In addition, absolute localization will take up limited resources per swarm UAV, which could have been avoided.

The model for UAV swarms is derived from the group behavior of flying creatures in nature [15]. They usually rely on organ functions such as vision and hearing to directly obtain information about their relative positions to each other. UAV swarms, as multi-intelligence systems, should also have the ability to achieve relative localization without relying on external facilities or information. Similar functions have already been implemented in the rapidly developing field of advanced driving assistance system (ADAS) research [16,17]. Based on the information provided by vision, laser, and other sensors, it has been possible to achieve accurate relative positioning of objects within a certain range while the vehicle is in motion. However, the environment in which vehicles are driven can be approximated as a two-dimensional space, whereas drones are in a more complex three-dimensional scenario.

Relative localization based on radio signals is a classical approach, currently represented by airborne UWB and relative localization based on carrier phase [18,19]. Although they are superior in terms of localization accuracy, they will significantly increase the cost, power consumption, and system complexity of each UAV, as well as taking into account mutual interference problems. While LIDAR has superior performance and proven applications, the same expensive price and high power consumption prevent it from being the first choice for swarm UAVs [20]. Millimeter-wave radar is less expensive, but it has lower localization accuracy and a smaller measurement range [21].

While relative localization achieved based on vision SLAM is not considered due to its indirectness and instability, vision sensors can also directly provide useful information for relative localization [22]. Wide-angle lenses, gimbals, camera scheduling algorithms, and target tracking algorithms [23] ensure flexible acquisition of environmental images [24]. Binocular cameras and depth cameras are the current mainstream vision solutions [25]. Binocular vision localization uses the principle of triangular geometric parallax to achieve relative localization. However, the co-processing of binocular data requires high computing resources and speed, and the accuracy and range of measurements are limited when the parallax is small. Depth cameras can obtain depth data based on the principle of structured light or time of flight (ToF), but they have a relatively small applicable distance and imaging field of view, making them unsuitable for the relative localization of drones in motion [26].

Monocular cameras are common onboard sensors for UAVs and have the advantage of being cheap and easy to deploy. However, information based solely on a single frame from a single camera can only measure direction but not distance unless more auxiliary information is introduced, which is also the core problem that needs to be solved for monocular visual localization [27]. The implementation of relative localization based on airborne monocular vision offers significant advantages in terms of cost, complexity, and hardware requirements compared to the other methods mentioned above, but there is a lack of mature solutions. Therefore, the development of a relative localization method based only on airborne monocular vision is of great practical importance to solve the relative localization problem of small multi-rotor UAV swarms.

In this research, we develop an airborne monocular-vision-based relative localization scheme using a small quadrotor UAV as an experimental platform. It achieves accurate real-time relative localization between UAVs based only on a single airborne camera's data and simple feature information of the quadrotor UAV. In summary, our contributions are as follows:

- We propose a new idea of directly using only the rotor motors as the basis for localization and use the deep-learning-based YOLOv8-pose keypoint detection algorithm to achieve fast and accurate detection of UAVs and their motors. Compared to other visual localization information sources, we do not add additional conditions and data acquisition is more direct and precise.
- A more suitable algorithm for solving the PnP (Perspective-n-Point) problem is derived based on the image plane 2D coordinates of rotor motors and the shape feature information of the UAV. Our algorithm is optimized for the application target, reduces the complexity of the algorithm by exploiting the geometric features of the UAV, and is faster and more accurate than classical algorithms.
- For the multi-solution problem of P3P, we propose a new scheme to determine the unique correct solution based on the pose information instead of the traditional reprojection method, which solves the problem of occluded motors during visual relative localization. The proposed method breaks the limitations of classical methods and reduces the amount of data necessary for visual localization.

A description of symbols and mathematical notations involved in this paper is shown in Table 1.

Table 1. Description of symbols and mathematical notations.

$\{A_i\}$	The set of points corresponding to all values of i.
(a, b)	Coordinates in the specified coordinate system.
$Oxyz$	The spatial coordinate system with O as the origin and Ox, Oy and Oz as the positive directions of the coordinate axes.
$\angle AOB$	The angle between the rays OA and OB with O as the vertex.
$\boldsymbol{A}$	Matrices, including vectors.
$\boldsymbol{AB}$	A vector with A as the starting point and B as the ending point.
$\boldsymbol{t}_n^m$	The displacement matrix of the O_m-coordinate system with respect to the O_n-coordinate system.
$\boldsymbol{R}_n^m$	The rotation matrix of the O_m-coordinate system with respect to the O_n-coordinate system.
$\boldsymbol{A} \times \boldsymbol{B}$	Multiply matrix $\boldsymbol{A}$ with matrix $\boldsymbol{B}$.
$[\cdot]^{\mathrm{T}}$	The transpose of the matrix.
$\lVert\cdot\rVert$	The modulus of the vector.

2. Related Work

2.1. Monocular Visual Localization

Currently, the main specific methods for monocular visual localization are feature point methods, direct methods, deep-learning-based methods, and semantic-information-based methods. References [28,29] both propose the use of deep learning target detection algorithms to classify and detect images from different angles of the UAV and then combine this with the corresponding dimensional information to estimate the relative position of the UAV. However, this places high demands on the detection model; an accurate detection model often means a larger amount of data collection for training as well as slower detection speeds, while simplifying the model will lead to a significant increase in error. Another idea is to artificially add features to the UAV to aid detection. In reference [30], Zhao et al. used the derived P4P algorithm to solve the relative position information of the target UAV based on the image positions of four LEDs pre-mounted on the UAV, but only semi-physical simulation experiments were carried out. Walter et al. obtained real-time relative position information of the UAV by detecting scintillating UV markers added to the UAV and using a 3D time-position Hough transform [31]. In reference [32], Saska et al. achieved relative localization in their study by deploying geometric patterns on the UAV and detecting them, with the study also incorporating inertial guidance information. Zhao et al. instead used the April Tag algorithm to achieve the acquisition of UAV position and attitude information by detecting and processing the onboard 2D code [33]. While these methods can achieve good results, the additional addition of features is not conducive to practical application and is not a preferred option. In reference [34], Pan et al. propose a learning-based correspondence point matching model to solve the position information of ground targets based on multiple frames from the UAV's onboard monocular camera. But this method is based less on real time and cannot adapt to the high-speed movement characteristics of UAVs. Reference [35] presents a method for obtaining UAV position and attitude information by inspecting the four rotor motors and other key components of the UAV and applying an improved PnP algorithm. However, we do not believe it is possible to detect so many characteristics of a UAV at the same time when detecting it in the air.

Based on the above analysis, harsh condition constraints, higher acquisition difficulty, and lower real-time and accuracy are the main problems in acquiring data sources for visual localization. We believe that relative localization based on the image feature information of the UAV itself is a feasible idea. Moreover, the number of feature points should be required to be as small as possible to facilitate detection and fast solving. The rotor motors are a necessary component of a quadcopter drone, and there are at least three of them visible when viewed from almost any angle. Therefore, we consider the motors as a reference point for visual localization and explore solving the PnP problem based on better parameters and computational effort.

2.2. Target and Keypoint Detection

Accurate detection of the UAV and its motors is the basis for visual localization. Deep-learning-based target detection algorithms are the current mainstream solution, with representative algorithms such as Faster R-CNN, YOLO, and SSD. Compared to other algorithms, the YOLO algorithm is based on the idea of one-off detection, which is faster to process and more suitable for applications in real-time scenarios [36]. Thanks to the simple network architecture and optimized algorithm design, the YOLO algorithm is simple to deploy and more conducive to deployment on lower-performance edge computers. Based on these advantages, the YOLO algorithm is widely used in ground-to-UAV and UAV-to-ground target detection in real time. However, detection accuracy, localization precision, and performance on small targets have been the relative disadvantages of the YOLO algorithm and have been the focus of its iteration and improvement [37].

The YOLO algorithm has now evolved to the latest v8 version, with many improvements referencing the strengths of previous versions. YOLOv8 improves on the FPN (feature pyramid networks) idea and the Darknet53 backbone network by replacing the C3 structure in YOLOv5 with the more gradient flow-rich C2f structure. This improves the multi-scale predictive capability and lightness of the algorithm. In the Head section, YOLOv8 uses the mainstream decoupled head structure and replaces Anchor-Base with Anchor-Free. in addition, YOLOv8 is optimized for multi-scale training, data enhancement, and post-processing optimization, making it easier to deploy and train [38]. The YOLOv8 development team has also released a pre-trained human pose detection model, YOLOv8-pose, as seen in reference [39]. Pose estimation is realized based on the detection and localization of specific parts and joints of the human body. Therefore, YOLOv8-pose can be considered as a method for keypoint detection [40].

Previous related work has focused on detecting UAV motors as area targets based on their additional characteristics [30,31,35]. In this study, we apply YOLOv8-pose, which is used for human posture detection, to the detection of the motors of UAVs. We hope to realize direct, accurate, and real-time access to localization data sources based on the advantages of YOLOv8-pose.

2.3. Solving the PnP Problem

The PnP problem is one of the classic problems in computer vision. It involves determining the position and orientation of a camera, given n points in three-dimensional space and their corresponding projection points on the camera image plane, combined with the camera parameters. Common solution methods include Gao's P3P [41], direct linear transformation (DLT) [42], EPnP (Efficient PnP) [43], UPnP (uncalibrated PnP) [44], etc. They have different requirements for the number of 2D–3D point pairs and are suitable for different scenarios. In practice, there are often errors in the coordinates of the projected points. More point pairs tend to help improve the accuracy and robustness of the results but increase the amount of work involved in matching and solving the point pairs. Due to the occurrence of occlusion, when photographing another quadcopter UAV with the onboard camera, often only three motors are detected. Three sets of point pairs are also the minimum requirement for solving the PnP problem, also known as the P3P problem.

Current solution methods for P3P problems can be divided into two-stage methods and single-stage methods. The classical Gao's method [41] mainly uses similar triangles, the cosine theorem, and Wu's elimination method to solve the problem. In reference [45], Li et al. proposed a geometric feature based on a perspective similar triangle (PST), reducing the unknown parameters, reducing the complexity of the equations, and showing a more robust performance. However, they all require the distance from the camera to the three points to be found first, and then use methods such as singular value decomposition (SVD) to obtain position and pose information. The single-stage method eliminates the intermediate process of solving for distance values, which is more in line with the application needs of this study. The method proposed by Kneip is representative of the single-stage method, which derives the solution for camera position and pose directly by introducing an intermediate camera and a series of geometrical treatments [46]. It offers a significant speed improvement over Gao's method, although at the cost of complex geometric transformations. Furthermore, all P3P solutions mention the need to deal with the non-uniqueness of the solution of the P3P problem by the reprojection method using the fourth set of point pairs. However, in reality, when viewed from a partial angle, only three motors are often observable due to the fuselage's shading.

Classical PnP solution methods are devoted to solving general problems and do not satisfy the special cases in this study. Meanwhile, more geometric features of rotor UAVs are not utilized in these methods. In this research, we follow the idea of the single-stage method and derive the position result of the P3P problem directly from an algebraic resolution perspective based on the dimensional characteristics of the quadrotor UAV. For the multi-solution problem of P3P, we propose a solution that does not require a fourth set of point pairs based on the attitude characteristics of the UAV.

3. Detection of UAVs and Motors

3.1. Detection Model Training

First, we simulate the perceptual behavior of on-board vision by photographing a quadrotor UAV hovering in the air from different angles and distances, as shown in Figure 1. We then label the captured images, where UAVs are labeled as detection targets with rectangles and motors are labeled as keypoints with dots. In order to correctly correspond to the 2D–3D point pairs, the motor labeling order is specified as clockwise from the first motor on the left, viewed from the bottom up. Obscured motors are not labeled. Finally, following the general steps of YOLOv8-pose model training, the labeled images and data were imported to generate the training model.

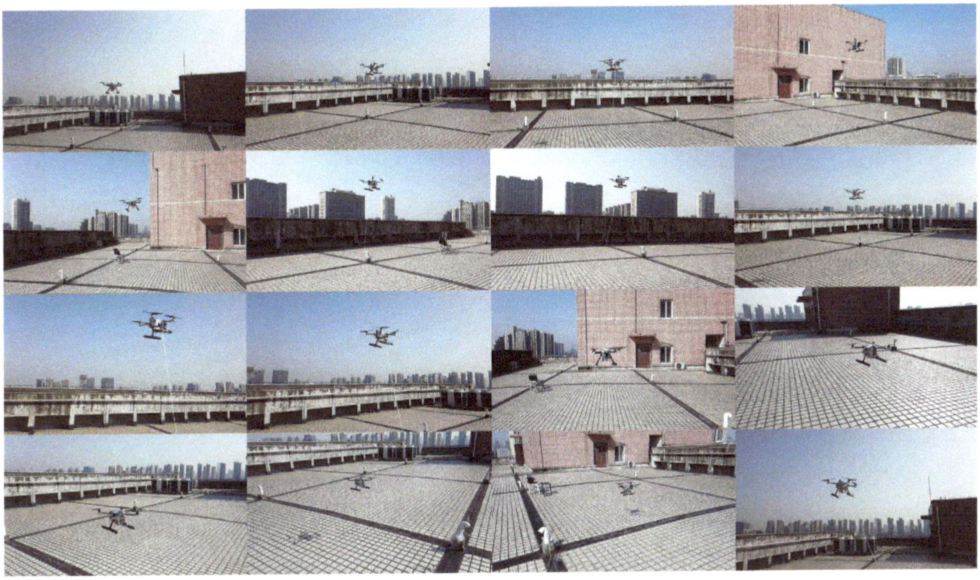

Figure 1. Acquisition of UAV images.

3.2. Sequencing of Motor Keypoints

Although the labeling order of the motors has been specified, the output order of the motor keypoints may still be wrong due to the complexity of the UAV's flight attitude and the multiple angles of detection. Therefore the sequence of keypoints of motors needs to be calibrated. Due to the presence of occlusion, two to four motors can be detected in one frame, as shown in Figure 2.

Figure 2. Three cases of the number of motors can be seen.

We set the pixel coordinates of the motors on the image plane to be $\{P_i^0 = (u_i^0, v_i^0)\}$ (i = 1,2,3,4), and the correct coordinates after sorting to be $\{P_i = (u_i, v_i)\}$. When two to three motors can be detected, we specify that the motors appearing on the screen are sorted from left to right. When all four motors are detected, we use the condition that the two midpoints of the lines connecting the non-adjacent motors should theoretically overlap to judge and correct the motor order. The specific algorithm for sorting is shown in Algorithm 1:

Algorithm 1 Sorting the four motors

Require: $\{P_i^0 = (u_i^0, v_i^0)\}, i \in \{1:n\}$
Ensure: $\{P_i = (u_i, v_i)\}$
if $n < 4$ **then**
  Sort $P_{1:n}^0$ by $u_1^0 < u_2^0\ (< u_3^0)$
else
  for $i, j \in \{1:n\}, i < j$ **do**
    $o_{ij} = [\frac{u_i^0 + u_j^0}{2}, \frac{v_i^0 + v_j^0}{2}]$
  end for
  $d_1 = \|o_{12}o_{34}\|, d_2 = \|o_{13}o_{24}\|, d_1 = \|o_{14}o_{23}\|$
  if $min\{d_{1:3}\} = d_1$ **then**
    Swap the values of P_2^0 and P_3^0
  else if $min\{d_{1:3}\} = d_3$ **then**
    Swap the values of P_3^0 and P_4^0
  end if
end if
$\{P_i\} = \{P_i^0\}$

4. Relative Position Solution Method

4.1. Problem Model

Typically, the onboard vision sensor can detect three to four motors of the UAV within the field of view. The solution of the relative position at this point is a P3P problem.

The model of the P3P problem is shown in Figure 3. Camera coordinate system, pixel coordinate system, and motor coordinate system are established separately. O_c is the optical centre of the camera and $O_p uv$ is the pixel coordinate system. The right-angle coordinate system $O_c x_c y_c z_c$ is established with O_c as the origin, where the x_c-axis is in the same direction as the u-axis, the z_c-axis is reversed with the v-axis, and the y_c-axis is on the optical axis. $\{M_i\}(i = 1,2,3,4)$ represents the four motors of the UAV and O_m is the intersection of the central axis of the UAV with the plane where the motors are located, here representing the spatial position of the UAV. We set up the right-angle coordinate system $O_m x_m y_m z_m$ with the point O_m as the origin, where the x_m-axis and y_m-axis are in the positive direction of $\boldsymbol{O_m M_3}$ and $\boldsymbol{O_m M_4}$, respectively, and the z_m-axis points above the top of the UAV.

In fact, the camera coordinate system and the motor coordinate system express the motion attitude of the camera gimbal and the UAV, which can be understood as the result of the transformation with respect to the Earth coordinate system or the inertial coordinate system. The pixel coordinate system is fixed with respect to the camera coordinate system and is determined by the internal parameters of the camera. Then, the P3P problem is converted to solving for the translation $\boldsymbol{t}_c^m$ and rotation $\boldsymbol{R}_c^m$ of the motors coordinate system with respect to the camera coordinate system, which are set as

$$\boldsymbol{t}_c^m = \begin{bmatrix} t_x \\ t_y \\ t_z \end{bmatrix}, \quad \boldsymbol{R}_c^m = \begin{bmatrix} r_{11} & r_{12} & r_{13} \\ r_{21} & r_{22} & r_{23} \\ r_{31} & r_{32} & r_{33} \end{bmatrix}, \tag{1}$$

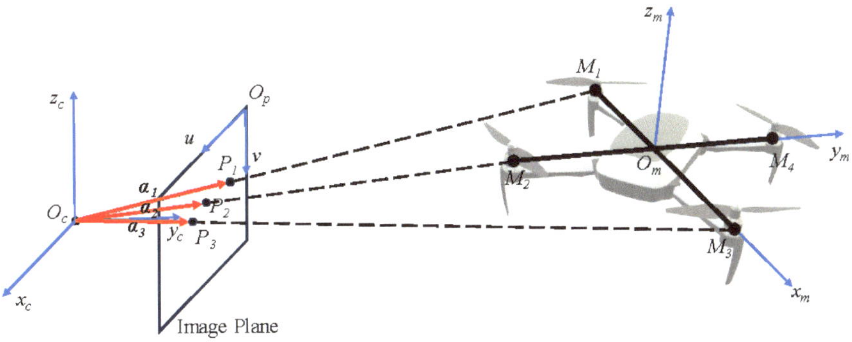

Figure 3. The model for the P3P problem.

4.2. Improved Solution Scheme for the P3P Problem

We first consider the general case where only three motors are detected. The pixel coordinates P_i of the motors and the camera focal length f are known. The vectors $\boldsymbol{\alpha}_i$ represent $\boldsymbol{O_cP_i}$. Obviously,

$$\boldsymbol{\alpha}_i = [u_i^c,\ f,\ v_i^c]^{\mathrm{T}},\ i = 1, 2, 3, \tag{2}$$

where

$$\begin{cases} u_i^c = \dfrac{u_i - \frac{W_p}{2}}{\frac{W_p}{2}} \cdot \dfrac{W_I}{2}, \\ v_i^c = -\dfrac{v_i - \frac{H_p}{2}}{\frac{H_p}{2}} \cdot \dfrac{H_I}{2}, \end{cases} \tag{3}$$

where W_p and H_p represent the pixel width and height of the image plane, and W_I and H_I represent the actual width and height of it.

Obviously, the point P_i is the projection on the image plane of the reflected rays from the point M_i when they strike the focal point O_c along a straight line. So, $\boldsymbol{O_cM_i}$ can be expressed as

$$\boldsymbol{O_cM_i} = k_i\boldsymbol{\alpha}_i,\ i = 1, 2, 3. \tag{4}$$

We set $\|\boldsymbol{O_mM_i}\| = d$, which can be obtained by measuring. Accordingly,

$$\begin{aligned} \boldsymbol{O_mM_1} &= [-d,\ 0,\ 0]^{\mathrm{T}}, \\ \boldsymbol{O_mM_2} &= [\ 0, -d,\ 0]^{\mathrm{T}}, \\ \boldsymbol{O_mM_3} &= [\ \ d,\ 0,\ 0]^{\mathrm{T}}. \end{aligned} \tag{5}$$

Based on the rules of vector transformation, $\boldsymbol{O_cM_i}$ can also be obtained from $\boldsymbol{O_mM_i}$ by the following transformation,

$$\boldsymbol{O_cM_i} = \boldsymbol{R_c^m} \times \boldsymbol{O_mM_i} + \boldsymbol{t_c^m},\ i = 1, 2, 3. \tag{6}$$

From (1), (4), (5), and (6), it follows that

$$
\begin{aligned}
k_1\boldsymbol{\alpha}_1 &= -\begin{bmatrix} r_{11} \\ r_{21} \\ r_{31} \end{bmatrix} d + \boldsymbol{t}_c^m, \\
k_2\boldsymbol{\alpha}_2 &= -\begin{bmatrix} r_{12} \\ r_{22} \\ r_{32} \end{bmatrix} d + \boldsymbol{t}_c^m, \\
k_3\boldsymbol{\alpha}_3 &= \begin{bmatrix} r_{11} \\ r_{21} \\ r_{31} \end{bmatrix} d + \boldsymbol{t}_c^m.
\end{aligned} \tag{7}
$$

To eliminate the unknown quantity k_i, the first and second rows of each equation in (7) are divided by the third row, respectively, and substitute (2), thus obtaining

$$
\begin{aligned}
&\frac{-r_{11}d + t_x}{-r_{31}d + t_z} = \frac{u_1^c}{v_1^c}, \quad \frac{-r_{21}d + t_y}{-r_{31}d + t_z} = \frac{f}{v_1^c}, \\
&\frac{-r_{12}d + t_x}{-r_{32}d + t_z} = \frac{u_2^c}{v_2^c}, \quad \frac{-r_{22}d + t_y}{-r_{32}d + t_z} = \frac{f}{v_2^c}, \\
&\frac{r_{11}d + t_x}{r_{31}d + t_z} = \frac{u_3^c}{v_3^c}, \quad \frac{r_{21}d + t_y}{r_{31}d + t_z} = \frac{f}{v_3^c}.
\end{aligned} \tag{8}
$$

Then, divide both the numerator and denominator on the left side of the Equation (8) by t_z, and we can obtain

$$
\begin{aligned}
&\frac{-r_{11}d/t_z + t_x/t_z}{-r_{31}d/t_z + 1} = \frac{u_1^c}{v_1^c}, \quad \frac{-r_{21}d/t_z + t_y/t_z}{-r_{31}d/t_z + 1} = \frac{f}{v_1^c}, \\
&\frac{-r_{12}d/t_z + t_x/t_z}{-r_{32}d/t_z + 1} = \frac{u_2^c}{v_2^c}, \quad \frac{-r_{22}d/t_z + t_y/t_z}{-r_{32}d/t_z + 1} = \frac{f}{v_2^c}, \\
&\frac{r_{11}d/t_z + t_x/t_z}{r_{31}d/t_z + 1} = \frac{u_3^c}{v_3^c}, \quad \frac{r_{21}d/t_z + t_y/t_z}{r_{31}d/t_z + 1} = \frac{f}{v_3^c}.
\end{aligned} \tag{9}
$$

For ease of expression, we make the following definitions:

$$
u_i^c = m_i v_i^c, \quad f = n_i v_i^c, \quad i = 1, 2, 3, \tag{10}
$$

$$
\begin{aligned}
&a_1 = t_x/t_z, \quad a_2 = t_y/t_z, \quad a_3 = r_{11}/t_z, \quad a_4 = r_{21}/t_z, \\
&a_5 = r_{31}/t_z, \quad a_6 = r_{12}/t_z, \quad a_7 = r_{22}/t_z, \quad a_8 = r_{32}/t_z.
\end{aligned} \tag{11}
$$

Substituting (10) and (11) into (9) gives

$$
\begin{aligned}
&\frac{-da_3 + a_1}{-da_5 + 1} = m_1, \quad \frac{-da_4 + a_2}{-da_5 + 1} = n_1, \\
&\frac{-da_6 + a_1}{-da_8 + 1} = m_2, \quad \frac{-da_7 + a_2}{-da_8 + 1} = n_2, \\
&\frac{da_3 + a_1}{da_5 + 1} = m_3, \quad \frac{da_4 + a_2}{da_5 + 1} = n_3.
\end{aligned} \tag{12}
$$

In (12), only $a_i(i = 1, 2, ..., 8)$ are unknown quantities, which can be simplified as

$$\begin{aligned} a_1 &= M_2 d^2 a_5 + M_1, \quad a_2 = N_2 d^2 a_5 + N_1, \\ a_3 &= M_1 a_5 + M_2, \quad a_4 = N_1 a_5 + N_2, \\ a_6 &= m_2 a_8 - M_2 d a_5 + M_3, \quad a_7 = n_2 a_8 - N_2 d a_5 + N_3, \end{aligned} \tag{13}$$

where

$$\begin{aligned} M_1 &= \frac{m_1 + m_3}{2}, \quad N_1 = \frac{n_1 + n_3}{2}, \\ M_2 &= \frac{m_1 - m_3}{2d}, \quad N_2 = \frac{n_1 - n_3}{2d}, \\ M_3 &= \frac{2m_2 - m_1 - m_3}{2d}, \quad N_3 = \frac{2n_2 - n_1 - n_3}{2d}. \end{aligned} \tag{14}$$

By the nature of the rotation matrix, we have

$$r_{11} r_{12} + r_{21} r_{22} + r_{31} r_{32} = 0, \tag{15}$$

$$r_{11}^2 + r_{21}^2 + r_{31}^2 = r_{12}^2 + r_{22}^2 + r_{32}^2 = 1. \tag{16}$$

Divide both sides of (15) and (16) by t_z^2, and substitute (11) and (13) into, and we can obtain

$$p_1 a_5^2 + p_2 a_5 a_8 + p_3 a_5 + p_4 a_8 + p_5 = 0, \tag{17}$$

$$q_1 a_8^2 + q_2 a_5^2 + q_3 a_5 a_8 + q_4 a_8 + q_5 a_5 + q_6 = 0. \tag{18}$$

where

$$\begin{aligned} p_1 &= -d(M_1 M_2 + N_1 N_2), \\ p_2 &= m_2 M_1 + n_2 N_1 + 1, \\ p_3 &= M_1 M_3 + N_1 N_3 - d(M_2^2 + N_2^2), \\ p_4 &= m_2 M_2 + n_2 N_2, \\ p_5 &= M_2 M_3 + N_2 N_3, \end{aligned} \tag{19}$$

$$\begin{aligned} q_1 &= m_2^2 + n_2^2 + 1, \\ q_2 &= d^2(M_2^2 + N_2^2) - M_1^2 - N_1^2 - 1, \\ q_3 &= -2d(m_2 M_2 + n_2 N_2), \\ q_4 &= 2m_2 M_3 + n_2 N_3, \\ q_5 &= -2d(M_2 M_3 + N_2 N_3) - 2(M_1 M_2 + N_1 N_2), \\ q_6 &= M_3^2 + N_3^2 - M_2^2 - N_2^2. \end{aligned} \tag{20}$$

From (17) we can also obtain

$$a_8 = -\frac{p_1 a_5^2 + p_3 a_5 + p_5}{p_2 a_5 + p_4}. \tag{21}$$

By substituting (21) into (18) and simplifying it, we can obtain

$$s_1 a_5^4 + s_2 a_5^3 + s_3 a_5^2 + s_4 a_5 + s_5 = 0, \tag{22}$$

where

$$
\begin{aligned}
s_1 &= p_1^2 q_1 + p_2^2 q_2 - p_1 p_2 q_3, \\
s_2 &= 2p_1 p_3 q_1 + 2p_2 p_4 q_2 - p_1 p_4 q_3 - p_2 p_3 q_3 - p_1 p_2 q_4 + p_2^2 q_5, \\
s_3 &= p_3^2 q_1 + 2p_1 p_5 q_1 + p_4^2 q_2 - p_3 p_4 q_3 - p_2 p_5 q_3 - p_1 p_4 q_4 + p_2^2 q_6 - p_2 p_3 q_4 + 2p_2 p_4 q_5, \\
s_4 &= 2p_3 p_5 q_1 - p_4 p_5 q_3 - p_3 p_4 q_4 - p_2 p_5 q_4 + p_4^2 q_5 + p_2 p_4 q_6, \\
s_5 &= p_5^2 q_1 - p_4 p_5 q_4 + p_4^2 q_6.
\end{aligned} \tag{23}
$$

Using the formula for the roots of an unary quartic equation, we can quickly obtain the value of a_5 by (22). The filtering of multiple solutions is described in the next subsection. The remaining value of a_i can then be solved for by (13) and (21).

From (11) and (16), we can obtain the value of t_z by

$$t_z = \frac{1}{\sqrt{a_3{}^2 + a_4{}^2 + a_5{}^2}}, \tag{24}$$

and solve for the values of t_x and t_y from (11). Here, we use the non-negativity of t_y to exclude the wrong solution of (24) and obtain the translation vector $\boldsymbol{t}_c^m$. Since rotation matrices are special orthogonal matrices, $\boldsymbol{R}_c^m$ also satisfies

$$r_{ij} = A_{ij}, \quad i,j = 1,2,3, \tag{25}$$

where A_{ij} stands for the algebraic cosine formula of r_{ij}. So, the rotation matrix $\boldsymbol{R}_c^m$ can be solved from (11) and (25). Due to the accuracy limitations of the actual calculations, Schmidt orthogonalization of $\boldsymbol{R}_c^m$ is also required.

4.3. Conversion of Coordinate Systems

The relative localization model of the two UAVs is shown in Figure 4. Multiple coordinate systems are established with O_b, O_c, and O_m as the origin, respectively. The definitions of O_c and O_m are given in the previous section, and O_b is determined in the same way as O_m. $O_{bi}x_{bi}y_{bi}z_{bi}$, $O_{ci}x_{ci}y_{ci}z_{ci}$, and $O_{ui}x_{ui}y_{ui}z_{ui}$ are three inertial coordinate systems, so each of their axes corresponds to parallel, respectively. $O_c x_c y_c z_c$ and $O_m x_m y_m z_m$ are defined in the previous section. $O_b x_b y_b z_b$ and $O_u x_u y_u z_u$ are the fuselage coordinate systems of the two UAVs, where the $x_b(x_u)$-axis points directly to the right of the fuselage, the $y_b(y_u)$-axis points directly in front, and the $z_b(z_u)$-axis is perpendicular to $O_b x_b y_b (O_u x_u y_u)$ and points above the fuselage. The difference between $O_u x_u y_u z_u$ and $O_m x_m y_m z_m$ is that unlike $O_m x_m y_m z_m$, which is set up to simplify calculations, $O_u x_u y_u z_u$ is a common coordinate system used when expressing UAV attitude. Due to the symmetry of the quadcopter UAV, we start by assuming that the positive direction of the y_u-axis is always in the first quadrant of the $O_m x_m y_m$.

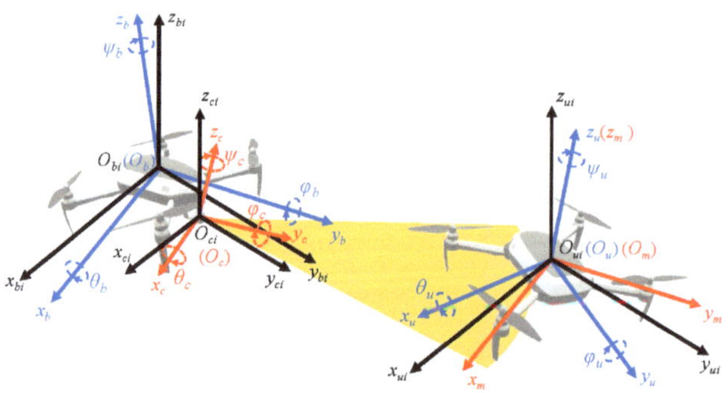

Figure 4. The coordinate system of interest for relative localization of the UAV.

Obviously, the relative position of the positioned UAV can be expressed as $\boldsymbol{t}_{bi}^{u} = \boldsymbol{O}_{bi}\boldsymbol{O}_{u}$. Due to the same orientation of the inertial coordinate systems, the attitude of the positioned UAV can be expressed as the rotation matrix $\boldsymbol{R}_{bi}^{u}$ of $O_u x_u y_u z_u$ with respect to $O_b x_b y_b z_b$. $\boldsymbol{R}_{bi}^{u}$ and $\boldsymbol{t}_{bi}^{u}$ can be considered as the result of a series of coordinate system transformations and the flexible kinematic properties of UAVs and gimbals increase the difficulty of solving them.

The solution scheme for $\boldsymbol{R}_{c}^{m}$ and $\boldsymbol{t}_{c}^{m}$ is given in the previous section. The attitude rotation matrices of the localization UAV and gimbal can be obtained based on their Euler angles acquired in real time. The Euler angle consists of roll angle φ, pitch angle θ, and yaw angle ψ, and the order of rotation is, based on an inertial coordinate system, first ψ degrees around the z-axis, then θ degrees around the transformed x-axis, and finally φ degrees around the transformed y-axis. The conversion formulas for Euler angles to the rotation matrix $\boldsymbol{R}$ in the right-handed coordinate system are

$$
\begin{aligned}
\boldsymbol{R}_x(\theta) &= \begin{bmatrix} 1 & 0 & 0 \\ 0 & \cos\theta & -\sin\theta \\ 0 & \sin\theta & \cos\theta \end{bmatrix}, \\
\boldsymbol{R}_y(\varphi) &= \begin{bmatrix} \cos\varphi & 0 & \sin\varphi \\ 0 & 1 & 0 \\ -\sin\varphi & 0 & \cos\varphi \end{bmatrix}, \\
\boldsymbol{R}_z(\psi) &= \begin{bmatrix} \cos\psi & -\sin\psi & 0 \\ \sin\psi & \cos\psi & 0 \\ 0 & 0 & 1 \end{bmatrix},
\end{aligned}
\tag{26}
$$

and

$$\boldsymbol{R} = \boldsymbol{R}_z(\psi) \cdot \boldsymbol{R}_x(\theta) \cdot \boldsymbol{R}_y(\varphi). \tag{27}$$

The attitude rotation matrices $\boldsymbol{R}_{bi}^{b}$ and $\boldsymbol{R}_{ci}^{c}$ can be obtained by substituting the Euler angles $\varphi_b, \theta_b, \psi_b$ and $\varphi_c, \theta_c, \psi_c$ of the localization UAV and the gimbal into (26) and (27), respectively.

Based on the above known information, we give the solution scheme for $\boldsymbol{R}_{bi}^{u}$ and $\boldsymbol{t}_{bi}^{u}$. Since the isotropy of inertial coordinate systems it follows that

$$\boldsymbol{R}_{bi}^{u} = \boldsymbol{R}_{ci}^{u}. \tag{28}$$

where $\boldsymbol{R}_{ci}^{u}$ denotes the rotation matrix of the positioned UAV relative to the camera inertial coordinate system. By the transitivity of the rotation matrix, $\boldsymbol{R}_{ci}^{u}$ can be expressed as

$$\boldsymbol{R}_{ci}^{u} = \boldsymbol{R}_{ci}^{c} \cdot \boldsymbol{R}_{c}^{m} \cdot \boldsymbol{R}_{m}^{u}, \tag{29}$$

where, according to the direction in which the coordinate system is set up, it is easy to know that

$$\boldsymbol{R}_{m}^{u} = \boldsymbol{R}_z(-\frac{\pi}{4}). \tag{30}$$

By the additive property of vectors, $\boldsymbol{t}_{bi}^{u}$ can be expressed as

$$\boldsymbol{t}_{bi}^{u} = \boldsymbol{t}_{bi}^{ci} + \boldsymbol{t}_{ci}^{u}, \tag{31}$$

where $\boldsymbol{t}_{bi}^{ci}$ can be obtained from

$$\boldsymbol{t}_{bi}^{ci} = \boldsymbol{R}_{bi}^{b} \cdot \boldsymbol{t}_0, \tag{32}$$

where $\boldsymbol{t}_0$ represents the initial value of $\boldsymbol{t}_{bi}^{ci}$ when $\varphi_b, \theta_b, \psi_b$=0, which can be easily obtained by measurement. And we can obtain $\boldsymbol{t}_{ci}^{u}$ by

$$\boldsymbol{t}_{ci}^{u} = \boldsymbol{R}_{ci}^{c} \cdot \boldsymbol{t}_{c}^{m}. \tag{33}$$

In summary, the relative position and attitude of the positioned UAV are finally given as

$$\begin{aligned} t_{bi}^{u} &= R_{bi}^{b} \cdot t_0 + R_{ci}^{c} \cdot t_c^m, \\ R_{bi}^{u} &= R_{ci}^{c} \cdot R_c^m \cdot R_m^u. \end{aligned} \tag{34}$$

4.4. Determination of Correct Solution

Theoretically, the quartic equation of one unknown of (22) has at most four different real roots. However, according to the conclusions of [47], in the P3P problem, the equation can be considered to have only two sets of real solutions, i.e., two sets of three-dimensional spatial points can be derived from one set of two-dimensional projected points. We verified this conclusion in simulation experiments, and the simulation model is shown in Section 5.

The two sets of solutions correspond to two sets of UAV positions and attitudes, as shown in Figure 5. $\{M_i'\}(i = 1,2,3,4)$ represents another set of erroneous motor positions derived from the projected points $\{P_i\}$, and O_m' is the erroneous position of the UAV. The degree of inclination of the UAV body corresponding to the two sets of solutions can be represented by the angle $\angle z_u O_m z_{ui}$ and angle $\angle z_u' O_m' z_{ui}'$, which are set as β_u and β_u', respectively.

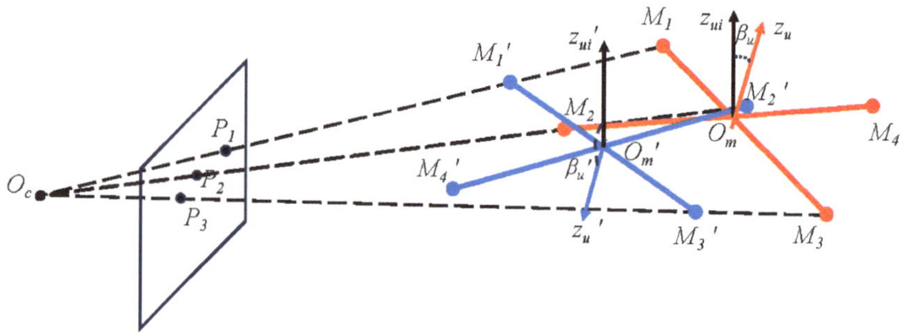

Figure 5. The position and attitude of the UAV corresponding to the two sets of solutions.

β_u is a result of the roll and pitch that occurs in the UAV, so the value of β_u should be within a limited range during normal flight. According to the vector angle formula, we can obtain

$$\cos \beta_u = \frac{\boldsymbol{w_3} \cdot \boldsymbol{z_{ci}}}{\|\boldsymbol{w_3}\| \|\boldsymbol{z_{ci}}\|}, \tag{35}$$

where $\boldsymbol{w_3}$ denotes the third row of $\boldsymbol{R_{bi}^u}$, which also represents the unit vector of the z_u-axis in the inertial coordinate system. Let $\boldsymbol{w_3} = [w_{31}, w_{32}, w_{33}]$ and $\boldsymbol{z_{bi}} = [0,\ 0,\ 1]$; β_u can be obtained from

$$\beta_u = \arccos w_{33}. \tag{36}$$

From (26) and (27), we have $w_{33} = \cos \varphi_u \cos \theta_u$. The roll and pitch angles of UAVs are usually finite, denoted as $\theta_u \in [\varphi_u^{min}, \varphi_u^{max}]$ and $\theta_u \in [\theta_u^{min}, \theta_u^{max}]$. And, due to the symmetry of quadrotor UAVs, usually $\varphi_u^{max} = \theta_u^{max} = -\varphi_u^{min} = -\theta_u^{min}$. Then, the range of β_u can be expressed as

$$\beta_u \in [\ 0,\ \cos^2 \varphi_u^{max}]. \tag{37}$$

We therefore set the maximum value of pitch and roll angles uniformly to α_u^{max}.

Since it is difficult to obtain the range of β_u' by mathematical derivation, we each obtained the approximate distribution of β_u' at $\varphi_u^{max} = \theta_u^{max} = \pi/6$ and $\varphi_u^{max} = \theta_u^{max} = \pi/4$ based on 10,000 simulation experiments, respectively, as shown in Figure 6.

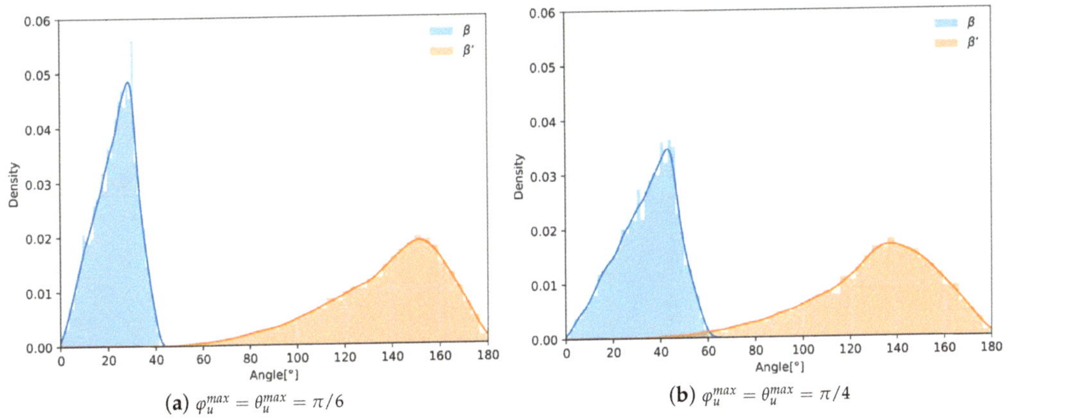

Figure 6. Distribution of UAV body tilt angles corresponding to the two sets of solutions.

It can be seen that the vast majority of the values of β'_u are greater than β_u^{max}, the maximum value of β_u, compared to the values of β_u that are strictly in the range shown in (37). In the two sets of experiments, the values of β'_u greater than β_u^{max} are approximately 99.8% and 98.8%, respectively. Therefore, in the vast majority of cases, the correct solution can be identified based on the value of β_u^{max}. Subject to errors in the projection points of motors, the value of β_u^{max} tends to be slightly larger than $\cos^2 \varphi_u^{max}$. Approximate values of β_u^{max} can be obtained based on a large number of simulation experiments.

When β'_u is also smaller than β_u^{max}, partially incorrect solutions can be further detected based on whether θ_u and φ_u corresponding to each set of solutions are simultaneously smaller than φ_u^{max} and θ_u^{max}, respectively. We set the maximum value of pitch and roll angles uniformly to α_u^{max}. Similar to β_u^{max}, the actual values obtained for α_u^{max} are slightly larger than φ_u^{max} and θ_u^{max}, and their approximations can be obtained through extensive randomized experiments.

For the mis-solutions that remain unfiltered, we find that their average error is much smaller than the measured distance and much lower than the average error of the full set of mis-solutions. When $\varphi_u^{max} = \theta_u^{max} = \pi/6$ and $\varphi_u^{max} = \theta_u^{max} = \pi/4$, simulation results show that the average errors of these incorrect solutions are only about $0.05\%\|\boldsymbol{t}_{bi}^{u}\|$ and $0.63\%\|\boldsymbol{t}_{bi}^{u}\|$, which are about 1/30 and 2/5 of the overall average error, respectively. We therefore take the average of these group solution pairs as the result.

In summary, the algorithm for determining the correct solution is shown in Algorithm 2:

4.5. Four Motors Detected

When all four motors are detected, positioning accuracy can be further improved. We divide the four projection points of motors into groups of three each in the order specified in Section 3.2. By substituting each of the four sets of projection points into the above solution scheme, four sets of localization results can be obtained. We set $\boldsymbol{t}_i$ to denote the relative position obtained based on the three points other than point P_i

The keypoint detection module gives the detection confidence for each motor, set to $c_{1:4}$. The weight W_i of $\boldsymbol{t}_i$ can be obtained based on c_i by

$$W_i = \frac{(\sum_{j=1}^{4} c_j) - c_i}{3 \sum_{j=1}^{4} c_j}. \tag{38}$$

Then $\boldsymbol{t}_{bi}^{u}$ can be given by

$$\boldsymbol{t}_{bi}^{u} = \sum_{i=1}^{4} W_i \boldsymbol{t}_i. \tag{39}$$

Algorithm 2 Determining the correct solution

Require: $T = \{\boldsymbol{t}_{bi1}^{u}, \boldsymbol{t}_{bi2}^{u}\}$, $\boldsymbol{B} = \{\beta_{u1}, \beta_{u2}\}$, $\boldsymbol{A} = \{\{\theta_{u1}, \varphi_{u1}\}, \{\theta_{u2}, \varphi_{u2}\}\}$
Ensure: $\boldsymbol{t}_{bi}^{u}$
if $min(\boldsymbol{B}) < \beta_u^{max}$ and $max(\boldsymbol{B}) > \beta_u^{max}$ **then**
    $idx = min(\boldsymbol{B})$'s index of $\boldsymbol{B}$
else if $max(abs(\boldsymbol{A}_1)) < \alpha_u^{max}$ and $max(abs(\boldsymbol{A}_2)) > \alpha_u^{max}$ **then**
    $idx = 1$
else if $max(abs(\boldsymbol{A}_1)) > \alpha_u^{max}$ and $max(abs(\boldsymbol{A}_2)) < \alpha_u^{max}$ **then**
    $idx = 2$
else if $max(abs(\boldsymbol{A}_1)) < \alpha_u^{max}$ and $max(abs(\boldsymbol{A}_2)) < \alpha_u^{max}$ **then**
    $idx = 0$
end if
if $idx = 0$ **then**
    $\boldsymbol{t}_{bi}^{u} = \frac{\boldsymbol{t}_{bi1}^{u} + \boldsymbol{t}_{bi2}^{u}}{2}$
else
    $\boldsymbol{t}_{bi}^{u} = T_{idx}$
end if

4.6. Two Motors Detected

Since the case where only two motors are detected rarely occurs, we give a transitional estimation scheme. The problem model at this point is shown in Figure 7.

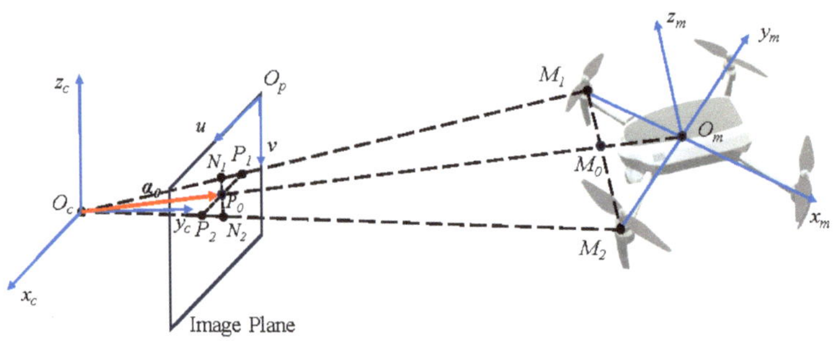

Figure 7. Schematic diagram when two motors are detected.

Taking into account the occlusion, we approximate that O_c is coplanar with $\{M_{1:4}\}$ and that $\|\boldsymbol{O_c M_1}\| = \|\boldsymbol{O_c M_2}\|$. So, $O_c O_m$ intersects $M_1 M_2$ at the midpoint of $M_1 M_2$ and the intersection is set to M_0. The projection point of O_m on the image plane is set to P_0 and $\boldsymbol{\alpha}_0$ represents the vector $\boldsymbol{O_c P_0}$. Then, the displacement vector $\boldsymbol{t}_c^m$ can be expressed as

$$\boldsymbol{t}_c^m = \frac{\|\boldsymbol{O_c M_0}\| + \|\boldsymbol{O_m M_0}\|}{\|\boldsymbol{\alpha}_0\|} \boldsymbol{\alpha}_0, \tag{40}$$

where $\|\boldsymbol{O_m M_0}\|$ is known to be $\frac{\sqrt{2}}{2} d$.

Make a parallel line of M_1M_2 through P_0, intersecting O_cM_1 and O_cM_2 at N_1 and N_2, respectively. From the properties of similar triangles we have

$$\frac{\|\boldsymbol{\alpha}_0\|}{\|\boldsymbol{O_cM_0}\|} = \frac{\|\boldsymbol{N_1N_2}\|}{\|\boldsymbol{M_1M_2}\|}, \tag{41}$$

where it is easy to see that $\|\boldsymbol{M_1M_2}\| = \sqrt{2}d$. Since P_1 and P_2 are known, the angles of $\angle P_1O_cP_2$, $\angle O_cP_1P_2$, and $\angle O_cP_2P_1$ can be obtained based on the vector pinch equations, which are set to η_1, η_2 and η_3, respectively. Here, it is specified that $\eta_2 < \pi/2 < \eta_3$. By the sine theorem, it can be obtained that

$$\begin{aligned} \frac{\|\boldsymbol{P_0N_1}\|}{\sin\eta_2} &= \frac{\|\boldsymbol{P_0P_1}\|}{\sin(\frac{\pi}{2}+\frac{\eta_1}{2})}, \\ \frac{\|\boldsymbol{P_0N_2}\|}{\sin(\pi-\eta_3)} &= \frac{\|\boldsymbol{P_0P_2}\|}{\sin(\frac{\pi}{2}-\frac{\eta_1}{2})}. \end{aligned} \tag{42}$$

It is also known that

$$\|\boldsymbol{P_0N_1}\| = \|\boldsymbol{P_0N_2}\|, \tag{43}$$

and

$$\|\boldsymbol{P_0P_1}\| + \|\boldsymbol{P_0P_2}\| = \|\boldsymbol{P_1P_2}\|. \tag{44}$$

From (42)–(44), we can obtain

$$\|\boldsymbol{N_1N_2}\| = 2\|\boldsymbol{P_1P_2}\|\frac{\sin\eta_2\sin(\pi-\eta_3)}{\sin(\frac{\pi}{2}+\frac{\eta_1}{2})\sin(\pi-\eta_3)+\sin\eta_2\sin(\frac{\pi}{2}-\frac{\eta_1}{2})}, \tag{45}$$

and

$$\|\boldsymbol{\alpha}_0\| = \frac{\frac{1}{2}\|\boldsymbol{N_1N_2}\|}{\tan\frac{\eta_1}{2}}. \tag{46}$$

Then, we can obtain $\|\boldsymbol{O_cM_0}\|$ first by (41) and then $\boldsymbol{t}_c^m$ by (40). Finally, after the coordinate transformation of Section 4.3, $\boldsymbol{t}_{bi}^u$ can be obtained.

5. Experimental Results and Analysis

Our experiment is divided into three parts. First, we obtained a self-training model of YOLOv8 by training based on the captured images and tested its effectiveness in detecting experimental UAVs and their motors. In the second part, we constructed the high-fidelity airborne gimbal camera model and localized UAV model based on the actual parameters, and examined the performance of the relative localization algorithm in various situations. Finally, we conducted system experiments based on two UAVs to verify the feasibility of our overall scheme using GPS-based relative localization data as a reference.

5.1. Experiment Platform

The hardware composition and operational architecture of the UAV experimental platform used to validate the proposed scheme is shown in Figure 8. We conduct secondary development and experiments based on two *Prometheus* 450 (*P*450) UAVs producted by *Amovlab*, Chengdu, China [48]. Each UAV is equipped with NVIDIA's Edge AI supercomputer Jeston Xavier NX and a Pixhawk 4 flight controller. The Jeston Xavier NX has a hexa-core NVIDIA Carmel ARM CPU, 6GB of LPDDR4x RAM and a GPU with 21TOPS of AI inference performance, which is capable of meeting the arithmetic requirements under Ubuntu 18.04. The Pixhawk 4 flight controller is the control hub of the UAV. We retrofitted the UAV with *amovlab*'s G1 gimbal camera to stream real-time images to the Jeston Xavier NX. The edge computer also obtains attitude data from the gimbal and flight controller through their ROS topics published in real time via the serial port. Based on the above data, the UAV achieves real-time detection and relative localization for other UAVs within its visual perception range on the Jeston Xavier NX. All experimental data were obtained based on this platform system. Key parameters of the UAV: $d = 21$ cm, $\boldsymbol{t}_0 = [0, 13, -6]$ cm.

5.2. Detection Performance Experiment

We labeled 1250 collected images of experimental UAVs and used them as a dataset to obtain a self-training model by training. We conducted UAV-to-UAV target detection experiments at distances ranging from 2 to 12 m. The experimental results show that the YOLOv8-pose target detection module based on the self-trained model is able to stably detect the target UAV and its visible motors. The motor's image plane positioning point can basically remain within the range of the motor's projected image. Screenshots of the detection results are shown in Figure 9, where the motors are marked by blue dots. The average detection time of the on-board target detection module for each image frame is about 43.5 ms.

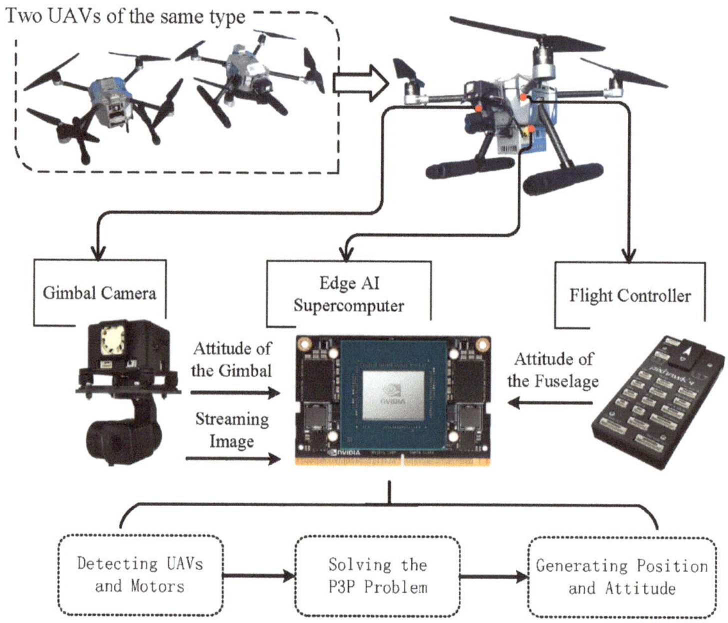

Figure 8. The hardware composition and operational architecture of the UAV experimental platform.

Figure 9. Detection effects of the UAV and its motors.

In summary, we verified the feasibility of realizing real-time detection of UAVs and their motors with an airborne camera based on YOLOv8.

5.3. Relative Localization Simulation Experiment

We tested the speed and accuracy of the proposed algorithm based on a self-built simulation model and compared it with three mainstream algorithms, which are Gao's, iterative method (IM) and AP3P. In order to increase the fidelity, all of our simulation experiments were performed on the edge computer of the *P*450 UAV.

5.3.1. Simulation Model

We constructed a virtual camera model based on the parameters of the *G*1 gimbal camera with an intrinsic matrix $\boldsymbol{K}$ of

$$\boldsymbol{K} = \begin{bmatrix} 640 & 0 & 640 \\ 0 & 640 & 360 \\ 0 & 0 & 1 \end{bmatrix}. \tag{47}$$

Based on the camera calibration work that has been performed, we assume that the camera's distortion is zero. The pitch angle of the gimbal $\theta_c \in [-\pi/3, \pi/3]$. The camera is capable of detecting drones from 2 to 12 m away from itself, which means that $D \in [2, 12]$ m, where $D = \|\boldsymbol{t}_{true}\|$.

In order to describe the situation where the motor is obscured, we designed a UAV model based on the *P*450, as shown in Figure 10. In the aforementioned $O_m x_m y_m z_m$ coordinate system, the body of the fuselage is represented by a sphere with O_f as the center and radius $R = 10$ cm, and the motors are represented by spheres with $M_{1:4}$ as the center and radius $r = 2$ cm. The coordinate of O_f is $[\,0,\ 0, -5]$ cm.

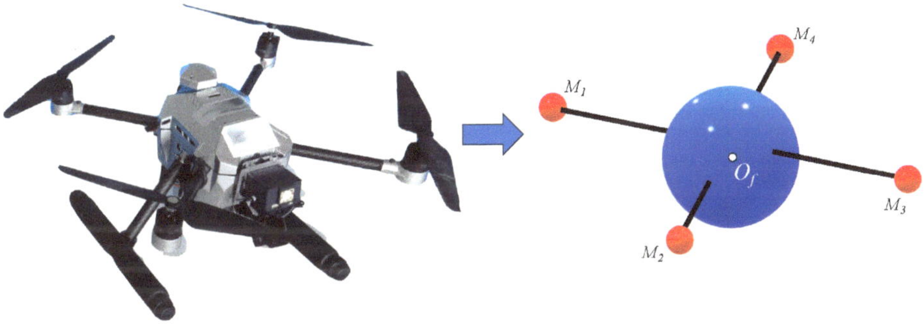

Figure 10. Simplification of the UAV.

The attitude of the UAV is determined by randomly generated Euler angles and Euler angles $\varphi_b, \theta_b, \psi_b \in [-\pi/4, \pi/4]$. The coordinates of O_f and $M_{1:4}$ in the Oxyz coordinate system can be obtained based on the Euler angles. Then, based on the projection relation, the projection points P_f and $P_{1:4}$ of O_f and $M_{1:4}$ on the image plane, and the radius R_p and $r_{p1:p4}$ of the projection circles of the fuselage and motors can be obtained.

According to the masking relation, the decision condition that three motors can be detected is expressed as

$$\|\boldsymbol{P}_f\boldsymbol{P}_4\| < R_p, \tag{48}$$

and the decision condition for detecting only two motors is

$$\|\boldsymbol{P}_1\boldsymbol{P}_4\| < r_{p1} \wedge \|\boldsymbol{P}_2\boldsymbol{P}_3\| < r_{p2}. \tag{49}$$

To simulate the error in motor detection, we add white noise obeying a two-dimensional Gaussian distribution to the image plane projection point $\{P_i(u_i, v_i)\}$ $(i = 1, 2, 3, 4)$ of motors, i.e., the actual projection point $P_i'(u_i', v_i')$ is denoted as

$$(u_i',\ v_i') \sim N(u_i,\ v_i,\ \sigma_{i1}^2,\ \sigma_{i2}^2,\ 0), \tag{50}$$

where

$$\sigma_{i1} = \sigma_{i2} = \sigma\frac{f}{y_i}. \tag{51}$$

σ is the standard deviation in centimeters of the 3D spatial point corresponding to the motor's localization point on the image plane and the position of the motor's true point. f represents the focal length and y_i denotes the coordinates of the motor M_i in the y-axis under the camera coordinate system, in meters.

We designed three values of σ, which are 0.5 cm, 1.0 cm, and 1.5 cm, based on the actual radius of the *P*450, which is 2 cm for the motor. The three values from small to large correspond to high to low accuracy and can be described as the localization point basically on the motor center, basically on the motor, and partially on the motor, respectively.

5.3.2. Execution Speed

The time taken to solve the P3P problem is the main factor affecting the speed of the relative localization algorithm. We performed execution time tests of the proposed algorithm as well as other classical algorithms at the same performance state of the edge computer. Each algorithm was run for 10,000 rounds. The distribution of single execution time is shown in Figure 11, and the average time taken is shown in Table 2.

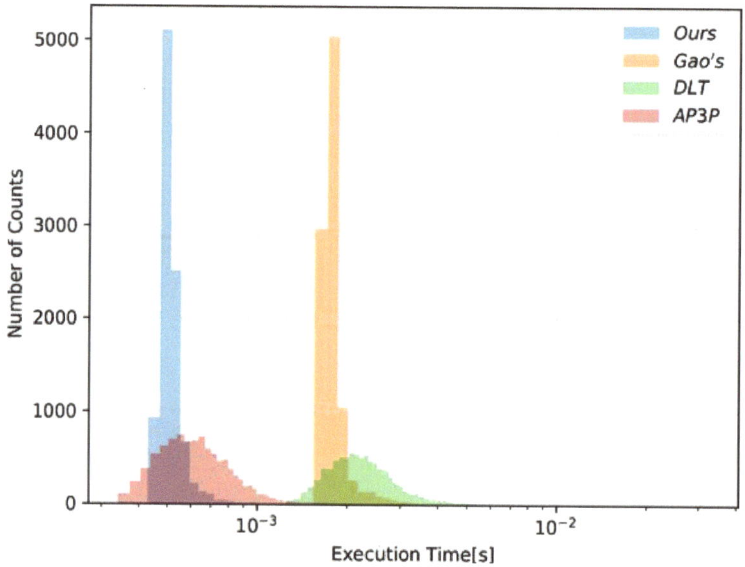

Figure 11. Distribution of single execution time for four algorithms.

Table 2. Average single execution time for the four algorithms.

Algorithms	Time [ms]	Proportionality
Ours	0.534	1
Gao's	1.845	3.46
IM	2.614	4.90
AP3P	0.722	1.35

It can be seen that our algorithm executes approximately 3.5 times faster than Gao's, 5 times faster than IM, and 35% faster than AP3P. Experimental results show that our proposed algorithm executes significantly faster than Gao's and IM. Compared to AP3P, we have a smaller but more consistent speed advantage. This is largely due to the fact that we have taken full advantage of the geometric characteristics of UAVs for targeted problem modeling. Our algorithm takes relative position as the unique objective and solves for it directly instead of obtaining it indirectly, reducing the accumulation and amplification of errors. Based on the results of the previous mathematical derivation, we only need to carry out simple algebraic calculations in the actual solution, which avoids the solution of the angle and the operation of the matrix and significantly reduces the computational complexity.

5.3.3. Computational Accuracy

In order to measure the accuracy of the relative localization and the correct choice of the solution, we denote the relative localization error as

$$e_t = \|\boldsymbol{t}_{est} - \boldsymbol{t}_{true}\|. \tag{52}$$

Following the approach of Section 4.4, we obtain reasonable values of β_u^{max} and α_u^{max} for three levels of detection accuracies with a sufficient number of randomized simulation experiments with known correct solutions. The values taken are shown in Table 3.

Table 3. Values of β_u^{max} and α_u^{max} for different detection accuracies.

σ [cm]	β_u^{max}	α_u^{max}
0.5	70°	52°
1.0	75°	58°
1.5	80°	62°

We randomly generated 10,000 sets of UAV position and attitude data in the simulation scenario. According to our occlusion model determination, there are 7871 sets of data where all four motors are detected, 2114 sets of data where three motors are detected, and 15 sets of data where only two motors are detected. This suggests that it is common for all four motors not to be detected. And given the simplified nature of the model and the fact that UAV swarms are often at similar altitudes during actual flight, the probability of detecting less than four motors should be greater. This supports the need for the study.

We first tested the overall accuracy of the proposed algorithm based on the simulation data and the experimental results are shown in Figure 12, and the vertical coordinate indicates the value of the kernel density estimate.

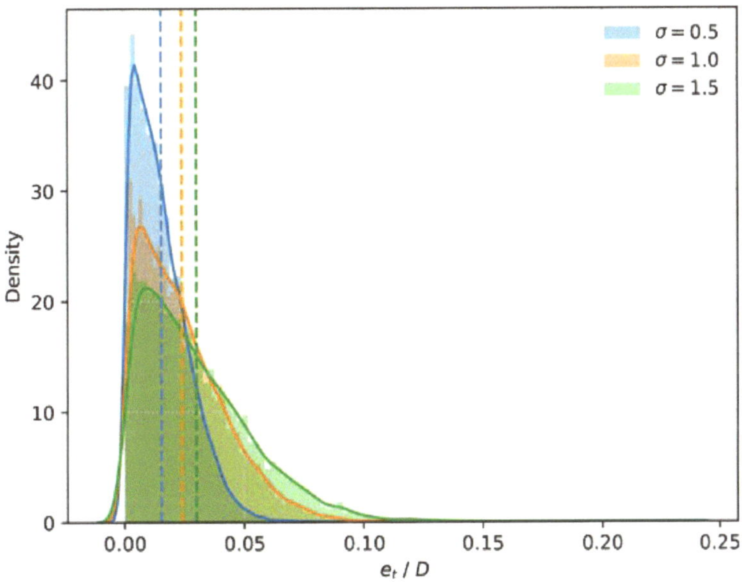

Figure 12. Error distributions of our algorithm under three levels of noise corresponding to σ = 0.5, 1.0 and 1.5, respectively.

The average localization errors at the three levels of noise are 1.53% D, 2.39% D, and 3.01% D, respectively, and are marked with vertical dashed lines in the figure (the same below). The data show that the localization accuracy of our algorithm has generally stabilized at a high level, and continues to provide less error-prone and stable localization data in the presence of increased noise. To further study the performance of the proposed algorithm, we analyze the specific performance of the algorithm when different numbers of motors are detected.

We solved the 7871 sets of data detected for the four motors by applying Gao's, IM, and AP3P methods, respectively, and compared them with the results of our algorithm. The error distribution of the four algorithms under different levels of noise is shown in Figure 13, and the corresponding average errors are shown in Table 4.

It is clear that the accuracy of IM and AP3P is significantly reduced when noise is present. The large error indicates that these two methods are not applicable to the solution of our research problem. The proposed algorithm is slightly more accurate than Gao's. We speculate that this advantage may stem from our weighting of the data based on the detection confidence of each motor. We speculate that this advantage may be the result of our multi-resolution solution as well as the regrouping weighting process. Therefore, we replaced our proposed post-processing scheme for the P3P solution with the reprojection method used by Gao and compared the experimental results with the results of our and Gao's schemes. The results of this experiment are shown in Figure 14.

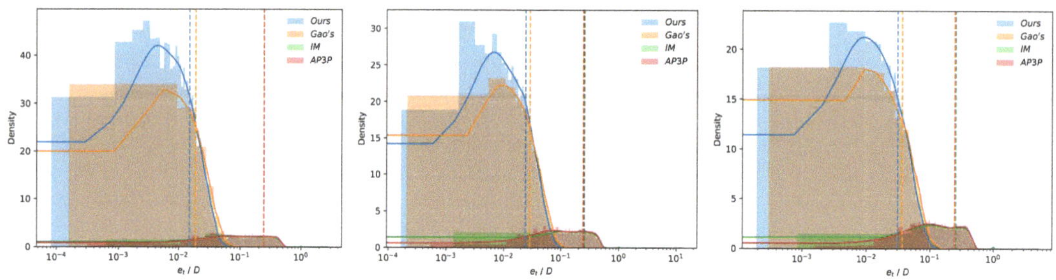

(a) $\sigma = 0.5$ **(b)** $\sigma = 1.0$ **(c)** $\sigma = 1.5$

Figure 13. Error distributions of the four algorithms for the three noise levels corresponding to σ = 0.5, 1.0 and 1.5.

Table 4. Localization errors of four algorithms with different detection accuracies.

σ [cm]	Ours	Gao's	IM	AP3P
0.5	0.015	0.019	0.242	0.239
1.0	0.024	0.029	0.251	0.239
1.5	0.030	0.036	0.252	0.240

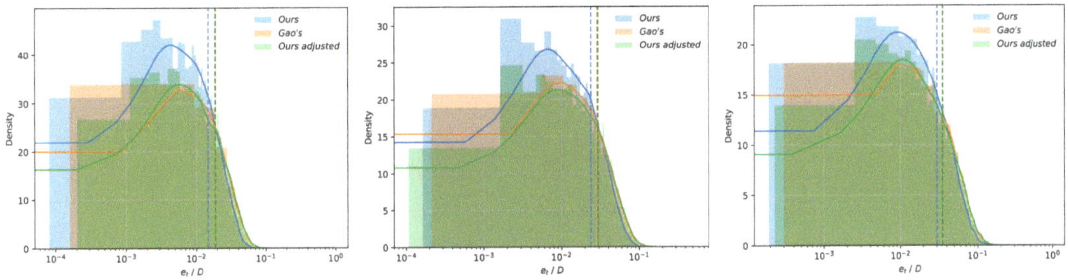

(**a**) $\sigma = 0.5$ (**b**) $\sigma = 1.0$ (**c**) $\sigma = 1.5$

Figure 14. Error distributions of our original, adjusted, and Gao's algorithm for three levels of noise corresponding to σ = 0.5, 1.0, and 1.5.

It can be seen that the accuracy of our algorithm is very close to that of Gao's after using the reprojection method instead of our post-processing scheme. This verifies the effectiveness of our post-processing scheme for accuracy improvement. By comparing the data in detail, we found that our post-processing algorithms are able to keenly detect outliers with large deviations and eliminate them or reduce their impact. Thus, our post-processing algorithm improves the robustness of the solution. However, our regrouping-weighted processing approach increases the computational cost, so we can choose to discard this part of the scheme when the arithmetic power is limited.

Due to the lack of other algorithms for obtaining the correct displacement based on the three key points, we can only compare the localization accuracy when three motors are detected with that when four motors are detected. Additional experiments were conducted, resulting in 7871 sets of localization data based on three motor points at each of the three levels of detection accuracy. The localization errors are shown in Figure 15.

As can be seen from the figure, our algorithm maintains a similar localization accuracy when only three motors are detected as when four motors are detected, specifically 1.68% D, 2.58% D, and 3.19% D. Localization errors still come mainly from detection errors. This shows that the performance of our pose-based multi-resolution determination scheme is robust. In the absence of a fourth motor point as a reprojection point, our method can effectively replace the reprojection method to obtain a stable and accurate solution.

We also tested the performance of the transitional solution when only two motors were detected. We obtained the results of 1,000 sets of experiments through a much larger number of randomized experiments, as shown in Figure 16.

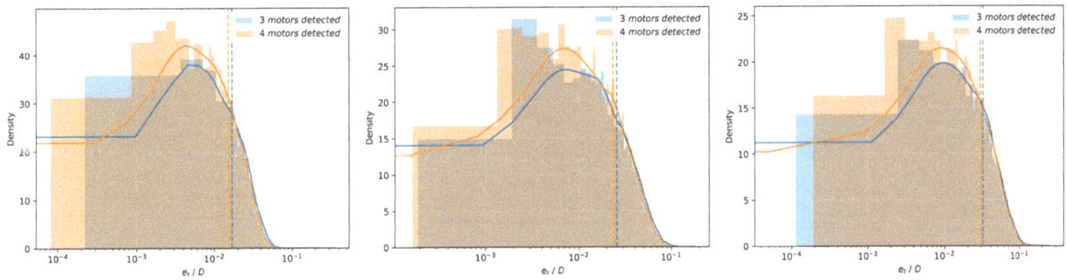

(**a**) $\sigma = 0.5$ (**b**) $\sigma = 1.0$ (**c**) $\sigma = 1.5$

Figure 15. Error distribution of our algorithm when only two motors are detected.

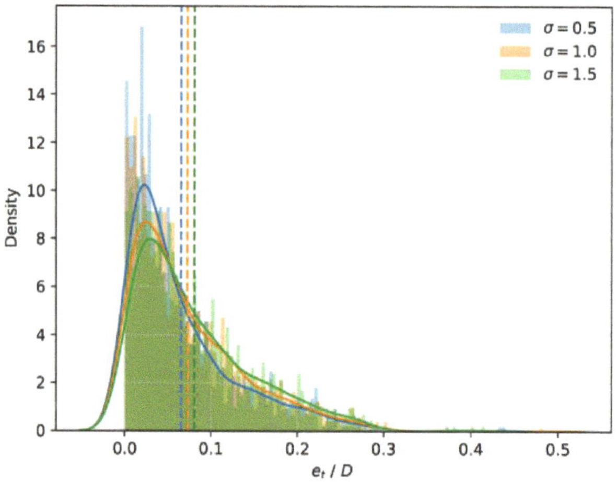

Figure 16. The localization error of our algorithm when two motors are detected.

It can be seen that the average error of our localization scheme when detecting two motors is controlled within 10% D, specifically 6.58% D, 7.33% D, and 8.10% D, respectively. Although some of the errors are large, given the small probability of the event occurring, we believe that its performance is acceptable as a transitional solution for special cases. In the process of processing data from consecutive frames, it is possible to combine the data from previous frames when more than two motors were detected and reduce the error by methods such as Kalman filtering.

5.3.4. System Experiment

Based on the demonstration of simulation experiments, we conducted real system experiments based on two *P*450 UAVs in a real environment. Due to the temporary lack of other more accurate means of localization, we generate the true relative position coordinates of the two UAVs based on GPS positioning data in an unobstructed environment. To minimize the increase in error due to other factors, we controlled the UAV used for localization to remain hovering in the air, and the localized UAV flew within the field of view of the camera for one minute, as shown in Figure 17. The real-time true relative position during the flight and the estimated relative position based on the proposed algorithm are shown in Figure 18, and Figure 19 illustrates the corresponding error distribution.

Figure 17. Real experimental scene diagram.

As shown in the figures, our scheme is generally able to achieve real-time vision-based relative localization between UAVs. The average relative error of the real experiment is 4.14%, which is slightly larger than the maximum average error of the simulation experiment. The error in the y-axis direction is significantly larger than that in the x-axis and z-axis directions, which is in line with the principle of our scheme. More outliers with larger deviations appear in the estimation results. By analyzing the data, we determined that this was the result of larger errors in the image plane coordinates of the motors. In addition, t_{true} itself, which is generated based on GPS and barometric altimeter data, actually has some error.

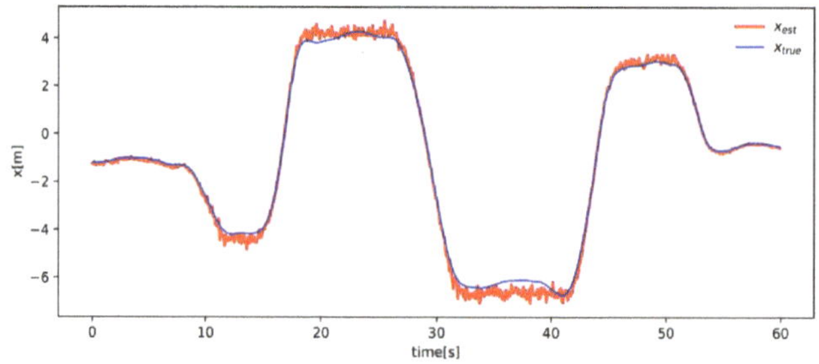

(**a**) True and estimated values in the x-axis direction

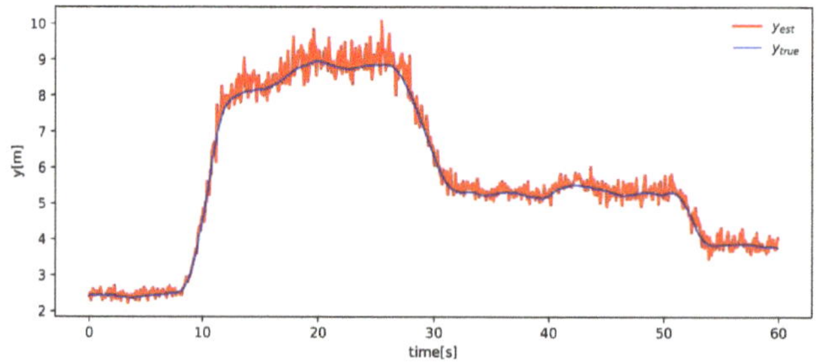

(**b**) True and estimated values in the y-axis direction

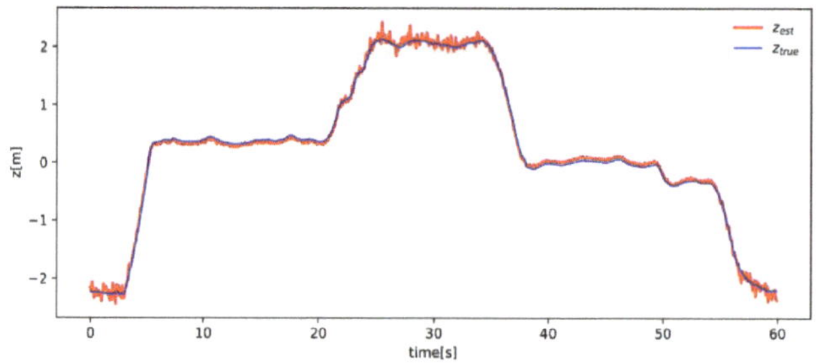

(**c**) True and estimated values in the z-axis direction

Figure 18. Comparison of true and estimated values of relative positions.

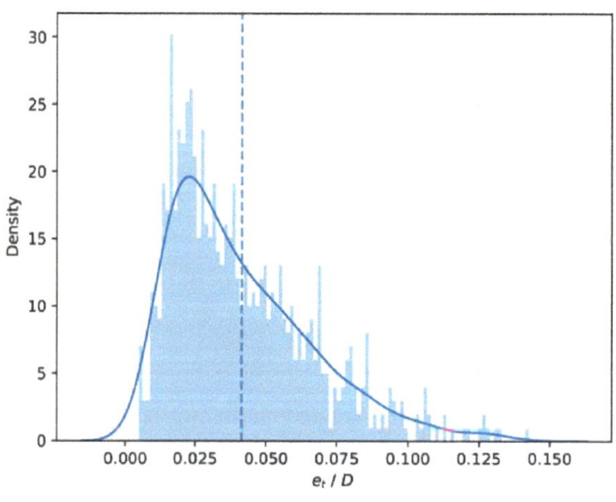

Figure 19. Error distribution in real experiments.

6. Conclusions

In order to realize real-time accurate relative localization within UAV swarms, we investigate a visual relative localization scheme based on onboard monocular sensing information. The conclusions of the study are as follows:

- Our study validates the feasibility of accurately detecting UAV motors in real time using the YOLOv8-pose attitude detection algorithm.
- Our PnP solution algorithm derived based on the geometric features of the UAV proved to be faster and more stable.
- Through the validation of a large number of stochastic experiments, we propose for the first time a fast scheme based on the rationality of UAV attitude to deal with the PnP multi-solution problem, which ensures the stability of the scheme when the visual information is incomplete.

Our scheme improves speed and accuracy while reducing data requirements, and the performance is verified in experiments.

However, there are limitations to our study. First, limited by the detection performance of the detection module for small targets, our relative localization can currently only be achieved at a distance of less than 12 m. Of course, with the improvement in the detection performance, the action distance will be larger. Second, our currently generated position data has not been filtered. So based on the experimental conclusions, our next research direction is to improve the detection performance of the detection module for the motors as small targets at long distances, and the second is to improve the overall stability of the estimation value under the time series through the filtering algorithm.

Author Contributions: Conceptualization, X.S., F.Q. and M.K.; methodology, X.S. and M.K.; software, X.S. and G.X.; validation, M.K. and H.Z.; formal analysis, F.Q.; investigation, K.T.; resources, K.T.; data curation, G.X.; writing—original draft preparation, X.S.; writing—review and editing, F.Q. and M.K.; visualization, X.S.; supervision, F.Q.; project administration, M.K.; funding acquisition, H.Z. All authors have read and agreed to the published version of the manuscript.

Funding: This work was supported by The Natural Science Foundation for Young Scholars of Anhui Province under Grant No. 2108085QF255, The Research Project of National University of Defense and Technology under Grant No. ZK21-45, The Military Postgraduate Funding Project under Grant No. JY2022A006, and in part by The 69th Project Funded by China Postdoctoral Science Foundation under Grant No. 2021M693977.

Data Availability Statement: The data are available from the corresponding author on reasonable request.

Conflicts of Interest: The authors declare no conflict of interest.

References

1. Yayli, U.C.; Kimet, C.; Duru, A.; Cetir, O.; Torun, U.; Aydogan, A.C.; Padmanaban, S.; Ertas, A.H. Design optimization of a fixed wing aircraft *Adv. Aircr. Spacecr. Sci.* **2017**, *1*, 65–80.
2. Wang, X.; Shen, L.; Liu, Z.; Zhao, S.; Cong, Y.; Li, Z.; Jia, S.; Chen, H.; Yu, Y.; Chang, Y.; et al. Coordinated flight control of miniature fixed-wing UAV swarms: methods and experiments. *Sci. China Inf. Sci.* **2019**, *62*, 134–150. [CrossRef]
3. Hellaoui, H.; Bagaa, M.; Chelli, A.; Taleb, T.; Yang, B. On Supporting Multiservices in UAV-Enabled Aerial Communication for Internet of Things. *IEEE Internet Things J.* **2023**, *10*, 13754–13768. [CrossRef]
4. Zhu, Q.; Liu, R.; Wang, Z.; Liu, Q.; Han, L. Ranging Code Design for UAV Swarm Self-Positioning in Green Aerial IoT. *IEEE Internet Things J.* **2023**, *10*, 6298–6311. [CrossRef]
5. Li, B.; Jiang, Y.; Sun, J.; Cai, L.; Wen, C.Y. Development and Testing of a Two-UAV Communication Relay System. *Sensors* **2016**, *16*, 1696. [CrossRef]
6. Ganesan, R.; Raajini, M.; Nayyar, A.; Sanjeevikumar, P.; Hossain, E.; Ertas, A. BOLD: Bio-Inspired Optimized Leader Election for Multiple Drones. *Sensors* **2020**, *11*, 3134. [CrossRef]
7. Zhou, L.; Leng, S.; Liu, Q.; Wang, Q. Intelligent UAV Swarm Cooperation for Multiple Targets Tracking. *IEEE Internet Things J.* **2022**, *9*, 743–754. [CrossRef]
8. Cheng, C.; Bai, G.; Zhang, Y.A.; Tao, J. Resilience evaluation for UAV swarm performing joint reconnaissance mission. *Chaos* **2019**, *29*, 053132. [CrossRef]
9. Luo, L.; Wang, X.; Ma, J.; Ong, Y. GrpAvoid: Multigroup Collision-Avoidance Control and Optimization for UAV Swarm. *IEEE Trans. Cybern.* **2023**, *53*, 1776–1789. [CrossRef]

10. Qi, Y.; Zhong, Y.; Shi, Z. Cooperative 3-D relative localization for UAV swarm by fusing UWB with IMU and GPS. *J. Phys. Conf. Ser.* **2020**, *1642*, 012028. [CrossRef]
11. Hu, J.; Hu, J.; Shen, Y.; Lang, X.; Zang, B.; Huang, G.; Mao, Y. 1D-LRF Aided Visual-Inertial Odometry for High-Altitude MAV Flight. In Proceedings of the 2022 International Conference on Robotics and Automation (ICRA), Philadelphia, PA, USA, 23–27 May 2022; pp. 5858–5864.
12. Masselli, A.; Hanten, R.; Zell, A. Localization of Unmanned Aerial Vehicles Using Terrain Classification from Aerial Images. In *Intelligent Autonomous Systems 13, Proceedings of the 13th International Conference IAS-13, Padova, Italy, 15–18 July 2014*; Springer: Cham, Switzerland, 2016; pp. 831–842.
13. Lin, H.; Zhan, J. GNSS-denied UAV indoor navigation with UWB incorporated visual inertial odometry. *Measurement* **2023**, *206*, 112256. [CrossRef]
14. Zhang, M.; Han, S.; Wang, S.; Liu, X.; Hu, M.; Zhao, J. Stereo Visual Inertial Mapping Algorithm for Autonomous Mobile Robot. In Proceedings of the 2020 3rd International Conference on Intelligent Robotic and Control Engineering (IRCE), Oxford, UK, 10–12 August 2020; pp. 97–104.
15. Jiang, Y.; Gao, Y.; Song, W.; Li, Y.; Quan, Q. Bibliometric analysis of UAV swarms. *J. Syst. Eng. Electron.* **2022**, *33*, 406–425. [CrossRef]
16. Mueller, F.d.P. Survey on Ranging Sensors and Cooperative Techniques for Relative Positioning of Vehicles. *Sensors* **2017**, *17*, 271. [CrossRef] [PubMed]
17. Dai, M.; Li, H.; Liang, J.; Zhang, C.; Pan, X.; Tian, Y.; Cao, J.; Wang, Y. Lane Level Positioning Method for Unmanned Driving Based on Inertial System and Vector Map Information Fusion Applicable to GNSS Denied Environments. *Drones* **2023**, *7*, 239. [CrossRef]
18. Garcia-Fernandez, M.; Alvarez-Lopez, Y.; Las Heras, F. Autonomous Airborne 3D SAR Imaging System for Subsurface Sensing: UWB-GPR on Board a UAV for Landmine and IED Detection. *Remote Sens.* **2019**, *11*, 2357. [CrossRef]
19. Fan, S.; Zeng, R.; Tian, H. Mobile Feature Enhanced High-Accuracy Positioning Based on Carrier Phase and Bayesian Estimation. *IEEE Internet Things J.* **2022**, *9*, 15312–15322. [CrossRef]
20. Song, H.; Choi, W.; Kim, H. Robust Vision-Based Relative-Localization Approach Using an RGB-Depth Camera and LiDAR Sensor Fusion. *IEEE Trans. Ind. Electron.* **2016**, *63*, 3725–3736. [CrossRef]
21. Liu, Z.; Zhang, W.; Zheng, J.; Guo, S.; Cui, G.; Kong, L.; Liang, K. Non-LOS target localization via millimeter-wave automotive radar. *J. Syst. Eng. Electron.* **2023**, 1–11. [CrossRef]
22. Arafat, M.Y.; Alam, M.M.; Moh, S. Vision-Based Navigation Techniques for Unmanned Aerial Vehicles: Review and Challenges. *Drones* **2023**, *7*, 89. [CrossRef]
23. Fan, H.; Wen, L.; Du, D.; Zhu, P.; Hu, Q.; Ling, H. VisDrone-SOT2020: The Vision Meets Drone Single Object Tracking Challenge Results. In Proceedings of the Computer Vision—ECCV 2020 Workshops, Glasgow, UK, 23–28 August 2020; pp. 728–749.
24. Zhao, X.; Yang, Q.; Liu, Q.; Yin, Y.; Wei, Y.; Fang, H. Minimally Persistent Graph Generation and Formation Control for Multi-Robot Systems under Sensing Constraints. *Electronics* **2023**, *12*, 317. [CrossRef]
25. Yan, J.; Zhang, Y.; Kang, B.; Zhu, W.P.; Lun, D.P.K. Multiple Binocular Cameras-Based Indoor Localization Technique Using Deep Learning and Multimodal Fusion. *IEEE Sens. J.* **2022**, *22*, 1597–1608. [CrossRef]
26. Yasuda, S.; Kumagai, T.; Yoshida, H. Precise Localization for Cooperative Transportation Robot System Using External Depth Camera. In Proceedings of the IECON 2021—47th Annual Conference of the IEEE Industrial Electronics Society, Toronto, ON, Canada, 13–16 October 2021; pp. 1–7.
27. Li, J.; Li, H.; Zhang, X.; Shi, Q. Monocular vision based on the YOLOv7 and coordinate transformation for vehicles precise positioning. *Connect. Sci.* **2023**, *35*, 2166903. [CrossRef]
28. Lin, F.; Peng, K.; Dong, X.; Zhao, S.; Chen, B.M. Vision-based formation for UAVs. In Proceedings of the 11th IEEE International Conference on Control and Automation (ICCA), Taichung, Taiwan, 18–20 June 2014; pp. 1375–1380.
29. Zhao, B.; Chen, X.; Jiang, J.; Zhao, X. On-board Visual Relative Localization for Small UAVs. In Proceedings of the 2020 Chinese Control And Decision Conference (CCDC), Hefei, China, 22–24 August 2020; pp. 1522–1527.
30. Zhao, H.; Wu, S. A Method to Estimate Relative Position and Attitude of Cooperative UAVs Based on Monocular Vision. In Proceedings of the 2018 IEEE CSAA Guidance, Navigation and Control Conference (CGNCC), Xiamen, China, 10–12 August 2018; pp. 1–6.
31. Walter, V.; Staub, N.; Saska, M.; Franchi, A. Mutual Localization of UAVs based on Blinking Ultraviolet Markers and 3D Time-Position Hough Transform. In Proceedings of the 2018 IEEE 14th International Conference on Automation Science and Engineering (CASE), Munich, Germany, 20–24 August 2018; pp. 298–303.
32. Li, S.; Xu, C. Efficient lookup table based camera pose estimation for augmented reality. *Comput. Animat. Virtual Worlds* **2011**, *22*, 47–58. [CrossRef]
33. Zhao, B.; Li, Z.; Jiang, J.; Zhao, X. Relative Localization for UAVs Based on April-Tags. In Proceedings of the 2020 Chinese Control And Decision Conference (CCDC), Hefei, China, 22–24 August 2020; pp. 444–449.
34. Pan, T.; Deng, B.; Dong, H.; Gui, J.; Zhao, B. Monocular-Vision-Based Moving Target Geolocation Using Unmanned Aerial Vehicle. *Drones* **2023**, *7*, 87. [CrossRef]
35. Jin, R.; Jiang, J.; Qi, Y.; Lin, D.; Song, T. Drone Detection and Pose Estimation Using Relational Graph Networks. *Sensors* **2019**, *19*, 1479. [CrossRef]

36. Zhao, Z.Q.; Zheng, P.; Xu, S.T.; Wu, X. Object Detection With Deep Learning: A Review. *IEEE Trans. Neural Netw. Learn. Syst.* **2019**, *30*, 3212–3232. [CrossRef]
37. Chen, C.; Zheng, Z.; Xu, T.; Guo, S.; Feng, S.; Yao, W.; Lan, Y. YOLO-Based UAV Technology: A Review of the Research and Its Applications. *Drones* **2023**, *7*, 190. [CrossRef]
38. Li, Y.; Fan, Q.; Huang, H.; Han, Z.; Gu, Q. A Modified YOLOv8 Detection Network for UAV Aerial Image Recognition. *Drones* **2023**, *7*, 304. [CrossRef]
39. Jocher, G.; Chaurasia, A.; Laughing, Q.; Kwon, Y.; Kayzwer; Michael, K.; Sezer, O.; Mu, T.; Shcheklein, I.; Boguszewski, A.; et al. Ultralytics YOLOv8. Available online: https://docs.ultralytics.com/tasks/pose/ (accessed on 25 September 2023)
40. Maji, D.; Nagori, S.; Mathew, M.; Poddar, D. YOLO-Pose: Enhancing YOLO for Multi Person Pose Estimation Using Object Keypoint Similarity Loss. In Proceedings of the 2022 IEEE/CVF Conference on Computer Vision and Pattern Recognition Workshops (CVPRW), New Orleans, LA, USA, 19–24 June 2022; pp. 2636–2645.
41. Gao, X.; Hou, X.; Tang, J.; Cheng, H. Complete solution classification for the Perspective-Three-Point problem. *IEEE Trans. Pattern Anal. Mach. Intell.* **2003**, *25*, 930–943.
42. Abdel-Aziz, Y.I.; Karara, H.M. Direct Linear Transformation from Comparator Coordinates into Object Space Coordinates in Close-Range Photogrammetry. *Photogramm. Eng. Remote Sens.* **2015**, *81*, 103–107. [CrossRef]
43. Lepetit, V.; Moreno-Noguer, F.; Fua, P. EPnP: An Accurate O(n) Solution to the PnP Problem. *Int. J. Comput. Vis.* **2009**, *81*, 155–166. [CrossRef]
44. Penate-Sanchez, A.; Andrade-Cetto, J.; Moreno-Noguer, F. Exhaustive Linearization for Robust Camera Pose and Focal Length Estimation. *IEEE Trans. Pattern Anal. Mach. Intell.* **2013**, *35*, 2387–2400. [CrossRef] [PubMed]
45. Li, S.; Xu, C. A Stable Direct Solution of Perspective-three-Point Problem. *Int. J. Pattern Recognit. Artif. Intell.* **2011**, *25*, 627–642. [CrossRef]
46. Kneip, L.; Scaramuzza, D.; Siegwart, R. A Novel Parametrization of the Perspective-Three-Point Problem for a Direct Computation of Absolute Camera Position and Orientation. In Proceedings of the IEEE Conference on Computer Vision and Pattern Recognition (CVPR), Colorado Springs, CO, USA, 20–25 June 2011; pp. 2969–2976.
47. Wolfe, W.; Mathis, D.; Sklair, C.; Magee, M. The perspective view of three points. *IEEE Trans. Pattern Anal. Mach. Intell.* **1991**, *13*, 66–73. [CrossRef]
48. Amovlab. Prometheus Autonomous UAV Opensource Project. Available online: https://github.com/amov-lab/Prometheus (accessed on 1 May 2023)

drones

Article

Drone Based RGBT Tracking with Dual-Feature Aggregation Network

Zhinan Gao [1], Dongdong Li [1,*], Gongjian Wen [1], Yangliu Kuai [2] and Rui Chen [1]

1 College of Electronic Science and Technology, National University of Defense Technology, Changsha 410073, China; gaozhinan22@nudt.edu.cn (Z.G.); wengongjian@sina.com (G.W.); 2019302130162@whu.edu.cn (R.C.)
2 College of Intelligent Science and Technology, National University of Defense Technology, Changsha 410073, China; kuaiyangliu09@nudt.edu.cn
* Correspondence: lidongdong12@nudt.edu.cn

Abstract: In the field of drone-based object tracking, utilization of the infrared modality can improve the robustness of the tracker in scenes with severe illumination change and occlusions and expand the applicable scene of the drone object tracking task. Inspired by the great achievements of Transformer structure in the field of RGB object tracking, we design a dual-modality object tracking network based on Transformer. To better address the problem of visible-infrared information fusion, we propose a Dual-Feature Aggregation Network that utilizes attention mechanisms in both spatial and channel dimensions to aggregate heterogeneous modality feature information. The proposed algorithm has achieved better performance by comparing with the mainstream algorithms in the drone-based dual-modality object tracking dataset VTUAV. Additionally, the algorithm is lightweight and can be easily deployed and executed on a drone edge computing platform. In summary, the proposed algorithm is mainly applicable to the field of drone dual-modality object tracking and the algorithm is optimized so that it can be deployed on the drone edge computing platform. The effectiveness of the algorithm is proved by experiments and the scope of drone object tracking is extended effectively.

Keywords: RGBT tracking; Drone based object tracking; transformer; feature aggregation

Citation: Gao, Z.; Li, D.; Wen, G.; Kuai, Y.; Chen, R. Drone Based RGBT Tracking with Dual-Feature Aggregation Network. *Drones* **2023**, *7*, 585. https://doi.org/10.3390/drones7090585

Academic Editor: Anastasios Dimou

Received: 7 August 2023
Revised: 5 September 2023
Accepted: 6 September 2023
Published: 18 September 2023

1. Introduction

Object tracking is one of the fundamental tasks in computer vision and has been widely used in robot vision, video analysis, autonomous driving and other fields [1]. Among them, the drone scene is an important application scenario for object tracking which assist drones in playing a crucial role in urban governance, forest fire protection, traffic management, and other fields. Given the initial position of a target, object tracking is to capture the target in subsequent video frames. Thanks to the availability of large datasets of visible images [2], visible-based object tracking algorithms have made significant progress and achieved state-of-the-art results in recent years. Currently, due to the diversification of drone missions, visible object tracking is unable to meet the diverse needs of drones in various application scenarios [3]. Due to the limitations of visible imaging mechanisms, object tracking heavily relies on optimal optical conditions. However, in realistic drone scenarios, UAVs are required to perform object tracking tasks in dark and foggy environments. In such situations, visible imaging conditions are inadequate, resulting in significantly noisy images. Consequently, object tracking algorithms based on visible imaging fail to function properly.

Infrared images are produced by measuring the heat emitted by objects. Compared with visible images, infrared images have relatively poor visual effects and complementary target location information [4,5]. In addition, infrared images are not sensitive to changes in scene brightness, and thus maintain good imaging results even in poor lightning environments. However, the imaging quality of infrared images is poor and the spatial resolution and grayscale dynamic range are limited, resulting in a lack of details and texture information in the images. In contrast, visible images are very rich in details and texture features. In summary, visible and infrared object tracking has received increasing attention as it can meet the mission requirements of drones in various scenarios, due to the complementary advantages of infrared and visible images (Figure 1).

Currently, two main kinds of methods in visual object tracking are deep learning (DL)-based methods and correlation filter (CF)-based approaches [1]. The methods based on correlation filtering utilize Fast Fourier Transform (FFT) to perform correlation operation in the frequency domain, which have a very fast processing speed and run in real-time. However, their accuracy and robustness are poor. The methods based on neural network mainly utilize the powerful feature extraction ability of neural network. Their accuracy is better than that of correlation filtering based methods while their speed is slower. With the proposal of Siamese networks [6,7], the speed of neural network-based tracking methods has been greatly improved. In recent years, the neural network-based algorithm has become the mainstream method for object tracking.

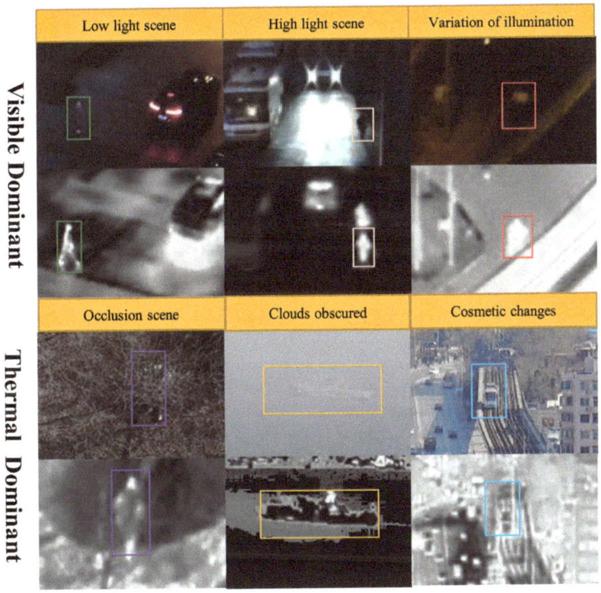

Figure 1. These are some visible-infrared image pairs captured by drones. In some scenarios, visible images may be difficult to distinguish different objects, while infrared images can continue to work in these scenarios. Therefore, information from visible and infrared modalities can complement each other in these scenarios. Introducing information from the infrared modality is very beneficial for achieving comprehensive object tracking in drone missions.

The transformer structure has achieved great success in the field of computer vision [8]. By introducing long-range attention mechanisms, it alleviates the problem of limited receptive fields in CNNs and has achieved State-of-the-art (SOTA) performance in multiple tasks in the field of computer vision [9,10]. Inspired by the remarkable success of transformer-based Siamese networks in object tracking on visible images, we propose a similar approach using a transformer-based Siamese network for RGB-Thermal (RGBT) tracking. Additionally, we drew inspiration from the transformer structure and used an attention mechanism to fuse visible and infrared image features in both spatial and channel dimensions. We visualized the effectiveness of this module through heat maps and achieved promising results on the drone dataset VTUAV [11] and drone hardware platform. The main contributions can be summarized as follows:

- A lightweight network applied to a drone visible-infrared object tracking mission is porposed with Swin transformer as the backbone network.
- A Dual-Feature aggregation module is integrated into this network, which aggregates visible-infrared image features from both spatial and channel dimensions using attention mechanisms. Ablation experiments are conducted to vefify the effectiveness of this module in fusing two modalities.
- Extensive experiments are conducted on the VTUAV dataset and drone hardware platform, the results showed that our tracker achieved good performance compared with other mainstream trackers.

2. Related Works

2.1. RGBT Tracking Algorithms

Many RGBT trackers have been proposed so far [12–15]. Due to the rapid development of RGB trackers, current RGBT trackers mainly consider the problem of dual-modal information fusion within mature trackers finetuned on the RGBT tracking task, where the key is to fuse visible and infrared image information. Several fusion methods are proposed, which are categorized as image fusion, feature fusion and decision fusion. For image fusion, the mainstream approach is to fusion image pixels based on weights [16,17], but the main information extracted from image fusion is the homogeneous information of the image pairs, and the ability to extract heterogeneous information from infrared-visible image pairs is not strong. At the same time, image fusion has certain requirements for registration between image pairs, which can lead to cumulative errors and affect tracking performance. Most trackers aggregate the representation by fusing features [18,19]. Feature fusion is a more advanced semantic fusion compared with image fusion. There are many ways to fuse features, but the most common way is to aggregate features using weighting. Feature fusion has the potential of high flexibility and can be trained with massive unpaired data, which is well-designed to achieve significant promotion. Decision fusion models each modality independently and the scores are fused to obtain the final candidate. Compared with image fusion and feature fusion, decision fusion is the fusion method on a higher level, which uses all the information from visible and infrared images. However, it is difficult to determine the decision criteria. Luo et al. [12] utilize independent frameworks to track in RGB-T data and then the results are combined by adaptive weighting. Decision fusion avoids the heterogeneity of different modalities and is not sensitive to modality registration. Finally, these fusion methods can also be used complementarily. For example, Zhang [11] used image fusion, feature fusion and decision fusion simultaneously for information fusion and achieved good results in multiple tests.

2.2. Transformer

Transformer originates from natural language processing (NLP) for machine translation and has been introduced to vision recently with great potential [8]. Inspired by the success in other fields, researchers have leveraged Transformer for tracking. Briefly, Transformer is an architecture for transforming one sequence into another one with the help of attention-based encoders and decoders. The attention mechanism can determine which parts of the sequence are important, breaking through the receptive field limitation of traditional CNN networks and capturing global information from the input sequence. However, the attention mechanism requires more training data to establish global relationships. Therefore, Transformer will have a lower effect than traditional CNN networks in some tasks with smaller sample size and more emphasis on regional relationships [20]. Additionally, the attention mechanism is able to replace correlation filtering operations in the Siamese network by finding the most relevant region to the template in the search area in a global scope. The method of [9] applies Transformer to enhance and fuse features in the Siamese tracking for performance improvement.

2.3. UAV RGB-Infrared Tracking

Currently, there are few visible-light-infrared object tracking algorithms available for drones, mainly due to two reasons. Firstly, there is a lack of training data for visible-light-infrared images of drones. Previously, models were trained using infrared images generated from visible images due to the difficulty in obtaining infrared images. With the emergence of datasets such as LasHeR [21], it is now possible to directly use visible and infrared images for training. In addition, there are also datasets such as GTOT [22], RGBT210 [23], RGBT234 [24], etc. available for evaluating RGBT tracking algorithm performance. However, in the field of RGBT object tracking for drones, only the VTUAV [11] dataset is available. Due to the different imaging perspectives of images captured by drones compared to normal images, training algorithms with other datasets does not yield good results. Secondly, existing algorithms have slow running speeds, making them difficult to use directly. Existing mainstream RGBT object tracking algorithms are based on deep learning, which have to deal with both visible and infrared images at the same time, with a large amount of data, a complex algorithmic structure and a low processing speed,such as JMMAC (4fps) [25], FANet (2fps) [18], MANnet (2fps) [26]. In drone scenarios, there is a high demand for speed in RGBT object tracking algorithms for drones. It is necessary to simplify the algorithm structure and improve its speed.

3. Material and Methods

This section introduces our visible-infrared object tracking algorithm. which is inspired by siamese-based RGB object tracking algorithms and adapted to RGBT tracking tasks based on SwinTrack [10]. The network is mainly divided into four parts: Feature Extraction Network, Dual-Feature Aggregation Network, Transformer-based Feature Fusion and Detection Head. The structure of network is shown as Figure 2.

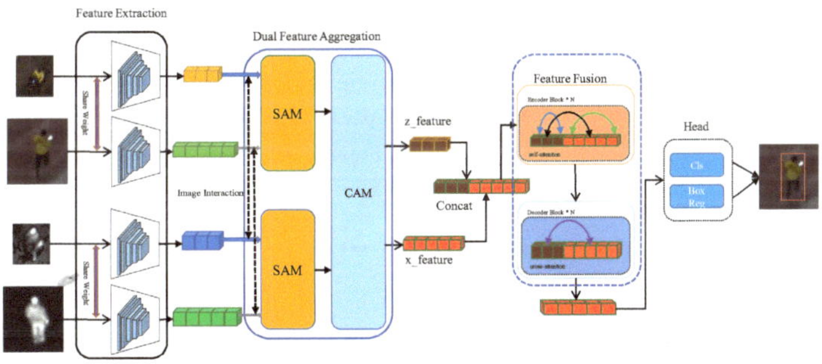

Figure 2. Architecture of our Transformer RGBT Tracking framework. This framework contains four fundamental components: Feature Extraction Network,Dual-Feature Aggregation Network, Feature Fusion Network and Detection Head.

3.1. Feature Extraction Network.

Traditionally, the ResNet network [27] is commonly used for feature extraction in computer vision tasks. However, with the development of the Transformer in the field of computer vision, feature extraction networks based on the Transformer have achieved better results than ResNet [10]. Moreover, our model is entirely based on the Transformer structure, using Swin-Transformer as the feature extraction network, which can provide a more compact feature representation and richer semantic information. This is very advantageous for the subsequent modules.The structure of Swin-Transformer is shown as Figure 3.

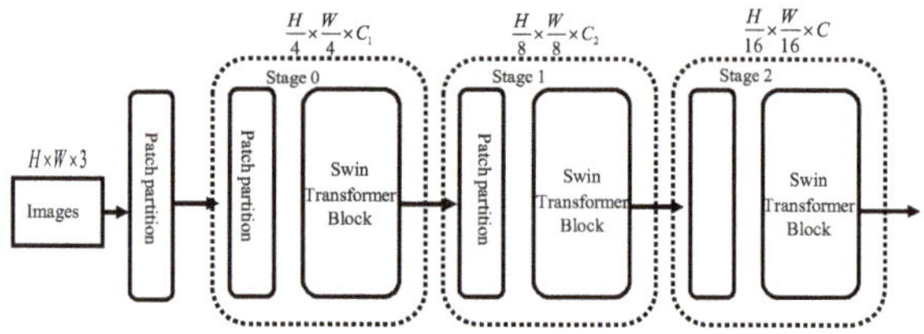

Figure 3. Architecture of Swin-Transformer. To accelerate the computation speed, images are divided into patches by Patch partition and fed into the network.

Our tracker follows the Siamese network structure, which requires a pair of image patches as inputs, i.e.,a template image $z \in \mathbf{R}^{H_z \times W_z \times 3}$and a search region image $x \in \mathbf{R}^{H_x \times W_x \times 3}$. The image pairs are firstly segmented into small patches and fed into the network. Attention operations are performed on these small patches, which significantly reduces the computation cost of the transform. Finally, template tokens $t_z \in \mathbf{R}^{\frac{H_z}{S} \times \frac{W_z}{S} \times C}$ and region tokens $t_x \in \mathbf{R}^{\frac{H_x}{S} \times \frac{W_x}{S} \times C}$ are generated, where s is the stride of the backbone network and C is the hidden dimension of the feature. To reduce model complexity, the lightweight Swin-Transform is used,where $s = 16$ and $C = 386$.

For single-modality tracking, a shared feature extraction network that shares parameters between the template and search images is sufficient. However, our task is RGB-T tracking with different imaging characteristics in each modality. Therefore, here visible and infrared features are extracted separately and two independent feature extraction modules are needed to extract features separately.

3.2. Dual-Feature Aggregation Network

Inspired by works such as Convolutional Block Attention Module (CBAM) [28] and Squeeze-and-Excitation Networks (SENet) [29], in the process of fusing visible and infrared information, we used attention mechanisms to enhance useful feature information in both spatial and channel dimensions. We proposed a Dual-Feature aggregation network and its structure diagram is shown in Figure 4.

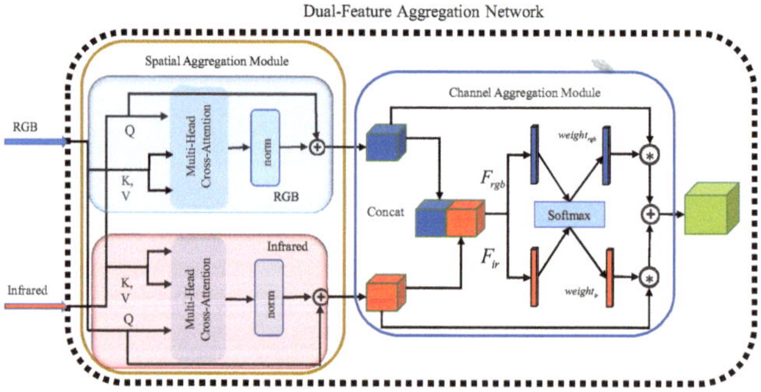

Figure 4. Architecture of Dual-Feature Aggregation Network. It mainly consists of two modules, namely Spatial Aggregation Module (SAM) and Channel Aggregation Module (CAM).

Different from other feature fusion methods that use a single modality to enhance another modality, we simultaneously enhance the visible-infrared dual-modality feature information by attention mechanism during the fusion process, which is very useful in some scenarios where single-modality algorithm is limited. The expression of the attention mechanism is:

$$Attention(Q,K,V) = Softmax(\frac{QK^T}{\sqrt{d_k}}V) \quad (1)$$

where Q, K and V stand for Query, Key and Value respectively, $\sqrt{d_k}$ means dimensionality normalization.

The attention mechanism is a feature weighting method. By selecting different Query, Key and Value, the image can be searched as a whole and the features of important areas in the image can be enhanced, so that the algorithm can pay more attention to important target areas in the image. For visible-infrared heterogeneous images, although their imaging methods are different, the target information is similar. For dual-modality object tracking, such similar target information is needed. By using the attention mechanism on heterogeneous images, by selecting different values of Query, Key and Value on heterogeneous images, one modality information can be used to enhance another modality information. In this case, the attention mechanism will strengthen the important target region in heterogeneous images and suppress the noise region in each modality. It is more conducive to the fusion of dual-modality features. The Dual-feature aggregation network consists of two modules: Spatial Aggregation Module (SAM) and Channel Aggregation Module (CAM). Below we will introduce these two modules respectively.

The Spatial Aggregation Module mainly focuses on the spatial features of the image. Inspired by the transformer structure, the attention mechanism searches for the regions of interest at any position in the image. By using the attention mechanism with an infrared image as the template and a visible image as the search image, the model can search for the most similar region in the visible image based on the information from the infrared image. This effectively aligns the information from the two modalities and highlights the parts of the visible image that are most similar to the infrared image. This approach allows for the incorporation of information from one modality into another while preserving the original information as much as possible. Similarly, by swapping the inputs of the visible and infrared images, the visible information can be used to calibrate the infrared image, resulting in calibrated features for both the infrared and visible images. This is why this module is called "Dual-feature". This process can be represented as:

$$SA_{RGB} = F_I + (MCA(F_I, Concat_s(F_{RGB}, F_I))) \tag{2}$$

$$SA_I = F_{RGB} + (MCA(F_{RGB}, Concat_s(F_I, F_{RGB}))) \tag{3}$$

where F_{RGB} and F_I are the visible-infrared feature information extracted by the backbone network. MCA is a multi-head Cross-Attention module, $Concat_s$ means concatenate in spatial domain.

The Channel Aggregation Module focuses on the features of image channels. The calibrated visible and infrared image features are obtained after spatial aggregation. If these features are directly fed into the encoder, the channel dimension will be twice that of the original features, which will seriously affect the efficiency of the encoder. Therefore, channel aggregation is needed to select the most important channel features from visible and infrared images, reduce the channel dimension and further fuse the information from visible and infrared images. The process can be expressed as:

$$weight_{RGB} = Softmax(\mathcal{F}_{RGB}(Concat_c(SA_{RGB}, SA_I))) \tag{4}$$

$$weight_I = Softmax(\mathcal{F}_I(Concat_c(SA_{RGB}, SA_I))) \tag{5}$$

$$F_{out} = weight_{RGB} \times SA_{RGB} + weight_I \times SA_I \tag{6}$$

where SA_{RGB} and SA_I represent the features that have been spatially aggregated. $\mathcal{F}_{RGB}$ and $\mathcal{F}_I$ are pooling layers used to generate weight parameters on channels. $Concat_c$ means concatenate in channel domain. F_{out} is the output of the Dual-Feature Aggregation Network.

3.3. Transformer-Based Feature Fusion Network

After the Dual-Feature aggregation network, fused template and search features are obtioned. During the fusion process, an encoder and decoder are utilized to fuse the template and search information and the resulting output is then refined. The encoder compresses the template and search features into a more compact representation. The decoder then decodes the compressed features back into the original feature space, achieving feature fusion. In the search regions, we utilize the fused features to locate the region that has the highest correlation with the template in the candidate region. Finally, these regions are fed into the detection head for detection. It is worth noting that we have employed concatenation-based fusion architecture in SwinTrack [10], which significantly reduces the model size and number of parameters compared to traditional methods.

The encoder consists of a sequence of Transformer blocks. Each block contains a Multi-Head Self-Attention (MSA) module and a Feed-Forward Network (FFN). The FFN contains a two-layer Multi-Layer Perceptron (MLP) with a GELU activation layer inserted after the first linear layer. To avoid overfitting, Layer Normalization (LN) is used. Moreover, residual connections are employed in both the MSA and FFN modules to facilitate gradient backpropagation. To reduce model complexity, four layers of Transformer blocks are used here as encoder. The process of encoder can be expressed as: Concatenate the features of template and search:

$$F = Concat(f_z, f_x) \tag{7}$$

Perform Attention operation on concatenated features in each encoder block:

$$F_{MSA} = F + MSA(LN(F)) \tag{8}$$

$$F_{FFN} = MLP(LN(F_{MSA})) + F_{MSA} \tag{9}$$

Separate the concatenated features into their original template and search features:

$$f_z, f_x = DeConcat(F_{FFN}) \tag{10}$$

Here f_x and f_z are respectively the template and search from Dual-Feature Aggregation Network. MSA is a multi-head Self-Attention module.

The decoder is composed of a Multi-Head Cross-Attention (MCA) module and the remaining parts are the same as the encoder and one layer of Transformer block is used here as decoder. The entire process of the decoder can be represented as follows:

$$F_{MCA} = f_x + MCA(LN(f_x), LN(Concat(f_x, f_z))) \tag{11}$$

$$F = MLP(LN(F_{MCA})) + F_{MCA} \tag{12}$$

Here f_x and f_z are produced by *DeConcat* in Encoder module. F will be fed to the Head network to generate a classification response map and a bounding box regression map.

3.4. Head and Loss

The Head network is split into two branches: classification and bounding box regression. Both are three-layer MLP networks that receive the feature map output from the decoder and respectively predict the classification response map $R_{cls} \in \mathbb{R}^{H_x \times W_x \times 2}$ and bounding box regression map $R_{reg} \in \mathbb{R}^{H_x \times W_x \times 4}$.

The classification Head receives the feature map output from the decoder and predicts the binary classification results of the search region. Only the annotated box is considered as a positive sample, while the rest are negative samples. As a result, the number of positive and negative samples is imbalanced. To alleviate this issue, we use a hyperparameter μ, which is set to 0.0625 based on experimental results, to reduce the loss from negative samples by a factor of μ. We use the standard binary cross-entropy loss for classification, which is defined as follows:

$$\mathcal{L}_{cls} = -\sum_{i} [y_i log(p_i) + \mu(1 - y_i) log(1 - p_i)] \tag{13}$$

Here, y_i denotes the ground-truth label of the i-th sample, $y_i = 1$ denotes foreground and p_i denotes the probability belong to the foreground predicted by the learned model.

For the bounding box regression, we use a linear combination of L1-norm loss $\mathcal{L}_1(.,.)$ and the generalized IoU loss $\mathcal{L}_{Giou}(.,.)$ [30]. The loss is calculated only for positive samples, while negative samples are ignored. The regression loss is defined as follows:

$$\mathcal{L}_{reg} = \sum_{i} [\lambda_G \mathcal{L}_{Giou}(b_i, \hat{b}) + \lambda_1 \mathcal{L}_1(b_i, \hat{b})] \tag{14}$$

where $\hat{b}$ denotes the normalized ground-truth bounding box. $\lambda_G = 3$ and $\lambda_1 = 4.3$ are hyperparameter weights determined through experiments.

4. Results

4.1. Implementation Details

Offline training. Experments are conducted on the VTUAV dataset and it is worth noting that the VTUAV dataset is annotated every ten frames. Therefore, there are only about 20,000 pairs of accurately usable training samples. To overcome this problem, the Lasher dataset is used for pre-training. The sizes of search region patch and template patch are 224×224 and 112×112, respectively. We trained the model using an AdamW optimizer with different initial learning rates for different modules. The backbone network used the Swin Transformer-Tiny pre-trained on Imagenet1K [31], with a stride of 16 and producing a feature map of size 14×14. The visible backbone network was most compatible with the pre-trained network and was set with a learning rate of 5×10^{-5}. The infrared backbone network required task fine-tuning, so its learning rate was set to 1×10^{-4}. The learning rates for all other modules were set to 5×10^{-4}, with a weight decay of 1×10^{-4}. Due to the use of concatenation-based Transform structures, our model has much lower GPU memory consumption compared to Transt [9]. We set the batchsize as 40, which can be trained on a single Nvidia Titan RTX GPU. We trained the model for 100 epochs and the learning rate decreased by a factor of 10 after 80 epochs.

Online Inference. We follow the inference steps of the Siamese network. First, we initialize the template based on the annotation results of the first frame. The target object is placed in the center of the image with a background area factor of 2. Then, we generate the search region based on the detection results. The background area factor for the search region is 5. During the inference process, the Detection Head outputs a 14×14 classification response feature map. We use a Hanning window to incorporate the prior information of the target's position into the tracking process, thereby suppressing sudden changes in the target's position. The process can be expressed as:

$$cls = (1 - \gamma) \times r_{cls} + \gamma \times h \tag{15}$$

Here r_{cls} is classification response feature map, γ is the weight parameter and h is the Hanning window with the same size as r_{cls}. And $\gamma = 0.49$ always get a very good result by experiments. After determining the target position based on the classification response feature map, the target's bounding box is estimated on the position response map. The new search region is then fed into the tracking network, completing a full target tracking inference process.

Evaluation metrics. In our experiment, all the trackers are run in one-pass evaluation (OPE) protocol and evaluated by Maximum Success Rate (MSR) and Maximum Precision Rate (MPR), which are widely used in RGB-T tracking [22–24]. A total of 175 short-term tracking video sequences from the VTUAV dataset were used, with each test sequence ranging from 2000 to 15,000 frames. The annotation results were labeled every ten frames, and the final evaluation results were tested using a sampling method.

4.2. Ablation Experiment

In order to verify the effectiveness of each component in the network, ablation experiments were conducted on the modules in the network and the results are shown as Table 1 and Figure 5.

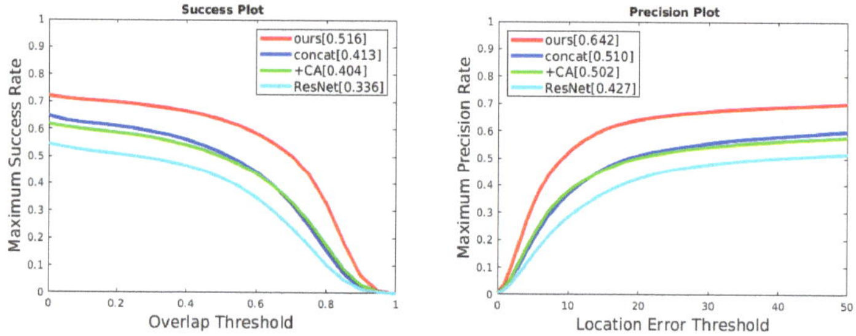

Figure 5. Ablation experiment result.

Table 1. Comparative Experiment Analysis Table.

RGBT Tracker	Maximum Success Rate(MSR)	Maximum Precision Rate (MPR)	Parameter
Concat	41.3	51.0	41.9 M
ResNet	33.6	42.7	41.3 M
+CAM	40.4	50.2	41.6 M
Ours	51.6	64.2	47.9 M

Firstly, the network without the Dual-Feature Aggregation Network (Concat) was tested as a baseline on the VTUAV dataset, and its MSR and MPR decreased by 10.3 and 13.2, respectively, which fully demonstrated the effectiveness of the Dual Feature Aggregation Network in RGB-T tracking. Secondly, experiments were conducted by replacing SwinTransform with ResNet as the backbone network (ResNet), and it was found that its performance decreased. This is because the stride of ResNet is 8, which requires the use of smaller template (56×56) and search (112×112) sizes, thus increasing the problem of limited receptive fields. Based on the baseline network (Concat), directly aggregating the visible and infrared channels after concatenation (+CAM) resulted in a relative decrease of 0.9 and 0.8 compared to the baseline. But performing spatial aggregation before channel aggregation has greatly improved its effectiveness (Ours).

Through analyzing the heatmap (Figure 6), it was found that when directly performing channel aggregation, multiple non-target areas with high confidence scores appeared in the heatmap. This would interfere with the network and easily lose the target, which is the main reason for the performance degradation. Although direct channel aggregation also complete the fusion of dual-modal information, this method has large errors and lead to tracking failure. However, by first using a spatial aggregation module to aggregate visible and infrared channels, the target information in the dual-modal can be selectively retained while suppressing noise. This reflects the effectiveness of the Dual Feature Aggregation Network in preserving bimodal information.

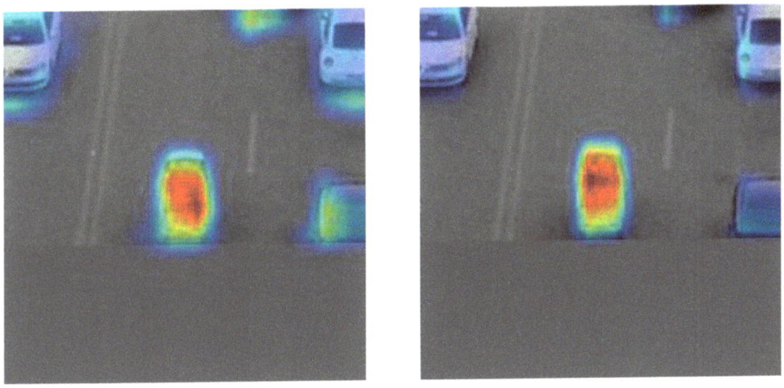

a. +CAM b. Dual-Feature Aggregation

Figure 6. Ablation experiment result.

4.3. Contrastive Experiment

Currently, the main mainstream deep-Learnning based RGBT object tracking algorithms are JMMAC [25], FANet [18], MANnet [26], DAFNet [32], ADRNet [14], FSRPN [15], etc., but some of these algorithms are too slow and not applicable to UAV RGBT object tracking tasks. We tested our algorithm on the VTUAV dataset and compared it with three additional fast trackers (DAFNet, ADRNet, FSRPN). Both DAFNet [32] and ADRNet [14] are the best performing multi-domain network trackers. The multi-domain network mainly performs classification and regression tasks in each domain such as visible and infrared and finally obtains the final tracking result of RGBT through competitive learning. FSRPN [15] is a Siamese-based tracker, which uses the pipeline of Siamese to improve the tracking accuracy and speed up the processing speed of the algorithm. The results are shown as Table 2 and Figure 7.

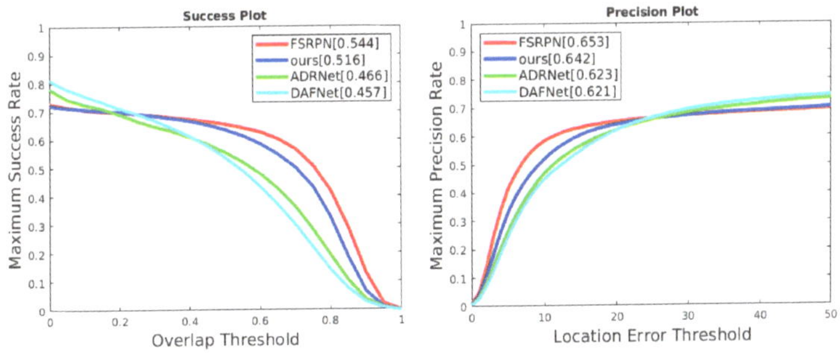

Figure 7. Comparison experiment result.

Table 2. Comparative Experiment Analysis Table.

RGBT Tracker	Maximum Success Rate(MSR)	Maximum Precision Rate (MPR)	FPS	Parameter
DAFNet	45.7	62.1	21.0	68.5 M
ADRNet	46.6	62.3	25.0	54.1 M
FSRPN	54.4	65.3	30.3	53.9 M
Ours	51.6	64.2	31.2	47.9 M

In the comparative experiments, our algorithm outperformed DAFNet and ADRNet in Maximum Success Rate and Maximum Precision Rate (MSR higher than DAFNet by 19% and ADRNet by 17%, MPR higher than DAFNet and ADRNet by 3%), but was slightly inferior to the FSPRN algorithm. Our algorithm has been specifically designed for drone missions, focusing on simplifying the network structure, reducing the number of network parameters and enhancing algorithm speed. As a result, our algorithm has successfully achieved a favorable balance between performance and efficiency on VTUAV in comparison to mainstream algorithms.

4.4. *Drone Hardware Platforms*

Our algorithms will eventually be deployed on a drone hardware platform, the components of which are described here, and the composition diagram is shown in Figure 8.

The DJI M300 drone is an industry-level unmanned aerial vehicle that can fly for 30 min in the air with a payload of up to 9 kg. It has a maximum flight altitude of 1500 m, making it suitable for most drone scenarios and tasks. The H20T camera is a visible-light and thermal infrared camera that can capture both visible-light and infrared images simultaneously. We use the H20T camera to complete our drone RGBT tracking tasks. We use Nvidia Orin NX as the on-board processing platform with the specific parameters shown in Table 3.

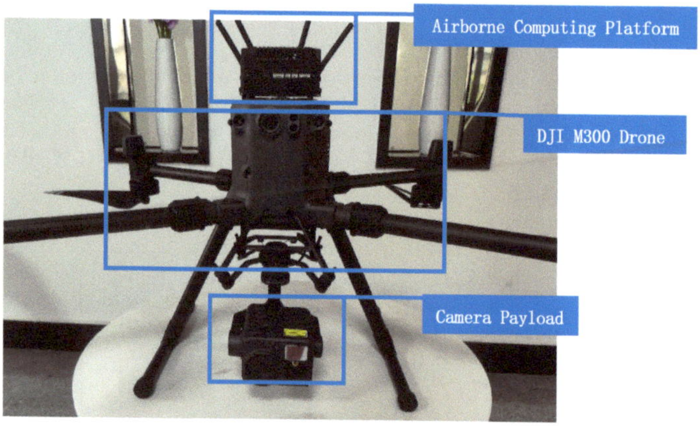

Figure 8. Structure of Drone Hardware Platforms. It mainly consists of three components, the DJI M300 drone, the H20T camera load, the Airborne Computing Platform Nvidia Orin NX.

Table 3. Nvidia Orin NX parameters Table *.

CPU	CPU Frequency	Display Memory	Computational Performance
Arm Cortex-A78AE	2 GHz	16 GB	100TOPS

* Parameters from official Nvidia documentation.

4.5. *Visualization and Analysis*

4.5.1. Algorithmic Effect of Drone Hardware Platform

We tested the images captured by the drone hardware platform with a processing speed of 13.1 fps when running the RGBT object tracking algorithm on the onboard computing platform, and the tracking results are shown in Figure 9.

From the tracking results, it can be observed that the infrared modality can solve the problem of tracking failure under conditions such as occlusion and lighting changes in complex scenes. Therefore, using visible-infrared object tracking can expand the scope of drone object tracking tasks and improve the environmental adaptability of drone object tracking tasks.

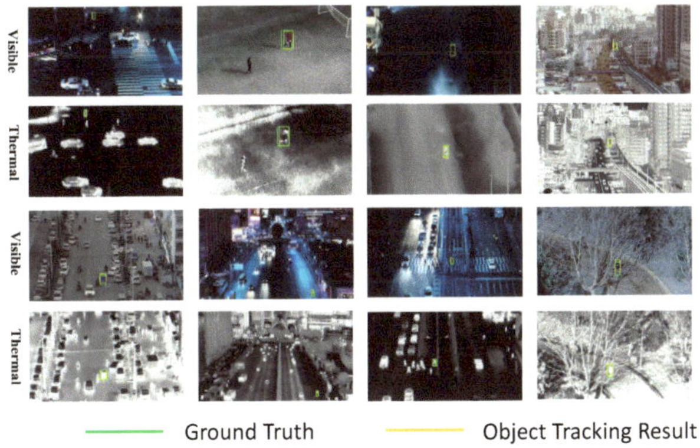

Figure 9. Graph of actual results of the algorithm.

4.5.2. Heatmap

We present the heatmaps generated from various main modules in the network, as shown in Figure 10. From the heatmaps, it can be observed that the backbone network extracts features separately from the visible and infrared images. Due to the differences between visible and infrared images, the features extracted by the network are also different. These features are then fed into the Dual Feature Aggregation Network, which combines the information from both modalities to obtain a fused feature map. In pedestrian-211 and Tricy-006 sequences, it can be seen more clearly that the aggregated feature map integrates all the features from both images. The fused feature map is then passed through the encoder and decoder modules, which focus the network's attention on the target. Finally, the maximum target response map is fed into the Head for classification and regression.

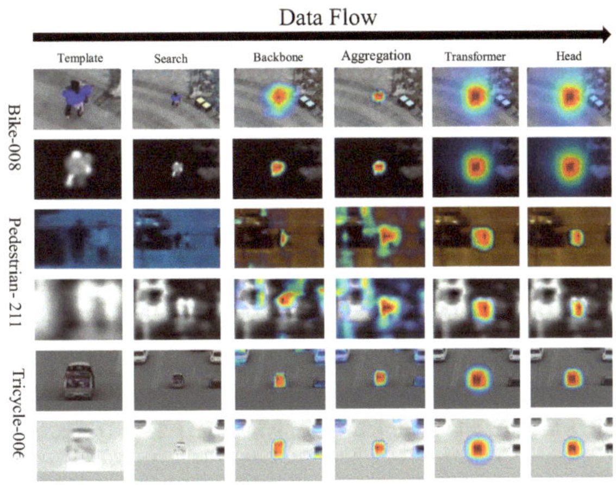

Figure 10. Heatmaps of Various Modules in the Network. The left two images are the visible-infrared template and search images input into the network. Following the direction of the network data flow, the heatmaps show the responses of the Backbone network, Feature Aggregation Network, Transform Fusion Network and Head modules, respectively.

4.5.3. Typical Failure Cases

We show some typical failure cases of our tracker in Figure 11.

Figure 11. Picture of the typical failure cases of our tracker.

The first image shows similar appearance, where the target is interfered with by similar targets in visible and infrared images. The Siamese network lacks the ability to distinguish between similar targets as its core is the matching of templates and searches. Although we introduce prior information about the target position by using the Hanning window and penalize sudden changes in the target's position, tracking similar targets at close range is still prone to failure. The second image shows shape changes. our tracker is non-temporal. When the appearance of the target in the search is significantly different from that in the template, the tracker will loses the target. Thus, temporal information is essential in some long sequence tracking processes. The third image shows imaging noise. In some environments, the infrared images obtained contain noise, severely affecting the quality of infrared images. When the single-mode noise is too strong, the tracker will be affected by interference from noise, affecting the tracking effect.

5. Discussion

This study primarily focuses on visible-infrared dual-modality object tracking in drone scenarios. We have conducted ablation experiments and visualization analysis to validate the effectiveness of the proposed dual-feature aggregation network in aggregating visible-infrared modality information. Our algorithm outperforms other mainstream algorithms in terms of tracking accuracy, while utilizing fewer parameters and achieving faster running speeds. Our algorithm is specifically designed for drone scenarios and can be seamlessly deployed and executed on the Nvidia Orin NX, a drone edge computing platform with limited computing resources. To evaluate the algorithm's performance, we conducted tests in an open scene using the aforementioned hardware platform. The results demonstrate that leveraging dual-modality information can significantly enhance the accuracy and robustness of object tracking, particularly in scenarios with illumination changes and occlusions. Additionally, we have analyzed the failure cases encountered during the experiments to identify potential areas for future research. The performance of our algorithm is degraded in scenes with similar appearance, shape changes and image noise. Furthermore, the algorithm's processing speed still falls short of meeting real-time requirements on edge computing devices. These challenges serve as important considerations for future improvements.

6. Conclusions

In this work, we mainly designed a visible-infrared object tracking network based on the Transformer architecture. It consists of four components, among which we focused on designing a Dual-Feature Aggregation Network to fuse visible and infrared information. Through ablation experiments and visualization analysis, we demonstrated the effectiveness of the Dual-Feature Aggregation Network. The algorithm is mainly for the task of RGBT object tracking in drone scenarios and the algorithm is simplified so that it can run on the drone edge computing platform. Compared with the mainstream RGBT object tracking algorithms, our algorithm still achieves better performance.

Author Contributions: Conceptualization, Z.G. and D.L.; methodology, Z.G.; software, Z.G.; validation, Z.G., D.L. and R.C.; formal analysis, Z.G.; investigation, Z.G.; resources, D.L.; data curation, Z.G.; writing—original draft preparation, Z.G.; writing—review and editing, Z.G., D.L., Y.K.; visualization, Z.G.; supervision, Z.G.; project administration, D.L., G.W.; funding acquisition, D.L. and Y.K. All authors have read and agreed to the published version of the manuscript.

Funding: This work is supported by the National Natural Science Foundation of China (NSFC) (NO.62102426), the natural science foundation of Hunan Province (NO.2021JJ40683), the Science and Technology Innovation Plan Project of Hunan Province (NO.2021RC2072) and the scientific research project of National University of Defense Technology (NO. ZK20-47, ZK21-29).

Institutional Review Board Statement: Not applicable.

Informed Consent Statement: Not applicable.

Data Availability Statement: The data presented in this study are available on request from the corresponding author.

Conflicts of Interest: The authors declare no conflict of interest. The funders had no role in the design of the study; in the collection, analyses, or interpretation of data; in the writing of the manuscript, or in the decision to publish the results.

References

1. Zhang, X.; Ye, P.; Leung, H.; Gong, K.; Xiao, G. Object Fusion Tracking Based on Visible and Infrared Images: A Comprehensive Review. *Inf. Fusion* **2020**, *63*, 166–187. [CrossRef]
2. Huang, L.; Zhao, X.; Huang, K. GOT-10k: A Large High-Diversity Benchmark for Generic Object Tracking in the Wild. *IEEE Trans. Pattern Anal. Mach. Intell.* **2021**, *43*, 1562–1577. [CrossRef] [PubMed]
3. Fan, H.; Ling, H.; Lin, L.; Yang, F.; Liao, C. LaSOT: A High-Quality Benchmark for Large-Scale Single Object Tracking. In Proceedings of the 2019 IEEE/CVF Conference on Computer Vision and Pattern Recognition (CVPR), Long Beach, CA, USA, 15–20 June 2019.
4. Liu, Q.; Li, X.; He, Z.; Fan, N.; Yuan, D.; Wang, H. Learning Deep Multi-Level Similarity for Thermal Infrared Object Tracking. *IEEE Trans. Multimed.* **2021**, *23*, 2114–2126. [CrossRef]
5. Liu, Q.; Li, X.; He, Z.; Fan, N.; Liang, Y. Multi-Task Driven Feature Models for Thermal Infrared Tracking. *arXiv* **2019**, arXiv:1911.11384.
6. Bertinetto, L.; Valmadre, J.; Henriques, J.F.; Vedaldi, A.; Torr, P.H.S. Fully-Convolutional Siamese Networks for Object Tracking. *arXiv* **2016**, arXiv:1606.09549.
7. Li, B.; Yan, J.; Wu, W.; Zhu, Z.; Hu, X. High Performance Visual Tracking with Siamese Region Proposal Network. In Proceedings of the 2018 IEEE/CVF Conference on Computer Vision and Pattern Recognition, Salt Lake City, UT, USA, 18–23 June 2018; pp. 8971–8980.
8. Dosovitskiy, A.; Beyer, L.; Kolesnikov, A.; Weissenborn, D.; Houlsby, N. An Image is Worth 16x16 Words: Transformers for Image Recognition at Scale. *arXiv* **2020**, arXiv:2010.11929v1.
9. Chen, X.; Yan, B.; Zhu, J.; Wang, D.; Yang, X.; Lu, H. Transformer Tracking. In Proceedings of the 2021 IEEE/CVF Conference on Computer Vision and Pattern Recognition (CVPR), Nashville, TN, USA, 2 November 2021; pp. 8122–8131.
10. Lin, L.; Fan, H.; Xu, Y.; Ling, H. SwinTrack: A Simple and Strong Baseline for Transformer Tracking. *arXiv* **2021**, arXiv:2112.00995v1.
11. Zhang, P.; Zhao, J.; Wang, D.; Lu, H.; Ruan, X. Visible-Thermal UAV Tracking: A Large-Scale Benchmark and New Baseline. In Proceedings of the 2022 IEEE/CVF Conference on Computer Vision and Pattern Recognition (CVPR), New Orleans, LA, USA, 18–24 June 2022; pp. 8876–8885.
12. Luo, C.; Sun, B.; Yang, K.; Lu, T.; Yeh, W.C. Thermal infrared and visible sequences fusion tracking based on a hybrid tracking framework with adaptive weighting scheme. *Infrared Phys. Technol.* **2019**, *99*, 265–276. [CrossRef]
13. Yun, X.; Sun, Y.; Yang, X.; Lu, N. Discriminative Fusion Correlation Learning for Visible and Infrared Tracking. *Math. Probl. Eng.* **2019**, *2019*, 2437521. [CrossRef]
14. Zhang, P.; Wang, D.; Lu, H.; Yang, X. Learning Adaptive Attribute-Driven Representation for Real-Time RGB-T Tracking. *Int. J. Comput. Vis.* **2021**, *129*, 2714–2729. [CrossRef]
15. Kristan, M.; Matas, J.; Leonardis, A.; Felsberg, M.; Pflugfelder, R.; Kämäräinen, J.K.; Zajc, L.C.; Drbohlav, O.; Lukezic, A.; Berg, A.; et al. The Seventh Visual Object Tracking VOT2019 Challenge Results. In Proceedings of the 2019 IEEE/CVF International Conference on Computer Vision Workshop (ICCVW), Seoul, Republic of Korea, 27–28 October 2019; pp. 2206–2241.
16. Jingchao, P.; Haitao, Z.; Zhengwei, H.; Yi, Z.; Bofan, W. Siamese Infrared and Visible Light Fusion Network for RGB-T Tracking. *arXiv* **2021**, arXiv:2103.07302v1.
17. Wu, Y.; Blasch, E.; Chen, G.; Bai, L.; Ling, H. Multiple source data fusion via sparse representation for robust visual tracking. In Proceedings of the 2011 Proceedings of the 14th Conference on Information Fusion, Chicago, IL, USA, 5–8 July 2011.

18. Zhu, Y.; Li, C.; Luo, B.; Tang, J. FANet: Quality-Aware Feature Aggregation Network for Robust RGB-T Tracking. *arXiv* **2018**, arXiv:1811.09855.
19. Li, C.; Wu, X.; Zhao, N.; Cao, X.; Tang, J. Fusing Two-Stream Convolutional Neural Networks for RGB-T Object Tracking. *Neurocomputing* **2017**, *281*, 78–85. [CrossRef]
20. Liu, W.; Quijano, K.; Crawford, M.M. YOLOv5-Tassel: Detecting Tassels in RGB UAV Imagery with Improved YOLOv5 Based on Transfer Learning. *IEEE J. Sel. Top. Appl. Earth Obs. Remote Sens.* **2022**, *15*, 8085–8094. [CrossRef]
21. Li, C.; Xue, W.; Jia, Y.; Qu, Z.; Luo, B.; Tang, J. LasHeR: A Large-scale High-diversity Benchmark for RGBT Tracking. *arXiv* **2021**, arXiv:2104.13202v2.
22. Li, C.; Cheng, H.; Hu, S.; Liu, X.; Tang, J.; Lin, L. Learning Collaborative Sparse Representation for Grayscale-Thermal Tracking. *IEEE Trans. Image Process.* **2016**, *25*, 5743–5756. [CrossRef]
23. Li, C.; Liang, X.; Lu, Y.; Zhao, N.; Tang, J. RGB-T Object Tracking: Benchmark and Baseline. *Pattern Recognit.* **2019**, *96*, 106977. [CrossRef]
24. Li, C.; Zhao, N.; Lu, Y.; Zhu, C.; Tang, J. Weighted Sparse Representation Regularized Graph Learning for RGB-T Object Tracking. In Proceedings of the Acm on Multimedia Conference, Bucharest, Romania, 6–9 June 2017; pp. 1856–1864. [CrossRef]
25. Zhang, P.; Zhao, J.; Wang, D.; Lu, H.; Yang, X. Jointly Modeling Motion and Appearance Cues for Robust RGB-T Tracking. *arXiv* **2020**, arXiv:2007.02041.
26. Wang, S.; Zhou, Y.; Yan, J.; Deng, Z. Fully Motion-Aware Network for Video Object Detection. In Proceedings of the European Conference on Computer Vision, Munich, Germany, 8–14 September 2018. [CrossRef]
27. He, K.; Zhang, X.; Ren, S.; Sun, J. Deep Residual Learning for Image Recognition. In Proceedings of the 2016 IEEE Conference on Computer Vision and Pattern Recognition (CVPR), Las Vegas, NV, USA, 27–30 June 2016; pp. 770–778.
28. Woo, S.; Park, J.; Lee, J.Y.; Kweon, I.S. *CBAM: Convolutional Block Attention Module*; Springer: Cham, Switzerland, 2018.
29. Hu, J.; Shen, L.; Sun, G. Squeeze-and-Excitation Networks. In Proceedings of the 2018 IEEE/CVF Conference on Computer Vision and Pattern Recognition, Salt Lake City, UT, USA, 18–23 June 2018; pp. 7132–7141.
30. Rezatofighi, H.; Tsoi, N.; Gwak, J.; Sadeghian, A.; Reid, I.; Savarese, S. Generalized Intersection Over Union: A Metric and a Loss for Bounding Box Regression. In Proceedings of the 2019 IEEE/CVF Conference on Computer Vision and Pattern Recognition (CVPR), Long Beach, CA, USA, 15–20 June 2019; pp. 658–666.
31. Russakovsky, O.; Deng, J.; Su, H.; Krause, J.; Satheesh, S.; Ma, S.; Huang, Z.; Karpathy, A.; Khosla, A.; Bernstein, M.a. ImageNet Large Scale Visual Recognition Challenge. *Int. J. Comput. Vis.* **2015**, *115*, 211–252. [CrossRef]
32. Gao, Y.; Li, C.; Zhu, Y.; Tang, J.; He, T.; Wang, F. Deep Adaptive Fusion Network for High Performance RGBT Tracking. In Proceedings of the 2019 IEEE/CVF International Conference on Computer Vision Workshop (ICCVW), Seoul, Republic of Korea, 27–28 October 2019; pp. 91–99.

drones

MDPI

Article

Implicit Neural Mapping for a Data Closed-Loop Unmanned Aerial Vehicle Pose-Estimation Algorithm in a Vision-Only Landing System

Xiaoxiong Liu *, Changze Li, Xinlong Xu, Nan Yang and Bin Qin

School of Automation, Northwestern Polytechnical University, Xi'an 710129, China;
cz_li@mail.nwpu.edu.cn (C.L.); xuxinlong@mail.nwpu.edu.cn (X.X.); yang_nan@mail.nwpu.edu.cn (N.Y.);
binq3638@mail.nwpu.edu.cn (B.Q.)
* Correspondence: liuxiaoxiong@nwpu.edu.cn

Abstract: Due to their low cost, interference resistance, and concealment of vision sensors, vision-based landing systems have received a lot of research attention. However, vision sensors are only used as auxiliary components in visual landing systems because of their limited accuracy. To solve the problem of the inaccurate position estimation of vision-only sensors during landing, a novel data closed-loop pose-estimation algorithm with an implicit neural map is proposed. First, we propose a method with which to estimate the UAV pose based on the runway's line features, using a flexible coarse-to-fine runway-line-detection method. Then, we propose a mapping and localization method based on the neural radiance field (NeRF), which provides continuous representation and can correct the initial estimated pose well. Finally, we develop a closed-loop data annotation system based on a high-fidelity implicit map, which can significantly improve annotation efficiency. The experimental results show that our proposed algorithm performs well in various scenarios and achieves state-of-the-art accuracy in pose estimation.

Keywords: vision-only landing system; runway-line detection; pose estimation; implicit neural mapping; data closed-loop

Citation: Liu, X.; Li, C.; Xu, X.; Yang, N.; Qin, B. Implicit Neural Mapping for a Data Closed-Loop Unmanned Aerial Vehicle Pose-Estimation Algorithm in a Vision-Only Landing System. *Drones* **2023**, *7*, 529. https://doi.org/10.3390/drones7080529

Academic Editors: Dongdong Li, Gongjian Wen, Yangliu Kuai and Runmin Cong

Received: 16 July 2023
Revised: 31 July 2023
Accepted: 9 August 2023
Published: 12 August 2023

1. Introduction

Safe and reliable flight is an important research topic in aircraft, and the process of approaching and landing is the phase with the highest accident rate during the flight of fixed-wing aircraft, so it is very important to guide the landing safely. traditional landing systems rely on landing systems with instruments, which are a proven landing solution, but the system requires expensive equipment and maintenance. For UAV (unmanned aerial vehicle) landing, typical ground-based landing systems include OPATS and SADA. With the continuous development of visual perception and positioning technologies, it has become possible to apply vision to guided landing systems in recent years. Vision sensors are resistant to interference and not easily detected compared to active sensors, such as radar and laser, so the application of vision sensors to guided landings has received a lot of attention [1].

Vision-based landing systems for fixed-wing aircraft are composed of ground-based visual landing systems and space-based visual landing systems according to the implementation principle. Ground-based visual landing systems place vision sensors around the runway to determine the position of the UAV through multi-point observation to achieve landing. The scheme has sufficient computing resources, but it needs to rely on communication links, and its autonomy and applicability are somewhat limited. Space-based visual landing systems use the information provided by vision to achieve navigation and positioning, which further completes the vision-guided landing. The C2Land project is a typical example of this solution [2].

The space-based visual landing system can be divided into image-based visual servoing (IBVS) and position-based visual servoing (PBVS). IBVS compares the image signal obtained from real-time measurements with a given image signal and uses the acquired image error for closed-loop control. However, PBVS uses the camera parameters to establish the relationship between the image signal and the aerial vehicle's attitude and utilizes the attitude information in the closed-loop control. IBVS does not need to rely on the camera model, but the scheme is more scene-dependent. PBVS achieves the decoupling of vision problem and control problem, but the scheme requires an accurate camera model [3].

This paper proposes a solution to the pose-estimation problem in vision-only landing systems. We use the PBVS strategy to make the whole pose-estimation system robust and interpretable. To achieve higher accuracy, we propose a novel pose-estimation algorithm in a visual landing system, which is an implicit neural mapping solution (refer to Figure 1). We use camera images as input and the pose estimation as output. The runway detection, initial pose estimation, and NeRF-inverting [4] modules are computed on the on-board device (blue color in Figure 1; implicit mapping and GT annotation modules are computed on the cloud device (red color in Figure 1). The detection algorithm proposed in this paper is abbreviated as FMRLD (flexible multi-stage runway-line detection) in the experiment.

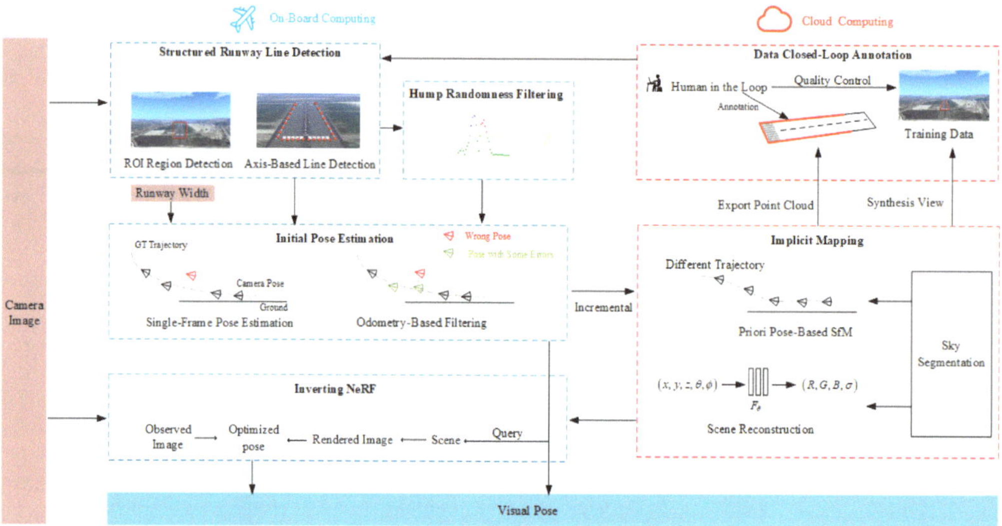

Figure 1. Our proposed implicit neural mapping pose estimation method in a vision-only landing system.

Our proposed algorithm follows the basic paradigm of pose estimation. Firstly, we perform feature extraction on the runway lines. The extracted features are then used for initial pose estimation, which is further optimized to obtain accurate estimation results. In the feature extraction phase, we use deep learning-based runway line detection methods to enhance accuracy and robustness (Section 3.1.1). These methods rely on high-quality datasets, so we utilize diverse data sources to construct datasets and perform data augmentation accordingly (Section 3.3.1). Since the accuracy of runway line detection directly affects the initial pose, we propose hump randomness filtering to refine the detection results (Section 3.1.2). During the initial pose estimation phase, we utilize the principle of multi-view geometry to estimate the pose. To ensure accuracy, we eliminate some incorrect estimation results (Section 3.2.1). The pose optimization is divided into two parts: on-board and cloud-based. On the on-board computing platform, the pose optimization results are obtained through inverting NeRF (Section 3.2.3). Meanwhile, on the cloud computing platform, the initial pose estimation results of the current trip are combined with the poses from historical trips for incremental pose optimization. The optimized results are then utilized for NeRF implicit mapping (Section 3.2.2). To address the challenges of expensive and inefficient runway data annotation, we propose a data closed-loop annotation strategy that leverages mapping results to assist in the annotation process. Specifically, we export the explicit point cloud of NeRF and allow annotators to annotate directly on the 3D point cloud. This approach significantly enhances the efficiency of data reuse compared to traditional image annotation methods. As a result, the entire algorithm operates in a closed-loop data flow (Section 3.3.2). The modules included in our proposed method are described below.

Runway detection: Accurate detection of runway lines is extremely important for navigation and positioning. Our structured runway line-detection and hump-randomness filtering modules provide consistent and reliable information on runway features. During the landing process, the visual features vary greatly among different runways, different weather conditions, and different landing phases, and these problems pose certain challenges to the accurate detection of runway lines. In this paper, our proposed coarse-to-fine accurate runway line-detection method fully considers the change in viewpoint during the landing of the aerial vehicle and the applicability of the algorithm to different scenarios. First, we use an object-detection algorithm to extract high-level semantic information about the runway, which ensures the uniform distribution of the runway in the image and facilitates the detection of subsequent runway lines. Then, we extract the left and right runway lines and the virtual start line in the focused image. We propose a column-anchor-based detection and parallel acceleration scheme for virtual start-line detection. Last, a runway line fine-tuning method based on clustering and optimization is proposed due to the randomness of detection arising from the width of the runway line. Our runway-detection module can provide good front-end detection information for pose estimation.

Initial pose estimation: The goal of our initial pose-estimation module is to estimate the UAV pose information with scales using a runway line feature. To obtain the scale, the module needs to input the runway width as a priori information. We use multi-view geometry, such as the vanishing point principle, to estimate the UAV's initial pose. However, the pose is generated from a single image and does not guarantee the stability of the pose. We adopt the results of the visual odometry pose estimation as a reference to fix the instability in the initial pose estimation.

Incremental implicit mapping: The incremental implicit mapping module provides map information to the initial pose estimation and improves the accuracy of the pose estimation. It also provides high-quality point-cloud maps due to the differentiability and high fidelity of the neural radiance field (NeRF [5]). Due to the limitations of NeRF [5] in pose optimization in large scale scenes, we have split the implicit mapping module into two sub-modules: offline pose optimization and NeRF mapping. In the offline pose optimization sub-module, we have adopted the standard structure from the motion (SfM) process. However, we have two modifications. One is that we introduce a sky segmentation sub-module, which ensures that SfM does not extract feature points from the sky during the feature-extraction stage, preventing the problem of poor pose-estimation results due to feature mismatch. The other point is that we use the results of the initial pose estimation as prior information for triangulation and bundle adjustment, thus preventing the failure of pose estimation caused by local optima that SfM may fall into in large-scale scenes. In NeRF mapping, a submodule and a grid-based NeRF approach [6] are adopted. We introduce appearance embedding to ensure robustness in different weather conditions. In addition, based on some characteristics of the runway itself, we introduce regularization losses (smoothness loss, sky loss, etc.) to improve the geometry of the NeRF mapping. Please refer to Section 4 for more details.

Inverting NeRF: Inverting NeRF aims to optimize the pose-estimation result based on the implicit map when a new initial pose arrives. We use the initial pose to query the NeRF map, and we can obtain a rendered image. Meanwhile, we can also obtain the camera image on that timestamp. Using the pose as an optimization variable, we optimize the pose by constructing a loss of the observed and rendered images.

GT annotation: The runway-detection network must be trained using annotated data, which is an extremely labor-intensive process. The GT annotation module reduces the annotation cost significantly by generating a 3D point-cloud map, annotating the runway in 3D space, and then projecting it into the 2D image. At the same time, due to the differentiable representation, NeRF can synthesize images with a novel view, thus providing true 3D data augmentation. The GT annotation module achieves a closed loop of data and enhances the iterative efficiency of the whole system.

Combining the above modules, we propose a complete algorithm for estimating the pose in a vision-only landing system. The proposed algorithm has been proven effective in simulation experiments.

The main contributions of our work are follows.

(1) A novel pose-estimation framework in a vision-only landing system is proposed, which introduces implicit mapping and ground-truth annotation modules to improve the pose-estimation accuracy and data-annotation efficiency.
(2) We build a runway-detection pipeline. The multi-stage detection framework proposed in this paper makes full use of the features of different stages, which can guarantee semantic features and positioning ability and therefore greatly improves the runway line detection accuracy.
(3) We present a NeRF-based mapping module in a visual landing system, whose high fidelity provides the possibility of reusing ground truth annotation, while its differentiability provides the basis for accurate pose estimation. Our NeRF-based mapping allows for the coding of different temporal styles, which is not possible with other mapping methods.

This paper is organized as follows: in Section 2, we introduce related work, including runway detection algorithms and neural radiance fields; in Section 3, we provide a detailed description of our algorithm, including implementation details of runway line detection, pose estimation, implicit mapping, and the data loop-closure module; in Section 4, we validate our proposed algorithm through experiments on runway line detection, pose estimation, and lightweight network; in Section 5, we discuss the advantages and disadvantages of our proposed algorithm, as well as future research directions; the conclusion is given in Section 6.

2. Related Work

2.1. Runway Detection

Runway detection methods can be roughly divided into three categories: detection based on a priori information, detection based on templates, and detection based on features. Feature-based detection methods have become the dominant detection method in recent years.

A Priori information-based runway detection: In a priori information-based methods, runway detection is achieved using known runway models and the aircraft attitude, and the upper limit of landing is considered in terms of safety and reliability, with the vision system primarily used as an auxiliary navigation system. The authors of [7] propose a model-based runway detection method that requires a known runway model (available through aeronautical information publication), the internal reference of the camera, and the rough pose provided by other sensors, and each line segment in the runway model can be mapped into the image using the above information. In [8], a camera model is also mapped to the image first, but unlike [7], the ROI given in this paper is the ROI of the smallest rectangle containing the left and right runway lines rather than the ROI of each segment of the runway model line. However, in tasks such as emergency landings, the initial attitude estimation is noisy and the sensor type is limited, and the model-based runway detection is less effective in this case.

Template-based runway detection: Template-based runway line detection uses the comparison of the query image and the template image to achieve detection. In [9], LSD is used for line feature extraction, and chamfer matching is later used to achieve runway search, but due to the limitations of template matching itself, the template often cannot adapt to the large changes in view during the landing process. The authors of [10] used a manually designed template to find the ROI and rotates the image in different directions after obtaining the binarized edge gradient map. Then, the sum of the pixel values in different columns is counted to find their peaks, and the peaks are clustered under different rotation angles. Finally, the clustering centers are mapped to straight lines in the original image to achieve runway line detection. Template-based detection methods are poorly generalized and often fail because they are more sensitive to runway geometry and light conditions.

Feature-based runway detection: Runway line detection based on image features is mainly achieved using visual images. Unlike remotely sensed runway detection [11], the proportion of the runway in the image changes continuously in the landing scenario, and the left and right runway edges no longer have parallel characteristics. In [12], the HSV color model and LSD algorithm were used to detect non-standard airfields, and the paper concluded that using the HSV color model could achieve better detection results than the RGB color model. In [13], ROIs are formed by corner-point detection and clustering, and then a neural network is used to classify these ROIs to determine the location of runway edges. However, it is a challenge to choose the number of clusters effectively. The authors of [14] use an end-to-end segmentation network for runway line detection and a self-attention module to enhance the segmentation, while a lightweight network is used to ensure real-time detection, but the paper does not give the impact of detection on subsequent tasks.

None of the above detection methods consider the effectiveness of detection under large viewpoint changes during landing, resulting in these methods only being effective when there is a small variation in perspective and therefore requiring different detection models to be set up for different landing stages (e.g., detection parameters need to be fine-tuned). Additionally, the detection of the starting line can enhance pose estimation performance; however, the above-mentioned methods often fail to detect the virtual start line as it often does not exist. Our proposed method overcomes these problems effectively and provides accurate and reliable runway line detection results.

2.2. Neural Radiance Field

NeRF is a recent breakthrough in the field of computer vision that allows for the generation of highly realistic 3D models of objects and scenes from 2D images. The method works by training a deep neural network to predict the radiance at any point in 3D space, given a set of images and corresponding camera poses. This allows for the creation of photorealistic renderings of objects and scenes from any viewpoint and even enables the synthesis of novel views that were not captured by the original images.

NeRF has been applied to a wide range of applications, including virtual reality, augmented reality, and robotics. It has also been used to generate 3D models of real-world objects and scenes, such as buildings, landscapes, and even human faces.

While NeRF has shown remarkable success in generating high-quality 3D models from a small number of images, it faces several challenges when applied to large-scale scenes.

Computation complexity: The continuity expression of NeRF and the weak assumption of spatial consistency result in slow convergence during training and while requiring large networks to compute the RGB and density of spatial sampling points, which also leads to the slow inference speed of the network. In large-scale scenes, a large number of points in the scene need to be calculated, so the computational requirements can become prohibitively large.

To address the challenge, several approaches have been proposed. It has been shown in recent research that grid-based representations can be used to speed up the training and inference of NeRF significantly. Plenoxels [15] store density values and colors directly on a voxel grid, rather than relying on an MLP network. Instant-NGP [6] greatly improves the training efficiency by utilizing hash encoding and multi-resolution mechanisms. F2NeRF [16] delves deep into the mechanism of space warping to handle unbounded scenes and achieves fast free-viewpoint rendering by allocating limited resources to highlight the necessary details.

Few shot: The original NeRF method requires a 360-degree view of the target object, allowing the network to effectively learn the geometric properties of the scene due to the large amount of co-visible areas. However, in some scenes, the number of input views is limited or the view directions are relatively uniform, which may deceive the network and prevent it from learning the correct geometric information from the images. RegNeRF [17] alleviates artifacts caused by the sparse input by adding regularizations on both geometry and appearance. DS-NeRF [18] and Urban-NeRF [19] improve the geometry of the scene by adding depth supervision.

During the visual landing process, the observation viewpoint is relatively uniform and falls into this category. To address these challenges, prior regularization constraints or depth supervision are often required to be added to the network.

Different resolutions: When there are multiple resolutions present in the input images, NeRF can exhibit blurring and aliasing. MipNeRF [20] solves this problem effectively by using cone sampling. In the process of visual landing, there is a significant difference in resolution between the early and late stages of landing. Therefore, our paper adopts a MipNeRF-based approach to address this issue.

3. Method

3.1. Multi-Stage Flexible Runway Detection

Our multi-stage runway-line-detection algorithm constructed in this paper follows the design principle from coarse to fine, which can largely improve the reliability and accuracy of runway-line detection. The first stage uses the object-detection algorithm, which can effectively extract the high-level semantic information of the runway. By extracting ROIs (regions of interest), it can ensure the uniform distribution of the runway in the image and facilitate the detection of subsequent runway lines. The second stage of the runway-line-detection algorithm is used to extract left and right runway lines and the virtual start line in the focused image, and the extracted runway lines are described in the form of point sets. The runway-line-detection algorithm does not use object segmentation techniques but rather row- and column-specific classification, which is able to reduce the computational effort and increase the inference performance. The third stage mainly adjusts the results of runway-line detection using pixel tuning and sub-pixel tuning to ensure that the detection results of runway lines are attached to the inner edges of the runway lines, thus effectively reducing the randomness of runway line detection.

3.1.1. Structured Runway-Line Detection

We adopt a row anchor-based [21] mechanism for runway-line detection, which samples the image in equally spaced rows, then uses the sampled rows as anchor rows and classifies several adjacent columns into the same grid. With these two processing techniques, the computational effort of the algorithm can be significantly reduced. Below, the network feature of the image is denoted as F, the runway-line-detection classifier is denoted as f, and the predicted results of the runway line are denoted as P.

For the i-th runway line and the j_1-th row anchor, the prediction result can be expressed as:

$$P_i^{j_1} = f_i^{j_1}(F), \tag{1}$$

where the number of row anchors is a_r the number of column grids is n_c, and $P_i^{j_1}$ is an $n_c + 1$ dimensional vector, where the extra dimension is used to indicate the presence or absence of the runway line.

However, the method does not allow for virtual start-line detection, as the slope of the start line is close to zero and the start line can not be effectively detected using the row anchor. In order to solve this problem, we design a column-anchor-based virtual start-line-detection method.

Similar to row anchors, column-anchor detection is defined as follows. For the i-th runway line and the j_2-th column anchor, the prediction can be expressed as:

$$P_i^{j_2} = f_i^{j_2}(F), \tag{2}$$

where the number of column anchors is a_c, the number of row grids is n_r, and $P_i^{j_2}$ is an $n_r + 1$ dimensional column vector.

Although such a design works in theory, the left and right runway lines and the virtual start line need to be predicted separately; i.e., two sets of models are required, which is detrimental to the reuse of network features, the management of the network model, and parallel GPU acceleration. Considering the pairwise characteristics between column and row, we propose an ingenious design approach to unify the left and right runway line and the start line in a unified detection framework.

The left runway line, right runway line, and the virtual start runway line are numbered i as 1–3, respectively, and then the prediction can be expressed in the following form:

$$\begin{cases} P_1^{j_1} = f_1^{j_1}(F) \\ P_2^{j_1} = f_2^{j_1}(F) \\ P_3^{j_2} = f_3^{j_2}(F) \end{cases} \tag{3}$$

To ensure matching dimensions, two merging methods can be generated, which are:

$$\begin{cases} a_r = n_r \\ a_c = n_c \end{cases} \tag{4}$$

or:

$$\begin{cases} a_r = a_c \\ n_c = n_r \end{cases} \tag{5}$$

If the form of Equation (4) is used, the $P_1^{j_1}$, $P_2^{j_1}$, and $P_3^{j_2}$ column vectors may have different dimensions, and in this case, if the different runway lines are processed uniformly, there will be invalid elements in the matrix P. If the form of Equation (5) is used, the $P_1^{j_1}$, $P_2^{j_1}$ and $P_3^{j_2}$ column vectors have the same dimension, in which case all the data in the matrix P are valid, and P can be expressed as:

$$P = \begin{bmatrix} P_1^{j_1} & P_2^{j_1} & P_3^{j_2} \end{bmatrix} = \begin{bmatrix} f_1^{j_1}(F_2) & f_2^{j_1}(F_2) & f_3^{j_2}(F_2) \end{bmatrix} \tag{6}$$

We use this combined form to unify the three runway lines and then interpret them differently in the post-processing stage.

Although the detection of three runway lines can be achieved using the above approach, in order to enable the runway-line-detection network to learn more essential features and achieve better generalizability, we add regular terms based on the geometric properties of the runway. There is a certain constraint relationship between the left and right runway lines. Due to perspective, the closer one gets to the top of the image, the closer the left and right runway lines are from the perspective of the image. For the i-th runway line and the j_1-th row anchor, the probability can be expressed as:

$$p_i^{j_1} = softmax\left(P_i^{j_1}\right) \tag{7}$$

So the location prediction results are:

$$L_i^{j_1} = \sum_{k=1}^{n_c} k \cdot p_i^{j_1}(k) \tag{8}$$

For the same anchor row, the difference between the predicted positions of the left and right runway lines is:

$$D^{j_1} = L_1^{j_1} - L_0^{j_1}, \text{ s.t. } j \in [0, h] \tag{9}$$

The difference in distance between the left and right runway lines between adjacent anchor rows is expressed as:

$$\Delta D^{j_1} = D^{j_1+1} - D^{j_1}, \text{ s.t. } j_1 \in [0,\ h-1] \tag{10}$$

Ideally, no loss is caused in the case of $\Delta D^{j_1} > 0$, while a loss is caused when $\Delta D^{j_1} < 0$. However, in the actual detection process, a certain tolerance threshold needs to be set. The reason for designing the threshold is mainly due to the fact that the constraint of a zero threshold is too strict, and the use of a zero threshold may reduce the performance of the detection. Assuming that the tolerance threshold is T, then, for each anchor row, the loss can be expressed as:

$$M^{j_1} = 0.5 \times \left(\left|\Delta D^{j_1} + T\right| - \Delta D^{j_1} - T\right), \text{ s.t. } j_1 \in [0, h-1] \tag{11}$$

Therefore, the correlation loss of the left and right runway lines can be expressed as:

$$Loss_{relation} = \sum_{j_1=1}^{h-1} \left\| M^{j_1} \right\|_1 \tag{12}$$

For the starting runway line, which is itself a virtual line, the distortion is prevented by adding a linear constraint-regularization term. The experimental results show that the structured loss of the left and right runway lines proposed in this paper and the linear loss of the starting line can improve the generalization.

3.1.2. Hump Randomness Filtering

The accuracy of the runway-line detection directly affects the subsequent position estimation. However, as the result of the width of the runway lines themselves, there is randomness in the location of the detection points in the structured runway-line detection. During the initial period of access to the visual landing system, the runway lines occupy fewer pixels in the image, but the positioning is more sensitive to small fluctuations in detection, and when the UAV is about to reach the ground, the runway lines occupy a certain width in the image, and if each runway is still considered as one edge in this case, it will produce great detection uncertainty. In this paper, the left and right runway line edges are absorbed toward the inner side of the runway, which effectively solves the problem of detection uncertainty.

Although there are off-the-shelf edge-detection algorithms, such as Sobel [22], Canny [23], etc., such generic edge-detection algorithms do not have direction selection characteristics and tend to introduce non-runway edge information, which causes some interference in the subsequent steps. In addition, since the general location of the runway line is already given in the second stage, there is no need to take the gradient of the whole map, but only to find the gradient at some specific locations, which can reduce the computational effort.

In order to enhance the gradient information of runway edges while suppressing the gradient information of non-runway edges, a directional gradient strategy is proposed in this section. The initial slope of the runway line k_{rough} can be determined from the detection points of the previous stage, so the directional gradient convolution kernel is determined based on the initial slope. Specifically, consider convolution kernel K_e as an $N \times N$ grid, given a straight line passing the center of the convolution kernel with slope k_{rough}. The straight line divides the original grid into three categories: grids on the top side of the straight line, grids on the bottom side of the straight line, and grids passing through the straight line. The values for these three types of grids are set to 1, −1, and 0, respectively. After the convolution kernel K is solved for, a directional gradient image can be obtained.

The detection point (x_0, y_0) is sampled in the directional gradient image along the orthogonal direction $y = n(x)$ to the left and right for N_0 pixels. The sampling sequence is defined as:

$$S_{(x_0,y_0)} = \begin{bmatrix} s_0 & \dots & s_{2\times N_0} \end{bmatrix}, \tag{13}$$

where $s_t(t = 0, \ldots, 2 \times N_0)$ denotes the gradient value on the specific sample point. $S_{(x_0,y_0)}$ is normalized to obtain $Sn_{(x_0,y_0)}$.

The values of $Sn_{(x_0,y_0)}$ are first filtered to remove the points whose amplitude is less than the specified threshold, e.g., 0.5, and then the remaining points are clustered by two-dimensional K-means++ [24]. The clustering results are shown in Figure 2. Assuming that the peaks of two categories are $r_i^{\max}$, where $i = 0, 1$, and then we can obtain the adjust points, and the result is used as initial value for subsequent optimization, which can be expressed as:

$$\begin{cases} x_i^{rough} = x_0 + \arg\max\limits_{\Delta x} r_i^{\max} \\ y_i^{rough} = n(x_i^{rough}) \end{cases} \tag{14}$$

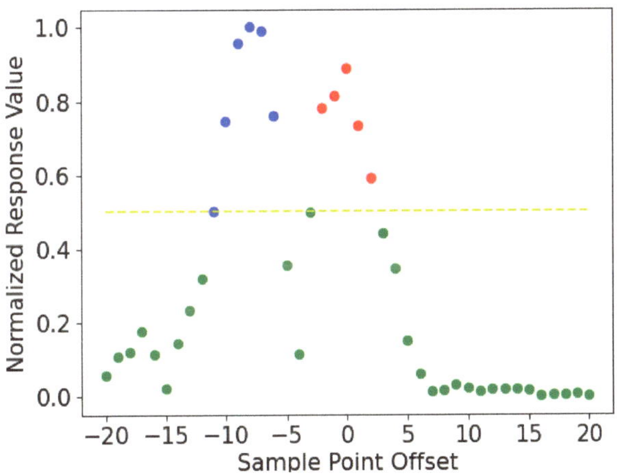

Figure 2. Sampling points and cluster. The blue and red points indicate different two types of data, and the green points are invalid. The yellow dashed line indicates the dividing line between the invalid and valid points.

Ideally, the edge gradient of an image is an impulse signal, but due to factors such as image blurring and filtering, the edges often do not conform to this model. We assume that the change in the image edge gradient conforms to the Gaussian model. Under this assumption, the horizontal coordinate corresponding to the peak of the Gaussian distribution is the edge position. For the runway line, there is a certain width itself, and the points on the runway line may be affected by both the left edge and the right edge. A hump model is proposed to deal with the sub-pixel fitting of the runway to obtain more accurate detection results. $B(\Delta x)$ is defined as:

$$B(\Delta x) = \frac{k_0}{\sqrt{2\pi\sigma^2}}\left(\exp\left(-\frac{(\Delta x - \mu_0)^2}{2\sigma^2}\right) + \exp\left(-\frac{(\Delta x - \mu_1)^2}{2\sigma^2}\right)\right), \tag{15}$$

where the horizontal coordinates of the two peaks, denoted μ_0 and μ_1, respectively, are the coordinates of the left and right edges of the runway to be sought. The standard deviation of the Gaussian function is denoted as σ, which is shared by two Gaussian functions.

We use the Levenberg–Marquardt method to optimize. In order to speed up the optimization and prevent the algorithm from falling into local optima, we add a regular term. Specifically, the distance of the peak points actually reflects the width of the runway line, which is actually relatively narrow. By constraining the distance between the two peaks, the result can be made to satisfy the actual physical scene. The experiments show that the regular term can effectively prevent the algorithm from falling into a local optimum, and the optimization problem can be described as follows:

$$\psi* = \arg\min_{\psi} \frac{1}{2N_0+1} \sum_{i=-N_0}^{N_0} \left(B_\psi(i) - Sn_{(x_0,y_0)}[i+N_0]\right)^2 + \lambda\|\mu_0 - \mu_1\|_2, \tag{16}$$

where the parameters are defined as $\psi = \{\mu_0, \mu_1, \sigma, k_0\}$, the model under a particular set of parameters p is defined as $B_\psi(\Delta x)$, and λ is the regularization coefficient. The initial values of $\mu_i^{initial}$ are selected as $\arg\max_{\Delta x} r_i^{\max}$, where $i = 0, 1$.

After obtaining μ_i, similarly to Equation (14), we can obtain (x_i^{fine}, y_i^{fine}). We then use the following criteria to check the optimized result, which is:

$$\left|\mu_i - \arg\max_{\Delta x} r_i^{\max}\right| < \tau \tag{17}$$

We set τ to 1 or less because the optimized model is fine-tuned sub-pixel and an adjustment value greater than 1 is considered unreasonable. When the optimization result does not satisfy this criterion, the initial value is used as the final result.

The hump randomness filtering algorithm is shown in Algorithm 1.

Algorithm 1 Hump Randomness Filter

Input: Detection Points Set **S**, Image **I**;
Output: Adjust Points Set **S**$'$
  Directional convolution kernel $K_e \leftarrow$ Detection points set **S**;
  for s in S **do**
    Sequence $Q \leftarrow$Sampling along the gradient direction for point s
    Sequence $S_n \leftarrow$ Get the directional gradient value of each point in Q using kernel K_e
    Initialize $\mu_i^{initial} \leftarrow$ Clustering with S_n to get two peak
    $\mu_i \leftarrow$ Using Sequence S_n and $\mu_i^{initial}$ to optimize the hump model
    if $\left|\mu_i - \mu_i^{initial}\right| < \tau$ **then**
      Use optimized parameters to get S'
    else
      Use init parameters to get S'
    end if
  end for

3.2. Implicit Reconstruction-Based Pose Estimation

3.2.1. Initial Pose Estimation

We use runway line features to initialize the UAV pose. The runway coordinate system and camera coordinate system are defined as shown in Figure 3. The origin O_r of the runway coordinate system is chosen as the midpoint of the runway start line, the x_r points from O_r to the front of the runway, and z_r starts from O_r and is perpendicular to the runway plane, and y_r can be determined according to the right-handed coordinate system.

The origin O_c of the camera coordinates system is located at the optical center of the camera, the z_c points from O_c to the camera directly in front, x_c points to the camera directly to the right, and y_c can be determined according to the right-handed coordinate system.

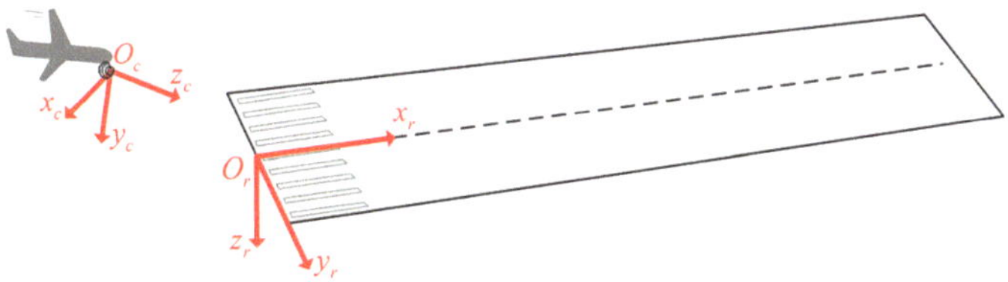

Figure 3. Runway coordinate system and camera coordinate system.

The positions of the 3D points in the runway coordinate system and the positions of the points in the pixel coordinate system are mathematically related as follows:

$$\vec{v}_p = \frac{1}{Z_c} K R_{cr} \left[I_{3\times3} | -\vec{t}_{cr} \right] \begin{bmatrix} \vec{v}_r \\ 1 \end{bmatrix} = \frac{1}{Z_c} P \begin{bmatrix} \vec{v}_r \\ 1 \end{bmatrix}, \tag{18}$$

where $P = KR_{cr}[I_{3\times3}| - \vec{t}_{cr}]$ is the projection matrix, R_{cr} denotes the rotation matrix from the runway coordinate system to the camera coordinate system, K is the camera's intrinsic matrix, and $\vec{t}_{cr}$ denotes the coordinates of the origin of the camera coordinate systems in the runway coordinate system. Assuming that the slope and bias of the runway line in the pixel coordinate system are denoted as k_i and b_i, where i is selected from $\{l, r, h\}$, which denote the left runway line, the right runway line, and the start runway line, respectively. The following location algorithm is given without proof [25,26]:

$$\vec{A}_i = \begin{bmatrix} k_i & -1 & b_i \end{bmatrix} K R_{cr} \tag{19}$$

The first three columns of $\vec{A}_i$ are denoted as a_i^1, a_i^2, a_i^3, respectively, and the width of the runway is W. The positioning result can be expressed as follows:

$$\begin{cases} \begin{bmatrix} y_{cr} \\ z_{cr} \end{bmatrix} = \begin{bmatrix} 1 & a_l^3/a_l^2 \\ 1 & a_r^3/a_r^2 \end{bmatrix}^{-1} \begin{bmatrix} -W/2 \\ W/2 \end{bmatrix} \\ x_{cr} = -a_l^3 z_{cr}/a_l^2 \end{cases} \tag{20}$$

With this method, the real width of the runway and the relative poses between the runway and the camera need to be given. The attitude of the camera can be obtained using an IMU and magnetometer, but the attitude of the runway is often difficult to obtain directly, so the relative attitude of the runway and the camera is more difficult to obtain. Below is a method to estimate the relative attitude of the UAV and the runway based on the runway line features [27].

The extinction points of the left and right runway lines are:

$$v_{lr}^p = \begin{pmatrix} (b_r - b_l)/(k_l - k_r) & (k_l b_r - k_r b_l)/(k_l - k_r) \end{pmatrix} \tag{21}$$

And the extinction point v_s^p of the starting line can be obtained from Equation (22):

$$\begin{bmatrix} \left(v_{lr}^p\right)^T \left(K^{-T}K^{-1}\right) \\ l_s^T \end{bmatrix} v_s^p = 0 \tag{22}$$

Based on the above, the rotation matrix R_{cr} can be expressed as follows:

$$R_{cr} = \begin{bmatrix} \frac{1}{\alpha_1} K^{-1} v_{lr}^p & \frac{1}{\alpha_2} K^{-1} v_s^p & \frac{1}{\alpha_3} \left(K^{-1} v_{lr}^p\right) \times \left(K^{-1} v_s^p\right) \end{bmatrix} \tag{23}$$

In Equation (23), α_1, α_2, and α_3 are normalized coefficients.

Based on the above, we obtain the single-frame pose of the UAV camera in the runway coordinate system. For the k-th frame, the pose estimated using above method can be abbreviated as Γ_k^{frame}.

However, a single-frame pose can suffer from unstable pose estimation. We used a visual odometer-based filtering method to remove the jumps. We initialized the scale of the monocular odometer using a single-frame pose. In feature extraction, we use the trained sky segmentation model to remove invalid feature points in the sky and increase the proportion of feature points in the runway area. Using the odometer's pose-estimation results, we are able to obtain pose transformation from k to $k+1$, denoted as $\Gamma_{k\to k+1}^{odom}$. Meanwhile, we are able to obtain $\Gamma_{k\to k+1}^{\text{frame}} = \Gamma_{k+1}^{\text{frame}}(\Gamma_k^{\text{frame}})^{-1}$. Then, we use the similarity metric $\Gamma_{k\to k+1}^{odom}(\Gamma_{k\to k+1}^{\text{frame}})^{-1}$ to determine whether to use a single-frame pose. If the threshold condition is not met, the odometer pose is used as the estimation pose. We denote the result as $\Gamma_k^{\text{initial}}$.

3.2.2. Implicit Mapping

Cloud computing, like offline computing, can see all the pose data of the previous flight trajectories in one batch, while cloud-based platforms have more computing resources, which is an advantage of cloud-based mapping and pose estimation over the onboard platform. We propose a scheme to reconstruct the implicit map using an implicit radiance field and optimize the UAV's pose attitude online.

We use Γ^{initial} as an a priori pose and use SfM for pose optimization [28]. During feature extraction, we use the segmentation model to remove sky feature points and moving objects such as birds. The feature matching is made more efficient by using a priori poses to guide this process. In the triangulation process, the SfM process itself has a real scale due to the a priori poses. During bundle adjustment, a priori poses are used as optimization regularization to prevent failure. In landing scenarios, there are often multiple trips with different flight paths, and it is important to merge them. One strategy is to optimize all trajectory poses together. However, this is computationally inefficient and often leads to optimization failure due to the high degree of freedom in the optimization process. We adopt a progressive merging strategy, where new trip data arrive and are first reconstructed separately and then merged with existing results. The experimental results show that the feature point extraction strategy, a priori poses, and incremental reconstruction can effectively improve the reconstruction accuracy of the SfM. We denote the pose optimization result as Γ^{opt}. This result is used for NeRF reconstruction.

Assume that the RGB color of a certain pixel is $\mathbf{C_t}$; to render this pixel in NeRF, a ray $\mathbf{r}(t) = \mathbf{o} + t\mathbf{d}$ is emitted from the camera's center of projection $\mathbf{o}$ in the direction $\mathbf{d}$ that passes through the pixel, and distance $t \in (t_n, t_f)$, where t_n and t_f are the predefined near and far distances. A sampling strategy is used to obtain the sampled t_k. For each distance $t_k \in t$, the 3D position can be expressed as $\mathbf{x} = \mathbf{r}(t_k)$. Then, a positional encoding strategy is used to improve rendering quality. The output of the specific sampling points k after passing through the neural radiance field are RGB colors $\mathbf{c_k}$ and a density σ_k, which can be expressed as:

$$[\sigma_k, \mathbf{c}_k] = \text{MLP}(\gamma(\mathbf{r}(t_k))), \forall t_k \in t, \tag{24}$$

where MLP represents the neural radiance network, while $\gamma(\cdot)$ denotes the positional encoding function.

The estimated densities and colors are utilized for approximating the volume-rendering integral through numerical quadrature, which is discussed in the volume-rendering review by Max [29]:

$$\mathbf{C}_p(\mathbf{r}) = \sum_k T_k(1 - \exp(-\sigma_k(t_{k+1} - t_k)))\mathbf{c}_k, \tag{25}$$

in which $T_k = \exp\left(-\sum_{k'<k} \sigma_{k'}(t_{k'+1} - t_{k'})\right)$, and $\mathbf{C}_p(\mathbf{r})$ is the final predicted color of the pixel. During the training of the NeRF network, the predicted pixel value $\mathbf{C}_p(\mathbf{r})$ is minimized with respect to the true pixel value $\mathbf{C}_t(\mathbf{r})$ using gradient descent.

Theoretically, we have the optimized poses Γ^{opt}, camera intrinsic matrix $\mathbf{K}$, and the corresponding images to perform the NeRF implicit reconstruction. However, there are some challenges in our specific scenario. Problem I is that the drastic changes in the viewpoint during landing and the large variation in runway resolution are also difficult problems for the NeRF model. Problem II is the problem of large-scale scenes: visual landing requires the representation of large scenes, which the original NeRF model is unable to handle. Problem III is the issue of appearance style: landing scene data may come from different times, and if this problem is not effectively addressed, it can affect the performance of the implicit map. Problem IV: due to the few viewpoints during landing, it is difficult to learn the geometric information of the scene in the mapping process.

To address Problem I, we adopt the approach proposed in MipNeRF [20], using conical frustum instead of rays in NeRF to alleviate the aliasing issues caused by multi-resolution.

To address Problem II, we utilize the scene parameterization mechanism from MipNeRF360 [30]. Specifically, we achieve coordinate transformation through defining contract$(\cdot)$:

$$\text{contract}(\mathrm{x}) = \begin{cases} \mathrm{x} & \|\mathrm{x}\| \leqslant 1 \\ (2 - 1/\|\mathrm{x}\|)(\mathrm{x}/\|\mathrm{x}\|) & \|\mathrm{x}\| > 1 \end{cases} \tag{26}$$

This approach allows us to compress the range of spatial points from $[0, +\infty)$ to $[0, 2)$. By choosing an appropriate unit scale, we can effectively represent unbounded scenes.

To address Problem III, we apply the appearance-embedding mechanism from NeRF-W [31] to our approach. NeRF-W assigns a unique appearance encoding to each image and obtains a corresponding word vector. This vector is then fed into a multilayer perceptron for backpropagation optimization, resulting in an appearance encoding that captures the style of the current image. However, unlike the "wild" images in NeRF-W, the images in the visual landing system maintain a consistent style throughout each trip. By setting the same appearance encoding for all images in a trip, we reduce the degree of freedom in appearance encoding and allow it to capture the essential features of the appearance. We denote the appearance encoding as $\mathbf{e}_i$, where i denotes the i-th trip. After the appearance embedding is incorporated, Equation (24) can be rewritten as:

$$[\sigma_k, \mathbf{c}_k] = \text{MLP}(\gamma(\mathbf{r}(t_k)), \mathbf{e}_i), \forall t_k \in t \tag{27}$$

To address the problem IV, we add some regularization constraints based on the physical properties of the real scene to limit the geometric degrees of freedom.

The sky's depth is considered to be infinite. Since the sky often lacks effective features, if the depth of the sky is not constrained, many floaters will appear in the scene. By adding regularization constraints to the sky, this problem can be alleviated [19]. All the rays belonging to the sky can be obtained based on the sky segmentation model, denoted as the set R_s, which contains n_s elements, and we define sky loss L_{sky} to encourage sky rays to have zero density:

$$L_{sky} = \frac{1}{n_s} \sum_{\mathbf{r} \in R_s} \sum_k [T_k(1 - \exp(-\sigma_k(t_{k+1} - t_k)))]^2 \tag{28}$$

The runway area, which is the focus of our attention, conforms to the assumption of planar smoothness. Therefore, we use the geometry regularization [17] to constrain the geometry of the runway. The depth of NeRF is generally represented as:

$$d_p(\mathbf{r}) = \sum_k T_k(1 - \exp(-\sigma_k(t_{k+1} - t_k)))t_k \tag{29}$$

All the rays belonging to the runway can be obtained based on the runway-detection model, denoted as the set R_r, which contains n_r elements, and we define runway loss L_{runway} as:

$$L_{runway} = \frac{1}{n_r} \sum_{\mathbf{r} \in R_r} \left(d\left(\mathbf{r}_{i,j}\right) - d\left(\mathbf{r}_{i+1,j}\right)\right)^2 + \left(d\left(\mathbf{r}_{i,j}\right) - d\left(\mathbf{r}_{i,j+1}\right)\right)^2 \tag{30}$$

In addition to the aforementioned losses, we propose a multi-view consistency loss. In this loss, we add a random rigid transformation T_r. To simplify the notation, we denote the function mapping from rays to rendered pixel colors as $M(\cdot)$. The set of common pixels between the images before and after the rigid transformation is denoted by R_c, which contains n_c elements, and we define consistency loss $L_{consistency}$ as:

$$L_{consistency} = \frac{1}{n_r} \sum_{\mathbf{r} \in R_c} \left(T_r^{-1}(M(T_r(\mathbf{r}), \mathbf{e}_i)) - M(\mathbf{r}, \mathbf{e}_i)\right)^2 \tag{31}$$

The loss function for our proposed method can be expressed as:

$$L_{total} = L_{rgb} + L_{sky} + L_{runway} + L_{consistency}, \tag{32}$$

in which $L_{rgb} = \frac{1}{n_i} \sum_{\mathbf{r} \in R_i} \|C_p(\mathbf{r}) - C_t(\mathbf{r})\|^2$, R_i represents all the rays that can be formed from the image, and n_i represents the number of rays.

3.2.3. Inverting NeRF

After training the implicit representation of the scene with NeRF, we use the scene map for online pose estimation. Unlike implicit mapping, the "inverting NeRF" module does not need to use subsequent frames of the current trip, so it is able to compute on an airborne platform. First, we perform appearance-style initialization after obtaining the image of this trip. We freeze all network parameters except for the appearance embedding and optimize it by minimizing the difference between the observed image and the predicted image; the appearance embedding of the new trip can be represented as $\mathbf{e}_{new}$.

For the k-th frame, we can obtain a set of rays R_k based on the initial pose $\Gamma_k^{initial}$ and intrinsic camera K. The mapping function from the rays to the RGB values for the k-th frame is denoted as $C_k(\cdot)$, and the optimization problem can be represented as:

$$T_k = \underset{T \in \mathrm{SE3}}{\arg\min} \sum_{\mathbf{r} \in R_k} \|T(M(\mathbf{r}, \mathbf{e}_{new})) - C_k(\mathbf{r})\|_2^2, \tag{33}$$

where T is the optimization variable. Then, we can obtain the optimized pose $\Gamma_k^{opt} = T_k \Gamma_k^{initial}$, which is a non-convex over the 6DoF space of SE(3). We used the optimization procedure from the paper [4].

3.3. Data Closed-Loop Strategy

3.3.1. Dataset

In order to achieve reliable runway-line detection, it is necessary to have a high-quality dataset [32]. However, to the best of our knowledge, there is no open-source dataset for this particular scenario of vision-based landing systems. Although runways exist in some remote sensing datasets, these runways are not directly applicable to the landing scenario as they are taken from a different perspective than in the vision-based landing system. Based on the information above, the dataset was produced for the landing system in this paper. We used four customized datasets. The first type was the Vega-Prime and X-plane runway image, which were directly generated by the simulator (Vega-Prime and X-Plane are simulators). The second type was the runway data collected from the real runway. The third type was the available data obtained using the perspective transformation of some remotely sensed runway data. The fourth type is the data collected from the internet. Runways come from different sources as shown in Figure 4.

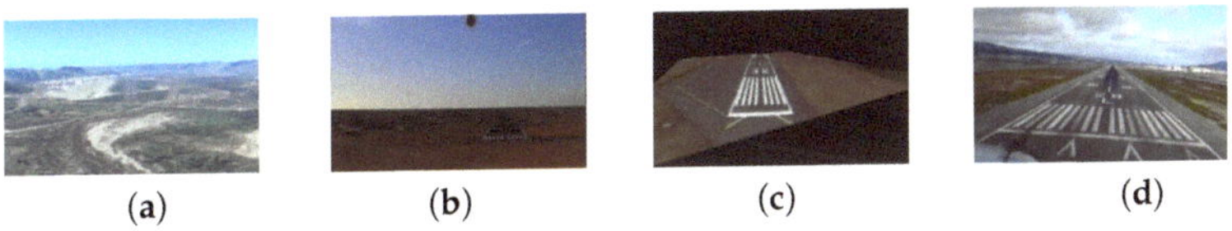

Figure 4. Different data sources of runways. (**a**) Vega-Prime runway. (**b**) Real runway. (**c**) Remote sensing dataset. (**d**) Internet collected.

The remote sensing runway dataset is transformed through perspective to form the view of the landing runway, which expands the diversity of the dataset. One remote sensing runway data point can simulate the runway of different landing stages through different perspective transformations, as shown in Figure 5.

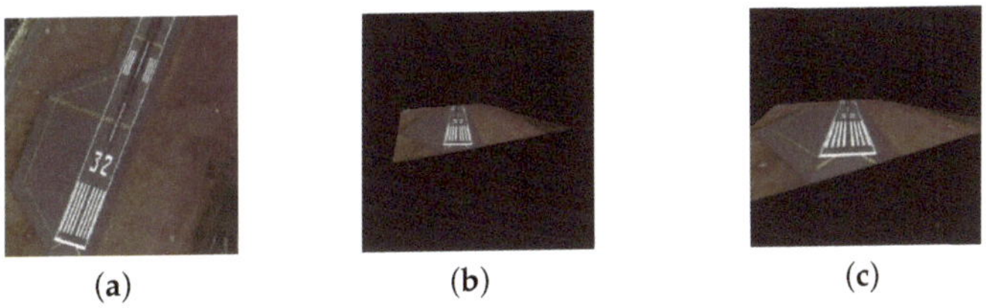

Figure 5. Different perspective transformations. (**a**) Original runway data. (**b**) Simulation landing angle 1. (**c**) Simulation landing angle 2.

The runway-line-detection framework proposed in this paper contains two deep network models, so two datasets are required. Directly labeling two datasets has a large labor overhead, and since there is some correlation between the two datasets, the datasets are only labeled with the runway lines in the images, and then the two datasets are automatically generated using the dataset preprocessing procedure. Considering the characteristic that the runway line itself is a straight line segment, in order to further reduce the annotation workload, the straight line segments are annotated instead of the point set in the program. For the bounding-box-localization dataset, the minimum axis-aligned rectangular box containing the runway can be generated according to the endpoints of the left and right runway lines and the start line marked in the original figure. In order to enhance the adaptability of the runway-line-detection algorithm to different scales of detection frames, the runway rectangular frames are scaled up and down, and three different scaling scales of 0.8, 1.0, and 1.2 are used in the experiment; this paper's strategy for dataset generation is shown in Figure 6.

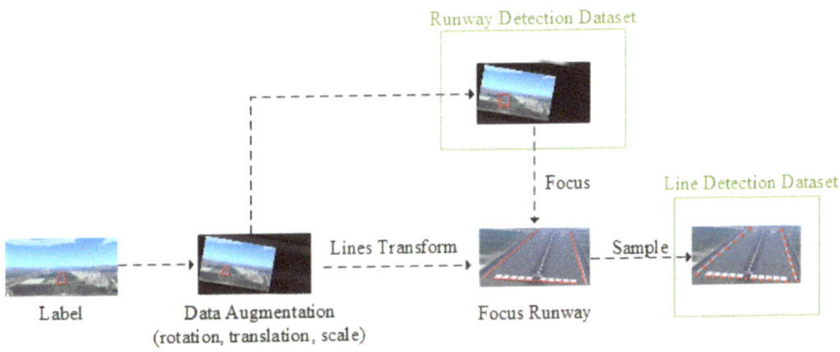

Figure 6. Dataset annotation strategy.

The annotation strategy used in this paper can effectively solve the image-rotation data-enhancement problem in the runway-detection process. In the generic object-detection task, the bounding boxes need to be rotated after image rotation. However, the problem of rotated data enhancement is often handled by means of the maximum frame due to the target-detection ground-truth axis-alignment target feature. Studies have shown that the maximum frame degrades the network performance [33]. In this paper, we adopt the method of labeling runway lines and rotate the runway lines in the process of rotating data enhancement before generating bounding boxes. This method is effective in avoiding the performance damage caused by the maximum frame, as shown in Figure 7.

(a) Maximum box (b) Rotating the minimum box

Figure 7. Different annotation boxes in the rotation-data augmentation. In Figure 7a, the red box shows the result obtained by rotating the conventional bounding box, while the blue box represents the bounding box generated from the red box after image rotation. The blue bounding box, known as the maximum box, includes a significant amount of irrelevant background information. In contrast, Figure 7b presents red line segments representing our proposed annotation method. The blue box represents the generated bounding box after image rotation, based on the annotated information, which contains less background information.

3.3.2. Data Closed-Loop Ground-Truth Annotation

Compared to other visual composition methods, NeRF has the advantage of high fidelity. At the same time, due to the continuity expression of the network, its point cloud export results can be infinitely densified.

The steps to export a point cloud from NeRF are described below. Firstly, we use the rays emitted by all effective pixels in training images as a set of rays. Secondly, for each ray, we use Equation (29) to calculate the mean depth, which is denoted as $d_p(\mathbf{r})$. Thirdly, we extract the appearance embedding of the most visually effective one from multiple trips, and then we can obtain the RGB value of that ray, which is denoted as $\mathbf{C_p}(\mathbf{r})$. Fourthly, by performing the aforementioned calculations, we are able to obtain the RGB values and depth values for all relevant pixels in the training images. By using the intrinsics and extrinsics of camera, we can then determine the coordinates of each 3D point within the runway coordinate system, ultimately forming a comprehensive point cloud. Finally, we divide the generated point cloud into blocks (with a size of 5 m $\times$ 5 m in our experiments) and calculate thickness mean and standard deviation statistics on each block. This process allows us to filter out outliers that fall outside of the 3σ range. By adding this step, we can significantly enhance the visualization of the point clouds.

Using the exported point cloud, manual 3D annotation can be performed on the left, right, and virtual-start runway lines in the point cloud. Then, using NeRF rendering with a new perspective, image annotation can be projected using the 3D annotation projection. By using different poses and appearance, labeled images can be generated, which can be used for the training of the runway line-detection network. The exported 3D point cloud is explained in Section 4.2.

4. Experiments

4.1. Runway Line Detection Experiments

To verify the effectiveness and accuracy of the algorithm proposed in this paper, we designed performance-evaluation metrics for runway-line detection. Unlike the evaluation metrics of general object detection and semantic segmentation, the runway-line-detection algorithm focuses on the error and accuracy of the slope and the bias of the runway lines.

The slope and bias of the detection results of a runway line are denoted as k_p and b_p, respectively, while the true labeling results are denoted as k_t and b_t. The detected angular error and bias error are expressed as follows:

$$\begin{cases} \Delta raw = \arctan k_p - \arctan k_t \\ \Delta angle = \min(180 - |\Delta raw|, |\Delta raw|) \\ \Delta bias = |b_p - b_t| \end{cases} \tag{34}$$

We denote the angle threshold as T_1(degree) and the distance threshold as T_2(pixel). The detection result under these thresholds is correct if the following conditions are met, abbreviated as $TA - T_1 - T_2$:

$$\begin{cases} \Delta angle < T_1 \\ \Delta bias < \sqrt{1 + k_t^2} T_2 \end{cases} \tag{35}$$

To ensure the reliability of the experimental results, all results in this section are the average of 1000 random experiments in which the PyTorch deep learning framework is used and the GPU used in the training and inference process is NVIDIA 3090.

The experimental results from Table 1 show continuous improvement in detection performance with the addition of different strategies with the exception of a few cases. FMRLD-basic represents the basic structured runway-line-detection algorithm.

Table 1. Detection algorithm performance. Bold indicates the best performing result. The strategy "correlation constraint" can be found in Section 3.1.1. The strategy "rotational data augmentation" can be found in Section 3.3.1. The strategy "hump filter" can be found in Section 3.1.2. The definition of $TA - T_1 - T_2$ can be found in Equation (34).

Methods	TA-1-5	TA-2-10	TA-3-20	TA-5-30	FPS
FMRLD-basic	42.0	63.3	80.2	87.1	40.7
+correlation constraint	42.4 (+0.4)	64.3 (+1.0)	81.6 (+1.4)	88.4 (+1.3)	40.7
+rotational data augmentation	45.5 (+3.1)	68.1 (+3.8)	84.9 (+3.3)	90.9 (+2.5)	40.7
+hump filter (rough)	51.0 (+5.5)	**70.8** (+2.7)	85.3 (+0.4)	91.5 (+0.6)	24.2
+hump filter (fine)	**52.3** (+1.3)	70.5 (−0.3)	**86.1** (+0.8)	**92.0** (+0.5)	10.6

The correlation constraint leads to an increase in detection accuracy. Specifically, the correlation loss has at least two effective gains to the algorithm. First, the addition of the correlation loss can improve the accuracy of the precision measurement of points, as shown in Figure 8. Second, the addition of correlation loss can reduce the missed detection, as shown in Figure 9.

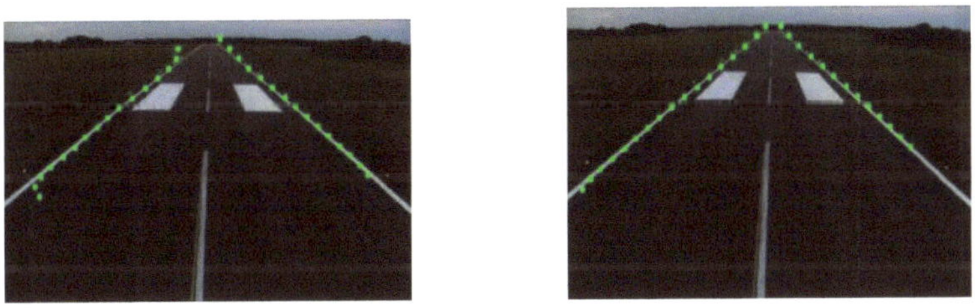

(a) (b)

Figure 8. Correlation loss improves detection accuracy. (**a**) Low detection accuracy (without correlation loss). (**b**) High detection accuracy (with correlation loss).

(a) (b)

Figure 9. Correlation loss eliminates missed detection. (**a**) Missed detection line (without correlation loss). (**b**) High detection accuracy (without correlation loss).

To further illustrate the performance of the FMRLD algorithm proposed in this paper, the algorithm is experimentally compared with other runway-line-detection algorithms [1,8,10,34–36]. To ensure the fairness of the comparison between different algorithms, the algorithm involving ROI extraction uses the same processing as in our paper, and the neural-network-based algorithm uses the same dataset as the FMRLD algorithm. As seen in Figure 10, the FMRLD-basic proposed in this paper has the highest accuracy of all comparison algorithms.

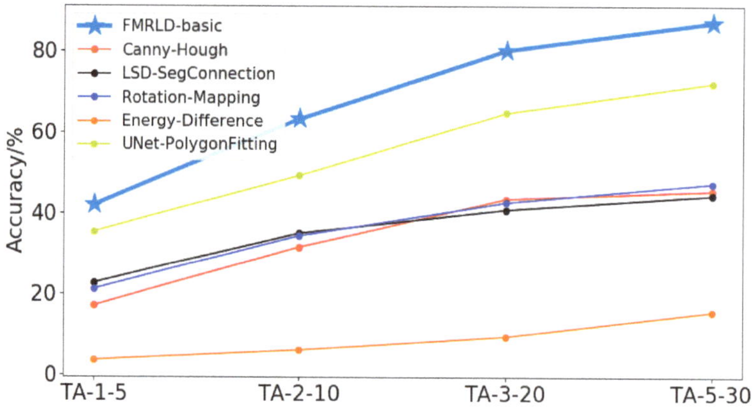

Figure 10. Performance comparison of different runway-line-detection algorithms. UNet-PolygonFitting's errors are directly transmitted, while FMRLD-basic prevents the transmission of errors through data enhancement in the runway detection stage.

Canny–Hough [8]: This method uses the Canny operator for edge detection and Hough transform for straight line extraction and then determines the slope and bias of the straight line according to the geometric constraints between the left and right lines.

LSD-SegConnection [34]: This method uses LSD linear detection to detect runway lines, and LSD is faster than Hough transform, but for images with low resolution, LSD will detect many small, discontinuous line segments. To address this issue, the method of pairing small line segments is adopted.

Rotation Mapping [10]: This method continuously rotates the image after extracting the edges, counts the average value of grayscale on each column of the image after each rotation, and records the column where the current rotation angle and the maximum grayscale average are located. After the image is rotated 180 degrees, the clustering is performed using the improved KMeans algorithm, and the cluster centers are used as the detection results and remapped back to the original image to obtain the detected runway lines.

Energy Difference [36]: The edge-based line-detection method is susceptible to interference, so a method to determine the runway line by maximizing the difference between the two sides of the runway line is proposed, and an iterative optimization strategy for determining the runway line is given.

UNet-PolygonFitting [37]: The runway is segmented using the segmentation network UNet to obtain the runway edge point set, and the quadrilateral is fitted using the edge point set to finally determine the slope and bias of the runway line.

The Canny–Hough and LSD-SegConnection algorithms based on edge detection can achieve better detection performance in specific landing scenarios by adjusting the parameters, but the datasets in this paper are more extensive, and the runway features and lighting conditions vary significantly, so the detection performance of such methods is poor. The energy difference method, based on energy difference, is more suitable for naturally formed runway edges (such as the edges formed by the concrete of the runway and the grass around the runway), but it is less effective in scenes where artificial runway lines exist. The difficulty of the rotation mapping method is to filter out the pseudo-peaks and determine the number of detection lines, so adjusting the parameters of this algorithm is also complicated.

Segmentation-based methods have higher detection accuracy than other methods. However, there is still a certain gap in detection performance compared to the method we proposed. To further investigate the reasons for the poor detection performance of the segmentation-based algorithm, we compare the differences in detection performance in detail between the FMRLD-basic algorithm and UNet-PolygonFitting. The analysis shows that the FMRLD-basic algorithm has a stable detection effect in all stages, while the UNet-PolygonFitting algorithm has a better detection effect in the early stage of landing when the runway occupies a relatively small proportion of the image. However, the detection result is poorer in the late stage of landing, which causes the overall performance of the algorithm to deviate. Our analysis show that the reason for this problem is the difference in the implementation of the two types of algorithms. The viewpoint changes rapidly in the late landing phase, and the dataset has relatively few samples of this type of data, which can lead to poor performance of both the UNet-PolygonFitting algorithm and the ROI phase of the FMRLD algorithm proposed in this paper. The segmentation result of the UNet-PolygonFitting is directly used for runway line detection, resulting in poor detection accuracy. However, the performance of the ROI phase detection frame in the FMRLD algorithm does not directly affect the performance of runway line detection. In addition, the rotation enhancement and scaling in the algorithm-design process make the FMRLD algorithm more fault-tolerant than the UNet-PolygonFitting algorithm.

4.2. Pose-Estimation Experiments

Unlike the evaluation method for runway detection, pose estimation evaluation requires a reference value for the pose. The Vega-Prime simulation environment can meet this requirement effectively.

Experiments were conducted using the FMRLD detection algorithm and the pose-initialization algorithm mentioned in Section 3.2.1. In order to avoid the randomness of the experiments, the localization algorithm was performed in 1000 random experiments, and the RMSE (root mean square error) was calculated as the final result. The error of the pose initialization algorithm is shown in Figure 11.

To further compare the effects of different detection algorithms on localization accuracy, we examined the position estimation of the FMRLD algorithm and the UNet-PolygonFitting algorithm. The Table 2 shows that FMRLD has a significant improvement in pose-estimation accuracy compared to UNet-PolygonFitting.

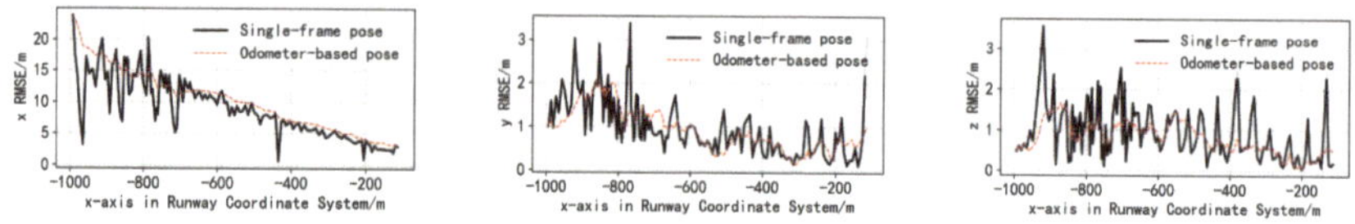

(**a**) x-axis RMSE. (**b**) y-axis RMSE. (**c**) z-axis RMSE.

Figure 11. Pose-initialization Result.By using visual odometry filtering (odometry-based pose), some jump points were effectively removed. The localization error in the x is relatively large, but it tends to decrease as the landing approaches. The estimation results in the y and z directions are relatively stable. Compared to the precise control in the y and z directions, the x direction's position only provides landing guidance. Therefore, an exact position is not required in the x direction.

Table 2. Initialization pose estimation RMSE using different runway-detection methods. Bold indicates the best performing result.

Method	x	y	z	Roll	Pitch	Yaw
FMRLD	**10.72 m**	**1.01 m**	**0.81 m**	**0.525°**	**0.338°**	**0.615°**
UNet-PolygonFitting	36.42 m	8.34 m	2.39 m	2.412°	3.183°	4.264°

We used SfM for pose optimization and compared the optimization results with and without the addition of the priors for initialization pose, as shown in Figure 12.

(**a**) (**b**)

Figure 12. The differences in adding prior pose constraints. The red area represents the camera pose, and the black points represent the constructed sparse point clouds. Figure 12a illustrates that without prior pose constraints, there are serious pose estimation errors. With the addition of prior pose constraints in Figure 12b, there is a significant improvement in the camera pose. (**a**) Initialization without pose priors. (**b**) Initialization with pose priors.

We constructed an implicit map and used inverse NeRF for pose estimation. Table 3 and Figure 13 show the experimental results for the one-trip reconstruction and progressive implicit reconstruction mentioned in Section 3.2.2. From the figures and tables, it can be seen that the accuracy of the one-trip reconstruction is higher than that of progressive implicit reconstruction, but this conclusion is only valid for the current trip. In a real landing scenario, each trip is different from the previous trip, and in such cases, the estimated RMSE is shown in Table 4. From the table, it can be seen that during the online pose-estimation process, progressive pose estimation has higher accuracy compared to one-trip pose estimation.

Table 3. SfM pose estimation RMSE.

Method	x	y	z	Roll	Pitch	Yaw
Initialized pose	10.75 m	1.04 m	0.96 m	0.542°	0.339°	0.617°
One trip pose (offline)	5.35 m	0.48 m	0.50 m	0.347°	0.284°	0.482°
Progressive implicit pose (offline)	6.94 m	0.56 m	0.54 m	0.425°	0.310°	0.535°

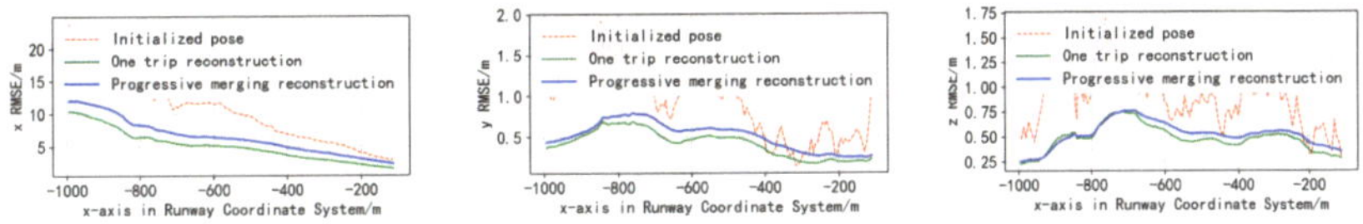

(**a**) x-axis RMSE. (**b**) y-axis RMSE. (**c**) z-axis RMSE.

Figure 13. Pose optimization result. The three subplots represent the RMSE for the three axes.

Table 4. Implicit pose estimation. Bold indicates the best performing result.

Method	x	y	z	Roll	Pitch	Yaw
Initialized pose	10.96 m	1.08 m	1.04 m	0.548°	0.346°	0.621°
One-trip pose (online)	9.32 m	1.01 m	0.63 m	0.492°	0.334°	0.587°
Progressive implicit pose (online)	**7.08 m**	**0.63 m**	**0.55 m**	**0.437°**	**0.315°**	**0.538°**

In order to demonstrate the effect of the regularization terms introduced in the implicit reconstruction, we performed implicit scene reconstruction using the original NeRF method and our proposed method and then exported the point clouds. As shown in Figure 14, our method effectively improves the reconstructed geometry by incorporating regularization terms.

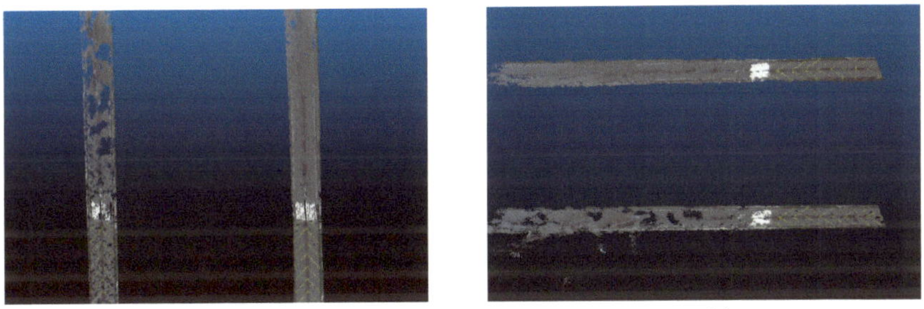

(**a**) (**b**)

Figure 14. Comparison between point clouds generated by original NeRF and our proposed method (both methods using the same pose input). In Figure 14a, the left point cloud shows the original method and the right one shows our proposed method. In Figure 14b, the bottom point cloud shows the original method and the top one shows our proposed method. (**a**) Bird's-eye view point cloud. (**b**) Side view point cloud.

We used the point cloud map generated by NeRF for annotation. As shown in Figure 14, the point clouds generated by our method contain complete geometric information and visual effects, making it easy to annotate from a bird's-eye view perspective. The improved point cloud quality in our method can be primarily attributed to the regularization terms mentioned in Equations (30) and (31) and the point cloud generation method discussed in Section 3.3.2.

By annotating on the point cloud and then projecting it back to the image, we can compare it with the ground truth manual annotation and obtain the accuracy of the projection. By statistics, the accuracy of TA1-5 after projection is 83.5 percent, and the accuracy of TA2-10 is 89.2 percent. On the other hand, we manually checked the annotated images after projection and found that 8% of the data needed to be modified and 25% needed to be fine-tuned, while the remaining annotated data could be used directly. By using our annotation tool, we were able to greatly improve the efficiency of data annotation.

4.3. Lightweight Neural Network Experiments

We have balanced the accuracy and time delay of the algorithm and designed a lightweight landing pose-estimation algorithm, FMRLD-Light, that can achieve real-time performance on edge computing devices. In this model, we have removed the steps of cloud-based mapping and pose optimization. The trained model is deployed on the Jetson Xavier NX platform, a low-power AI computer developed by NVIDIA. To fully exploit the performance of the platform, the TensorRT network model is used for inference in the experiments, and Numba is used for acceleration in the more time-consuming operations. In addition, the algorithm ensures minimal environmental dependencies, including only the image processing library OpenCV and the matrix processing library Numpy, in addition to the library functions necessary for TensorRT model inference. The average time of the algorithm running is 55.3 ms, which can meet the real-time requirements for detection and positioning during the landing process. The histogram of the time distribution of the FMRLD-Light algorithm is shown in Figure 15, with the quantization sampling at one second intervals in the histogram. The results of the histogram show that the detection time of the algorithm is relatively stable after the normal start-up of the system, and there is no systematic risk caused by too long of a detection time.

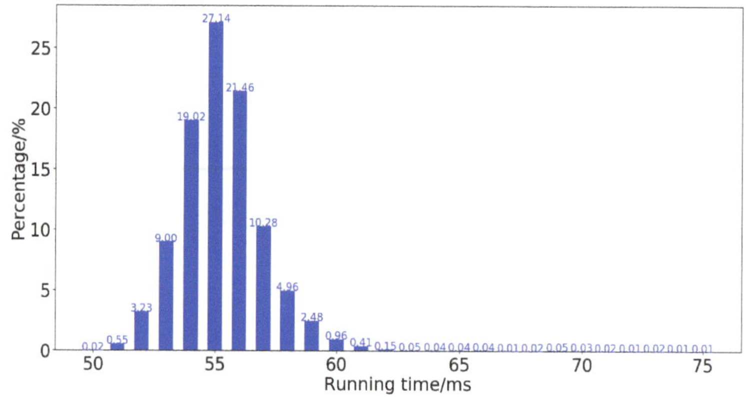

Figure 15. FMRLD–Light algorithm time distribution histogram.

The detection and localization accuracy of the FMRLD–Light algorithm is shown in Table 5. The results indicate that the lightweight algorithm has limited accuracy loss.

Table 5. Accuracy metrics for lightweight methods.

Method	TA-1-5	TA-2-10	TA-3-20	TA-5-30	x	y	z
FMRLD-Light	43.5	65.2	82.8	90.3	13.17 m	1.44 m	1.32 m
FMRLD	52.3	70.5	86.1	92.0	7.08 m	0.63 m	0.55 m
UNet-PolygonFitting	35.7	47.1	60.5	71.2	36.42 m	8.34 m	2.39 m

5. Discussion

Compared with conventional pose-estimation algorithms for landing systems, the pose-estimation algorithm proposed in this paper uses the runway coordinate system as the reference coordinate system, which naturally compensates for the runway slope. This paper proposes a new method for pure visual landing systems, aiming to explore the accuracy limit of the landing system in unfamiliar or complex environments and the accuracy limit of pure visual landing when other sensors are lost. In an engineered visual landing system, the pure visual solution proposed in this paper can function as a robust subsystem and provide more reliable pose data through multi-sensor fusion. However, there are still some limitations in this algorithm, specifically, the requirement for prior knowledge of the runway width, which can be removed through joint optimization of visual landing and IMU after introducing IMU. We are currently focusing on and exploring this research direction.

Next, the issue of real-time is discussed. The real-time aspect of the on-board detection algorithms has been thoroughly validated. The real-time performance of the pose estimation algorithm primarily depends on the speed of NeRF inference and inversion. With the widespread application of NeRF in various fields, acceleration schemes have been extensively studied. In the near future, this issue will no longer be a problem.

6. Conclusions

This paper proposes a novel pose-estimation algorithm for vision-based landing that achieves an accuracy level suitable for guidance and control using visual sensors. On the on-board computing platform, the algorithm first performs runway line detection and fine-tuning. It utilizes the detection results to estimate the initial pose, followed by pose optimization through NeRF inversion. On the cloud computing platform, we propose a multi-trip incremental reconstruction approach for pose estimation. And then we use the optimized pose for NeRF mapping. The lightweight algorithm presented in this paper can achieve real-time pose estimation on board and has strong engineering value. In addition, this paper proposes a closed-loop labeling scheme, which effectively improves labeling efficiency. Compared with previous runway line detection algorithms, this paper improves the detection accuracy by more than 10 points compared to previous runway-line-detection algorithms, and the position estimation accuracy can also achieve state-of-the-art performance.

Author Contributions: Conceptualization, X.L. and C.L.; methodology, C.L.; software, C.L.; validation, C.L. and X.X.; formal analysis, C.L.; investigation, C.L.; resources, X.L.; data curation, X.L., C.L., B.Q. and X.X.; writing—original draft preparation, C.L.; writing—review and editing, C.L., X.X. and B.Q.; visualization, C.L., X.X. and N.Y.; supervision, X.L.; project administration, X.L.; funding acquisition, X.L. All authors have read and agreed to the published version of the manuscript.

Funding: This research was funded by the National Natural Science Foundation of China, grant number No. 62073266, and the Aeronautical Science Foundation of China, grant number No. 201905053003.

Institutional Review Board Statement: Not applicable.

Informed Consent Statement: Not applicable.

Data Availability Statement: Not applicable.

Acknowledgments: Gratitude is extended to the Shaanxi Province Key Laboratory of Flight Control and Simulation Technology.

Conflicts of Interest: The authors declare that they have no known competing financial interest or personal relationships that could have appeared to influence the work reported in this paper.

References

1. Kong, W.; Zhou, D.; Zhang, D.; Zhang, J. Vision-based autonomous landing system for unmanned aerial vehicle: A survey. In Proceedings of the 2014 International Conference on Multisensor Fusion and Information Integration for Intelligent Systems (MFI), Beijing, China, 28–29 September 2014; pp. 1–8.
2. Kügler, M.E.; Mumm, N.C.; Holzapfel, F.; Schwithal, A.; Angermann, M. Vision-augmented automatic landing of a general aviation fly-by-wire demonstrator. In Proceedings of the AIAA Scitech 2019 Forum, San Diego, CA, USA, 7–11 January 2019; p. 1641.
3. Tang, C.; Wang, Y.; Zhang, L.; Zhang, Y.; Song, H. Multisource fusion UAV cluster cooperative positioning using information geometry. *Remote Sens.* **2022**, *14*, 5491. [CrossRef]
4. Yen-Chen, L.; Florence, P.; Barron, J.T.; Rodriguez, A.; Isola, P.; Lin, T.Y. inerf: Inverting neural radiance fields for pose estimation. In Proceedings of the 2021 IEEE/RSJ International Conference on Intelligent Robots and Systems (IROS), Prague, Czech Republic, 27 September–1 October 2021; pp. 1323–1330.
5. Mildenhall, B.; Srinivasan, P.P.; Tancik, M.; Barron, J.T.; Ramamoorthi, R.; Ng, R. Nerf: Representing scenes as neural radiance fields for view synthesis. *Commun. ACM* **2021**, *65*, 99–106. [CrossRef]

6. Müller, T.; Evans, A.; Schied, C.; Keller, A. Instant neural graphics primitives with a multiresolution hash encoding. *ACM Trans. Graph.* **2022**, *41*, 1–15. [CrossRef]
7. Tang, Y.L.; Kasturi, R. Runway detection in an image sequence. In *Image and Video Processing III*; SPIE: Bellingham, WA, USA, 1995; Volume 2421, pp. 181–190.
8. Angermann, M.; Wolkow, S.; Schwithal, A.; Tonhäuser, C.; Hecker, P. High precision approaches enabled by an optical-based navigation system. In Proceedings of the ION 2015 Pacific PNT Meeting, Honolulu, HA, USA, 20–23 April 2015; pp. 694–701.
9. Wang, J.; Cheng, Y.; Xie, J.; Niu, W. A real-time sensor guided runway detection method for forward-looking aerial images. In Proceedings of the 2015 11th International Conference on Computational Intelligence and Security (CIS), Shenzhen, China, 19–20 December 2015; pp. 150–153.
10. Guan, Z.; Li, J.; Yang, H. Runway extraction method based on rotating projection for UAV. I*n Proceedings of the 6th International Asia Conference on Industrial Engineering and Management Innovation: Innovation and Practice of Industrial Engineering and Management (Volume 2)*; Springer: Berlin/Heidelberg, Germany, 2016; pp. 311–324.
11. Akbar, J.; Shahzad, M.; Malik, M.I.; Ul-Hasan, A.; Shafait, F. Runway detection and localization in aerial images using deep learning. In Proceedings of the 2019 Digital Image Computing: Techniques and Applications (DICTA), Perth, WA, Australia, 2–4 December 2019; pp. 1–8.
12. Lin, C.E.; Chou, W.Y.; Chen, T. *Visual-Assisted UAV Auto-Landing System*; DEStech Transactions on Engineering and Technology Research: Lancaster, PA, USA, 2018.
13. Hiba, A.; Zsedrovits, T.; Heri, O.; Zarandy, A. Runway detection for UAV landing system. In Proceedings of the CNNA 2018, the 16th International Workshop on Cellular Nanoscale Networks and Their Applications, Budapest, Hungary, 28–30 August 2018; pp. 1–4.
14. Wang, Y.; Jiang, H.; Liu, C.; Pei, X.; Qiu, H. An airport runway detection algorithm based on Semantic segmentation. *Navig. Position. Timing CSTPCD* **2021**, *8*, 97–106.
15. Fridovich-Keil, S.; Yu, A.; Tancik, M.; Chen, Q.; Recht, B.; Kanazawa, A. Plenoxels: Radiance fields without neural networks. In Proceedings of the IEEE/CVF Conference on Computer Vision and Pattern Recognition, New Orleans, LA, USA, 19–20 June 2022; pp. 5501–5510.
16. Wang, P.; Liu, Y.; Chen, Z.; Liu, L.; Liu, Z.; Komura, T.; Theobalt, C.; Wang, W. F2-NeRF: Fast Neural Radiance Field Training with Free Camera Trajectories. In Proceedings of the IEEE/CVF Conference on Computer Vision and Pattern Recognition, Vancouver, BC, Canada, 18–22 June 2023; pp. 4150–4159.
17. Niemeyer, M.; Barron, J.T.; Mildenhall, B.; Sajjadi, M.S.; Geiger, A.; Radwan, N. Regnerf: Regularizing neural radiance fields for view synthesis from sparse inputs. In Proceedings of the IEEE/CVF Conference on Computer Vision and Pattern Recognition, New Orleans, LA, USA, 18–24 June 2022; pp. 5480–5490.
18. Deng, K.; Liu, A.; Zhu, J.Y.; Ramanan, D. Depth-supervised nerf: Fewer views and faster training for free. In Proceedings of the IEEE/CVF Conference on Computer Vision and Pattern Recognition, New Orleans, LA, USA, 18–24 June 2022; pp. 12882–12891.
19. Rematas, K.; Liu, A.; Srinivasan, P.P.; Barron, J.T.; Tagliasacchi, A.; Funkhouser, T.; Ferrari, V. Urban radiance fields. In Proceedings of the IEEE/CVF Conference on Computer Vision and Pattern Recognition, New Orleans, LA, USA, 18–24 June 2022; pp. 12932–12942.
20. Barron, J.T.; Mildenhall, B.; Tancik, M.; Hedman, P.; Martin-Brualla, R.; Srinivasan, P.P. Mip-nerf: A multiscale representation for anti-aliasing neural radiance fields. In Proceedings of the IEEE/CVF International Conference on Computer Vision, Montreal, BC, Canada, 11–17 October 2021; pp. 5855–5864.
21. Qin, Z.; Wang, H.; Li, X. Ultra fast structure-aware deep lane detection. In Proceedings of the Computer Vision–ECCV 2020: 16th European Conference (Proceedings, Part XXIV 16), Glasgow, UK, 23–28 August 2020; Springer: Berlin/Heidelberg, Germany, 2020; pp. 276–291.
22. Sobel, I.; Feldman, G. A 3 × 3 isotropic gradient operator for image processing. In *A Talk at the Stanford Artificial Project*; 1968; pp. 271–272. Available online: https://www.researchgate.net/publication/285159837_A_33_isotropic_gradient_operator_for_image_processing (accessed on 15 July 2023).
23. Canny, J. A computational approach to edge detection. In *IEEE Transactions on Pattern Analysis and Machine Intelligence*; IEEE: New York, NY, USA, 1986; pp. 679–698.
24. Arthur, D.; Vassilvitskii, S. K-means++ the advantages of careful seeding. In Proceedings of the Eighteenth Annual ACM-SIAM Symposium on Discrete Algorithms, Philadelphia, PA, USA, 7–9 January 2007; pp. 1027–1035.
25. Liu, C.; Liu, L.; Hu, G.; Xu, X. A P3P problem solving algorithm for landing vision navigation. *Navig. Position. Timing* **2018**, *5*, 58–61.
26. Tang, C.; Wang, C.; Zhang, L.; Zhang, Y.; Song, H. Multivehicle 3D cooperative positioning algorithm based on information geometric probability fusion of GNSS/wireless station navigation. *Remote Sens.* **2022**, *14*, 6094. [CrossRef]
27. Zhou, L.; Zhong, Q.; Zhang, Y.; Lei, Z.; Zhang, X. Vision-based landing method using structured line features of runway surface for fixed-wing unmanned aerial vehicles. *J. Natl. Univ. Def. Technol.* **2016**, *9*, 38.
28. Schonberger, J.L.; Frahm, J.M. Structure-from-motion revisited. In Proceedings of the IEEE Conference on Computer Vision and Pattern Recognition, Las Vegas, NV, USA, 27–30 June 2016; pp. 4104–4113.
29. Max, N. Optical models for direct volume rendering. *IEEE Trans. Vis. Comput. Graph.* **1995**, *1*, 99–108. [CrossRef]

30. Barron, J.T.; Mildenhall, B.; Verbin, D.; Srinivasan, P.P.; Hedman, P. Mip-nerf 360: Unbounded anti-aliased neural radiance fields. In Proceedings of the IEEE/CVF Conference on Computer Vision and Pattern Recognition, New Orleans, LA, USA, 18–24 June 2022; pp. 5470–5479.
31. Martin-Brualla, R.; Radwan, N.; Sajjadi, M.S.; Barron, J.T.; Dosovitskiy, A.; Duckworth, D. Nerf in the wild: Neural radiance fields for unconstrained photo collections. In Proceedings of the IEEE/CVF Conference on Computer Vision and Pattern Recognition, Montreal, BC, Canada, 11–17 October 2021; pp. 7210–7219.
32. Lindén, J.; Forsberg, H.; Haddad, J.; Tagebrand, E.; Cedernaes, E.; Ek, E.G.; Daneshtalab, M. Curating Datasets for Visual Runway Detection. In Proceedings of the 2021 IEEE/AIAA 40th Digital Avionics Systems Conference (DASC), San Antonio, TX, USA, 3–7 October 2021; pp. 1–9.
33. Kalra, A.; Stoppi, G.; Brown, B.; Agarwal, R.; Kadambi, A. Towards rotation invariance in object detection. In Proceedings of the IEEE/CVF International Conference on Computer Vision, Montreal, BC, Canada, 11–17 October 2021; pp. 3530–3540.
34. Dong, Y.; Yuan, B.; Wang, H.; Shi, Z. A runway recognition algorithm based on heuristic line extraction. In Proceedings of the 2011 International Conference on Image Analysis and Signal Processing, Wuhan, China, 21–23 October 2011; IEEE: Piscataway, NJ, USA, 2011; pp. 292–296.
35. Abu-Jbara, K.; Alheadary, W.; Sundaramorthi, G.; Claudel, C. A robust vision-based runway detection and tracking algorithm for automatic UAV landing. In Proceedings of the 2015 International Conference on Unmanned Aircraft Systems (ICUAS), Denver, CO, USA, 9–12 June 2015; IEEE: Piscataway, NJ, USA, 2015; pp. 1148–1157.
36. Zhou, Z.; Rahman Siddiquee, M.M.; Tajbakhsh, N.; Liang, J. Unet++: A nested u-net architecture for medical image segmentation. In *Deep Learning in Medical Image Analysis and Multimodal Learning for Clinical Decision Support: Proceedings of the 4th International Workshop, DLMIA 2018, and 8th International Workshop, ML-CDS 2018, Held in Conjunction with MICCAI 2018, Granada, Spain, 20 September 2018*; Proceedings 4; Springer: Berlin/Heidelberg, Germany, 2018; pp. 3–11.
37. Ronneberger, O.; Fischer, P.; Brox, T. U-net: Convolutional networks for biomedical image segmentation. In *Medical Image Computing and Computer-Assisted Intervention—Proceedings of the MICCAI 2015: 18th International Conference, Munich, Germany, 5–9 October 2015, Proceedings, Part III 18*; Springer: Berlin/Heidelberg, Germany, 2015; pp. 234–241.

drones

MDPI

Article

TAN: A Transferable Adversarial Network for DNN-Based UAV SAR Automatic Target Recognition Models

Meng Du [1], Yuxin Sun [2], Bing Sun [3], Zilong Wu [1], Lan Luo [4], Daping Bi [1] and Mingyang Du [1,*]

1 College of Electronic Engineering, National University of Defense Technology, Hefei 230037, China
2 Science and Technology on Electro-Optical Information Security Control Laboratory, Tianjin 300308, China
3 China Satellite Maritime Tracking and Control Department, Jiangyin 214430, China
4 College of Communication Engineering, Lanzhou University, Lanzhou 730030, China
* Correspondence: dumingyang17@nudt.edu.cn

Abstract: Recently, the unmanned aerial vehicle (UAV) synthetic aperture radar (SAR) has become a highly sought-after topic for its wide applications in target recognition, detection, and tracking. However, SAR automatic target recognition (ATR) models based on deep neural networks (DNN) are suffering from adversarial examples. Generally, non-cooperators rarely disclose any SAR-ATR model information, making adversarial attacks challenging. To tackle this issue, we propose a novel attack method called Transferable Adversarial Network (TAN). It can craft highly transferable adversarial examples in real time and attack SAR-ATR models without any prior knowledge, which is of great significance for real-world black-box attacks. The proposed method improves the transferability via a two-player game, in which we simultaneously train two encoder–decoder models: a generator that crafts malicious samples through a one-step forward mapping from original data, and an attenuator that weakens the effectiveness of malicious samples by capturing the most harmful deformations. Particularly, compared to traditional iterative methods, the encoder–decoder model can one-step map original samples to adversarial examples, thus enabling real-time attacks. Experimental results indicate that our approach achieves state-of-the-art transferability with acceptable adversarial perturbations and minimum time costs compared to existing attack methods, making real-time black-box attacks without any prior knowledge a reality.

Keywords: unmanned aerial vehicle (UAV); synthetic aperture radar (SAR); automatic target recognition (ATR); deep neural network (DNN); adversarial example; transferability; encoder–decoder; real-time attack

Citation: Du, M.; Sun, Y.; Sun, B.; Wu, Z.; Luo, L.; Bi, D.; Du, M. TAN: A Transferable Adversarial Network for DNN-Based UAV SAR Automatic Target Recognition Models. *Drones* **2023**, *7*, 205. https://doi.org/10.3390/drones7030205

Academic Editor: Sanjay Sharma

Received: 1 March 2023
Revised: 10 March 2023
Accepted: 13 March 2023
Published: 16 March 2023

1. Introduction

The ongoing advances in unmanned aerial vehicle (UAV) and synthetic aperture radar (SAR) technologies have enabled the acquisition of high-resolution SAR images through UAVs. However, unlike visible light imaging, SAR images reflect the reflection intensity of imaging targets to radar signals, making it difficult for humans to extract effective semantic information from SAR images without the aid of interpretation tools. Currently, deep learning has achieved excellent performance in various scenarios [1–3], and SAR automatic target recognition (SAR-ATR) models based on deep neural networks (DNN) [4–8] have become one of the most popular interpretation methods. With their powerful representation capabilities, DNNs outperform traditional approaches in image classification tasks. However, recent studies have shown that DNN-based SAR-ATR models are susceptible to adversarial examples [9].

The concept of adversarial examples was first proposed by Szegedy et al. [10], which suggests that a carefully designed tiny perturbation can cause a well-trained DNN model to misclassify. This finding has made adversarial attacks one of the most serious threats to artificial intelligence (AI) security. To date, researchers have proposed a variety of adversarial attack methods, which can be mainly divided into two categories from the perspective of prior knowledge: the white-box and black-box attacks. In the first case, attackers can utilize a large amount of prior knowledge, such as the model structure and gradient information, etc., to craft adversarial examples for victim models. Examples of white-box methods include gradient-based attacks [11,12], boundary-based attacks [13], and saliency map-based attacks [14], etc. In the second case, attackers can only access the output information or even less, making adversarial attacks much more difficult. Examples of black-box methods include probability label-based attacks [15,16] and decision-based attacks [17], etc. We now consider an extreme situation, where attackers have no access to any feedback from victim models, such that existing attack methods are unable to craft adversarial examples until researchers discover that adversarial examples can transfer among DNN models performing the same task [18]. Recent relevant studies focused on improving the basic FGSM [11] method to enhance the transferability of adversarial examples, such as gradient-based methods [19,20], transformation-based methods [20,21], and variance-based methods [22], etc. However, the transferability and real-time performance of the above approaches are still insufficient to meet realistic attack requirements. Consequently, adversarial attacks are pending further improvements.

With the wide application of DNNs in the field of remote sensing, researchers have embarked on investigating the adversarial examples of remote sensing images. Xu et al. [23] first investigated the adversarial attack and defense in safety-critical remote sensing tasks, and proposed the mixup attack [24] to generate universal adversarial examples for remote sensing images. However, the research on the adversarial example of SAR images is still in its infancy. Li et al. [25] generated abundant adversarial examples for CNN-based SAR image classifiers through the basic FGSM method and systematically evaluated critical factors affecting the attack performance. Du et al. [26] designed a Fast C&W algorithm to improve the efficiency of generating adversarial examples by introducing an encoder–decoder model. To enhance the universality and feasibility of adversarial perturbations, the work in [27] presented a universal local adversarial network to generate universal adversarial perturbations for the target region of SAR images. Furthermore, the latest research [28] has broken through the limitations of the digital domain and implemented the adversarial example of SAR images in the signal domain by transmitting a two-dimensional jamming signal. Despite the high attack success rates achieved by the above methods, the problem of transferable adversarial examples in the field of SAR-ATR has yet to be addressed.

In this paper, a transferable adversarial network (TAN) is proposed to improve the transferability and real-time performance of adversarial examples in SAR images. Specifically, during the training phase of TAN, we simultaneously trained two encoder–decoder models: a generator that crafts malicious samples through a one-step forward mapping from original data, and an attenuator that weakens the effectiveness of malicious samples by capturing the most harmful deformations. We argue that if the adversarial examples crafted by the generator are robust to the deformations produced by the attenuator, i.e., the attenuated adversarial examples remain effective to DNN models, then they are capable of transferring to other victim models. Moreover, unlike traditional iterative methods, our approach can one-step map original samples to adversarial examples, thus enabling real-time attacks. In other words, we realize real-time transferable adversarial attacks through a two-player game between the generator and attenuator.

The main contributions of this paper are summarized as follows.

(1) For the first time, this paper systematically evaluates the transferability of adversarial examples among DNN-based SAR-ATR models. Meanwhile, our research reveals that there may be potential common vulnerabilities among DNN models performing the same task.
(2) We propose a novel network to enable real-time transferable adversarial attacks. Once the proposed network is well-trained, it can craft adversarial examples with high transferability in real time, thus attacking black-box victim models without resorting to any prior knowledge. As such, our approach possesses promising applications in AI security.

(3) The proposed method is evaluated on the most authoritative SAR-ATR dataset. Experimental results indicate that our approach achieves state-of-the-art transferability with acceptable adversarial perturbations and minimum time costs compared to existing attack methods, making real-time black-box attacks without any prior knowledge a reality.

The rest parts of this paper are arranged as follows. Section 2 introduces the relevant preparation knowledge, and Section 3 describes the proposed method in detail. The experimental results and conclusions are given in Sections 4 and 5, respectively.

2. Preliminaries

2.1. Adversarial Attacks for DNN-Based SAR-ATR Models

Suppose $x_n \in [0, 255]^{W \times H}$ is a single channel SAR image from the dataset $\mathcal{X}$ and $f(\cdot)$ is a DNN-based K-class SAR-ATR model. Given a sample x_n as input to $f(\cdot)$, the output is a K-dimensional vector $f(x_n) = [f(x_n)_1, f(x_n)_2, \cdots, f(x_n)_K]$, where $f(x_n)_i \in \mathbb{R}$ denotes the score of x_n belonging to class i. Let $C_p = \arg\max_i (f(x_n)_i)$ represent the predicted class of $f(\cdot)$ for x_n. The adversarial attack is to fool $f(\cdot)$ with an adversarial example $\tilde{x}_n$ that only has a minor perturbation on x_n. The detail process can be expressed as follows:

$$\arg\max_i f(\tilde{x}_n)_i \neq C_p, \quad \text{s.t.} \|\tilde{x}_n - x_n\|_p \leq \xi \tag{1}$$

where the L_p-norm is defined as $\|v\|_p = (\sum_i |v_i|^p)^{\frac{1}{p}}$, and ξ controls the magnitude of adversarial perturbations. The common L_p-norm includes the L_0-norm, L_2-norm, and L_∞-norm. Attackers can select different norm types according to practical requirements. For example, the L_0-norm represents the number of modified pixels in $\tilde{x}_n$, the L_2-norm measures the mean square error (MSE) between $\tilde{x}_n$ and x_n and the L_∞-norm denotes the maximum variation for individual pixels in $\tilde{x}_n$.

Meanwhile, adversarial attacks can be mainly divided into two modes. The first basic mode is called the non-targeted attack, making DNN models misclassify. The second one is more stringent, called the targeted attack, which induces models to output specified results. There is no doubt that the latter poses a higher level of threat to AI security. In other words, the non-targeted attack is to minimize the probability of models correctly recognizing samples; conversely, the targeted attack maximizes the probability of models identifying samples as target classes. Thus, (1) can be transformed into the following optimization problems:

- For the non-targeted attack:

$$minimize(\frac{1}{N}\sum_{n=1}^{N} D(\arg\max_i f(\tilde{x}_n)_i == C_{tr})), \quad \text{s.t.} \|\tilde{x}_n - x_n\|_p \leq \xi \tag{2}$$

- For the targeted attack:

$$maximize(\frac{1}{N}\sum_{n=1}^{N} D(\arg\max_i f(\tilde{x}_n)_i == C_{ta})), \quad \text{s.t.} \|\tilde{x}_n - x_n\|_p \leq \xi \tag{3}$$

where the discriminant function $D(\cdot)$ equals one if the equation holds; otherwise, it equals zero. C_{tr} and C_{ta} represent the true and target classes of the input. N is the number of samples in the dataset. Obviously, the above optimization problems are exactly the opposite of a DNN's training process, and the corresponding loss functions will be given in the next chapter.

2.2. Transferability of Adversarial Examples

We consider an extreme situation where attackers have no access to any feedback from victim models, in which existing white-box and black-box attacks are unable to craft adversarial examples. In this case, attackers can utilize the transferability of adversarial examples to attack models. Specifically, the extensive experiments in [18] have demonstrated that adversarial examples can transfer among models, even if they have different architectures or are trained on different training sets, so long as they are trained to perform the same task. Details about the transferability are shown in Figure 1.

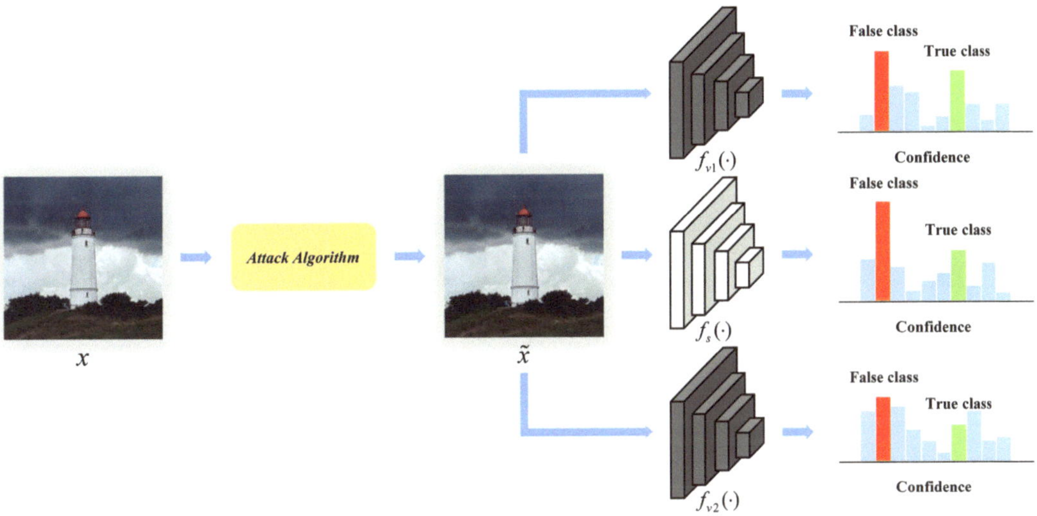

Figure 1. Transferability of adversarial examples.

As shown in Figure 1, for an image classification task, we have trained three recognition models. Suppose that only the surrogate model $f_s(\cdot)$ is a white-box model, and victim models $f_{v1}(\cdot)$, $f_{v2}(\cdot)$ are black-box models. Undoubtedly, given an sample x, attackers can craft an adversarial example $\tilde{x}$ to fool $f_s(\cdot)$ through attack algorithms. Meanwhile, given the transferability of adversarial examples, $\tilde{x}$ can also fool $f_{v1}(\cdot)$ and $f_{v2}(\cdot)$ successfully. However, the transferability generated by existing algorithms is very weak, so this paper is dedicated to crafting highly transferable adversarial examples.

3. The Proposed Transferable Adversarial Network (TAN)

In this paper, the proposed Transferable Adversarial Network (TAN) utilizes the encoder–decoder model and data augmentation technology to improve the transferability and real-time performance of adversarial examples. The framework of our network is shown in Figure 2. As we can see, compared to traditional iterative methods, TAN introduces a generator $G(\cdot)$ to learn the one-step forward mapping from the clean sample x to the adversarial example $\tilde{x}$, thus enabling real-time attacks. Meanwhile, to improve the transferability of $\tilde{x}$, we simultaneously trained an attenuator $A(\cdot)$ to capture the most harmful deformations, which are supposed to weaken the effectiveness of $\tilde{x}$ while still preserving the semantic meaning of x. We argue that if $\tilde{x}$ is robust to the deformations produced by $A(\cdot)$, i.e., $\tilde{x}^*$ remains effective to DNN models, then $\tilde{x}$ is capable of transferring to the black-box victim model $f_v(\cdot)$. In other words, we achieve real-time transferable adversarial attacks through a two-player game between $G(\cdot)$ and $A(\cdot)$. This chapter will introduce our method in detail.

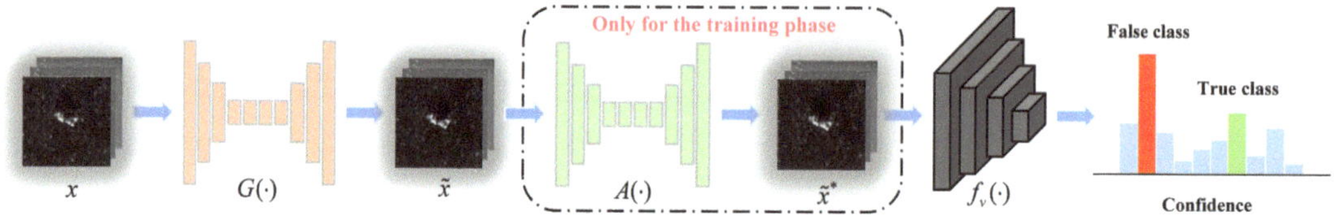

Figure 2. Framework of TAN.

3.1. Training Process of the Generator

For easy understanding, Figure 3 shows the detailed training process of the generator. Note that a white-box model is selected as the surrogate model $f_s(\cdot)$ during the training phase.

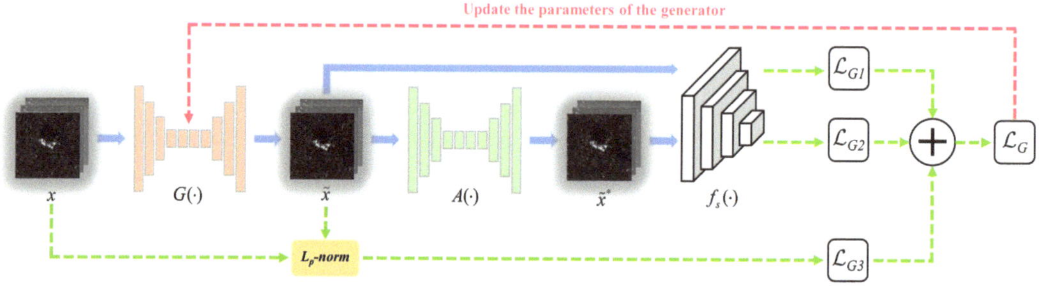

Figure 3. Training process of the generator.

As we can see, given a clean sample x, the generator $G(\cdot)$ crafts the adversarial example $\tilde{x}$ through a one-step forward mapping, as follows:

$$\tilde{x} = G(x) \tag{4}$$

Meanwhile, the attenuator $A(\cdot)$ takes $\tilde{x}$ as input and outputs the attenuated adversarial example $\tilde{x}^*$:

$$\tilde{x}^* = A(\tilde{x}) \tag{5}$$

Since $\tilde{x}$ has to fool $f_s(\cdot)$ with a minor perturbation, and $\tilde{x}^*$ needs to remain effective against $f_s(\cdot)$, the loss function of $G(\cdot)$ consists of three parts. Next, we will give the generator loss $\mathcal{L}_G$ of non-targeted and targeted attacks, respectively.

For the non-targeted attack: First, according to (2), $\tilde{x}$ is to minimize the classification accuracy of $f_s(\cdot)$, which means that it has to decrease the confidence of being recognized as the true class C_{tr}, i.e., to increase the confidence of being identified as others. Thus, the first part of $\mathcal{L}_G$ can be expressed as:

$$\begin{aligned}\mathcal{L}_{G1}(f_s(\tilde{x}), C_{tr}) &= -\log\left(\frac{\sum_{i \neq C_{tr}} \exp(f_s(\tilde{x})_i)}{\sum_i \exp(f_s(\tilde{x})_i)}\right) \\ &= -\log\left(1 - \frac{\exp(f_s(\tilde{x})_{C_{tr}})}{\sum_i \exp(f_s(\tilde{x})_i)}\right)\end{aligned} \tag{6}$$

Second, to improve the transferability of $\tilde{x}$, we expect that $\tilde{x}^*$ remains effective to $f_s(\cdot)$, so the second part of $\mathcal{L}_G$ can be derived as:

$$\begin{aligned}\mathcal{L}_{G2}(f_s(\tilde{x}^*), C_{tr}) &= -\log\left(\frac{\sum_{i \neq C_{tr}} \exp(f_s(\tilde{x}^*)_i)}{\sum_i \exp(f_s(\tilde{x}^*)_i)}\right) \\ &= -\log\left(1 - \frac{\exp(f_s(\tilde{x}^*)_{C_{tr}})}{\sum_i \exp(f_s(\tilde{x}^*)_i)}\right)\end{aligned} \tag{7}$$

Finally, the last part of $\mathcal{L}_G$ is used to limit the perturbation magnitude. We introduce the traditional L_p-norm to measure the degree of image distortion as follows:

$$\begin{aligned}\mathcal{L}_{G3}(x, \tilde{x}) &= \|\tilde{x} - x\|_p \\ &= (\sum_i |\Delta x_i|^p)^{\frac{1}{p}}\end{aligned} \tag{8}$$

In summary, we apply the linear weighted sum method to balance the relationship among $\mathcal{L}_{G1}$, $\mathcal{L}_{G2}$, and $\mathcal{L}_{G3}$. As such, the complete generator loss for the non-targeted attack can be represented as:

$$\mathcal{L}_G = \omega_{G1} \cdot \mathcal{L}_{G1}(f_s(\tilde{x}), C_{tr}) + \omega_{G2} \cdot \mathcal{L}_{G2}(f_s(\tilde{x}^*), C_{tr}) + \omega_{G3} \cdot \mathcal{L}_{G3}(x, \tilde{x}) \tag{9}$$

where $\omega_{G1} + \omega_{G2} + \omega_{G3} = 1$. $\omega_{G1}, \omega_{G2}, \omega_{G3} \in [0, 1]$ are the weight coefficients of $\mathcal{L}_{G1}$, $\mathcal{L}_{G2}$, and $\mathcal{L}_{G3}$, respectively. The weight coefficients represent the relative importance of each loss term during the training process. A larger weight implies that the corresponding loss will decrease more rapidly and significantly, allowing attackers to adjust the parameters flexibly according to their actual needs.

For the targeted attack: According to (3), $\tilde{x}$ is to maximize the probability of being recognized as the target class C_{ta}, i.e., to increase the confidence of C_{ta}. Thus, $\mathcal{L}_{G1}$ here can be expressed as:

$$\mathcal{L}_{G1}(f_s(\tilde{x}), C_{ta}) = -\log\left(\frac{\exp(f_s(\tilde{x})_{C_{ta}})}{\sum_i \exp(f_s(\tilde{x})_i)}\right) \tag{10}$$

To maintain the effectiveness of $\tilde{x}^*$ against $f_s(\cdot)$, $\mathcal{L}_{G2}$ here is derived as:

$$\mathcal{L}_{G2}(f_s(\tilde{x}^*), C_{ta}) = -\log\left(\frac{\exp(f_s(\tilde{x}^*)_{C_{ta}})}{\sum_i \exp(f_s(\tilde{x}^*)_i)}\right) \tag{11}$$

The perturbation magnitude is still limited by the $\mathcal{L}_{G3}$ shown in (8). Therefore, the complete generator loss for the targeted attack can be represented as:

$$\mathcal{L}_G = \omega_{G1} \cdot \mathcal{L}_{G1}(f_s(\tilde{x}), C_{ta}) + \omega_{G2} \cdot \mathcal{L}_{G2}(f_s(\tilde{x}^*), C_{ta}) + \omega_{G3} \cdot \mathcal{L}_{G3}(x, \tilde{x}) \tag{12}$$

3.2. Training Process of the Attenuator

According to Figure 2, during the training phase of TAN, an attenuator $A(\cdot)$ was introduced to weaken the effectiveness of $\tilde{x}$ while still preserving the semantic meaning of x. We show the detailed training process of $A(\cdot)$ in Figure 4.

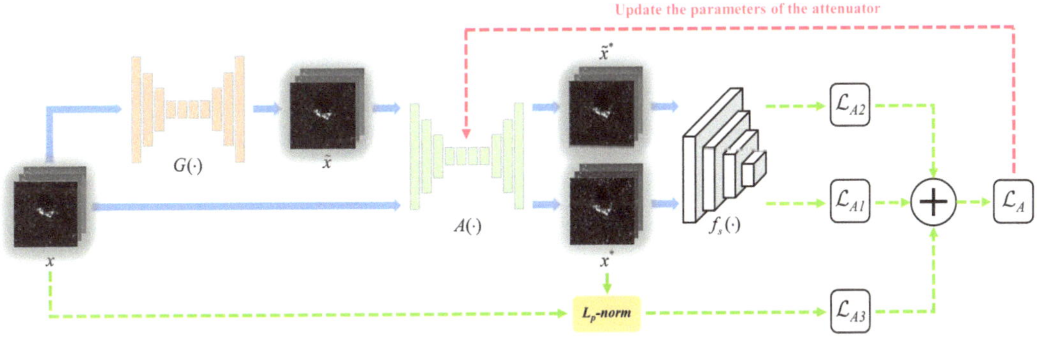

Figure 4. Training process of the attenuator.

As we can see, the attenuator loss $\mathcal{L}_A$ also consists of three parts. First, to preserve the semantic meaning of x, $f_s(\cdot)$ has to keep a basic classification accuracy on the following attenuated sample x^*:

$$x^* = A(x) \tag{13}$$

It means that the first part of $\mathcal{L}_A$ should increase the confidence of x^* being recognized as the true class C_{tr}, as follows:

$$\mathcal{L}_{A1}(f_s(x^*), C_{tr}) = -\log\left(\frac{\exp(f_s(x^*)_{C_{tr}})}{\sum_i \exp(f_s(x^*)_i)}\right) \tag{14}$$

Meanwhile, to weaken the effectiveness of $\tilde{x}$, $A(\cdot)$ also need to improve the confidence of the attenuated adversarial example $\tilde{x}^*$ being identified as C_{tr}, so the second part of $\mathcal{L}_A$ can be expressed as:

$$\mathcal{L}_{A2}(f_s(\tilde{x}^*), C_{tr}) = -\log\left(\frac{\exp(f_s(\tilde{x}^*)_{C_{tr}})}{\sum_i \exp(f_s(\tilde{x}^*)_i)}\right) \tag{15}$$

Finally, to avoid excessive image distortion caused by $A(\cdot)$, the third part of $\mathcal{L}_A$ is used to limit the deformation magnitude, which can be expressed by the traditional L_p-norm, as follows:

$$\begin{aligned}\mathcal{L}_{A3}(x, x^*) &= \|x^* - x\|_p \\ &= (\sum_i |\Delta x_i|^p)^{\frac{1}{p}}\end{aligned} \tag{16}$$

As with the generator loss, we utilize the linear weighted sum method to derive the complete attenuator loss as follows:

$$\mathcal{L}_A = \omega_{A1} \cdot \mathcal{L}_{A1}(f_s(x^*), C_{tr}) + \omega_{A2} \cdot \mathcal{L}_{A2}(f_s(\tilde{x}^*), C_{tr}) + \omega_{A3} \cdot \mathcal{L}_{A3}(x, x^*) \tag{17}$$

where $\omega_{A1} + \omega_{A2} + \omega_{A3} = 1$. $\omega_{A1}, \omega_{A2}, \omega_{A3} \in [0, 1]$ are the weight coefficients of $\mathcal{L}_{A1}$, $\mathcal{L}_{A2}$, and $\mathcal{L}_{A3}$, respectively.

3.3. Network Structure of the Generator and Attenuator

According to Sections 3.1 and 3.2, the generator and attenuator are essentially two encoder–decoder models, so the choice of a suitable model structure is necessary. We mainly consider two factors. First, as the size of original samples and adversarial examples should be the same, the model has to keep the input and output sizes identical. Second, to prevent our network from overfitting while saving computational resources, a lightweight model will be a better choice. In summary, we applied ResNet Generator proposed in [29] as the encoder–decoder model of TAN. The structure of ResNet Generator is shown in Figure 5.

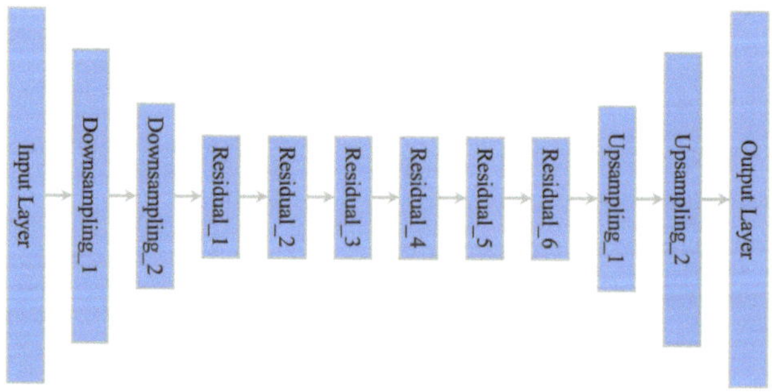

Figure 5. Structure of ResNet Generator.

As we can see, ResNet Generator mainly consists of downsampling, residual, and upsampling modules. For a visual understanding, given an input data of size $1 \times 128 \times 128$, the input and output sizes of each module are listed in Table 1.

Obviously, the input and output sizes of ResNet Generator are the same. Meanwhile, to ensure the validity of the generated data, we added a *tanh* function after the output module, which restricts the generated data to the interval $[0, 1]$. The total number of parameters in ResNet Generator has been calculated to be approximately $7.83M$, which is a fairly lightweight network. For more details, please refer to the literature [29].

Table 1. Input–output relationships for each module of ResNet Generator.

Module	Input Size	Output Size
Input	$1 \times 128 \times 128$	$64 \times 128 \times 128$
Downsampling_1	$64 \times 128 \times 128$	$128 \times 64 \times 64$
Downsampling_2	$128 \times 64 \times 64$	$256 \times 32 \times 32$
Residual_1 $\sim$ 6	$256 \times 32 \times 32$	$256 \times 32 \times 32$
Upsampling_1	$256 \times 32 \times 32$	$128 \times 64 \times 64$
Upsampling_2	$128 \times 64 \times 64$	$64 \times 128 \times 128$
Output	$64 \times 128 \times 128$	$1 \times 128 \times 128$

3.4. Complete Training Process of TAN

As we described earlier, TAN improves the transferability of adversarial examples through a two-player game between the generator and attenuator, which is quite similar to the working principle of generative adversarial networks (GAN) [30]. Therefore, we also adopted an alternating training scheme to train our network. Specifically, given the dataset $\mathcal{X}$ and batch size S, we first randomly divided $\mathcal{X}$ into M batches $\{b_1, b_2, \cdots, b_M\}$ at the beginning of each training iteration. Second, we set a training ratio $R \in N^*$, which means that TAN trains the generator R times and then trains the attenuator once, i.e., once per batch for the former and only once per R batch for the latter. In this way, we can prevent the attenuator from being so strong that the generator cannot be optimized. Meanwhile, to shorten training time, we set an early stop condition *Esc* so that training can be ended early when certain indicators meet the condition. Note that the generator and attenuator are trained alternately, i.e., the attenuator's parameters are fixed when the generator is trained, and vice versa. More details of the complete training process for TAN are shown in Algorithm 1.

4. Experiments

4.1. Data Descriptions

To date, there is no publicly available dataset for UAV SAR-ATR, thus this paper experiments on the most authoritative SAR-ATR dataset, i.e., the moving and stationary target acquisition and recognition (MSTAR) dataset [31]. MSTAR is collected by a high-resolution spotlight SAR and published by the U.S. Defense Advanced Research Projects Agency (DARPA) in 1996, which contains SAR images of Soviet military vehicle targets at different azimuth and depression angles. In standard operating conditions (SOC), MSTAR includes ten classes of targets, such as self-propelled howitzers (2S1); infantry fighting vehicles (BMP2); armored reconnaissance vehicles (BRDM2); wheeled armored transport vehicles (BTR60, BTR70); bulldozers (D7); main battle tanks (T62, T72); cargo trucks (ZIL131); and self-propelled artillery (ZSU234). The training dataset contains 2747 images collected at a depression angle of 17°, and the testing dataset contains 2426 images captured at a depression angle of 15°. More details about the dataset are given in Table 2, and Figure 6 shows the optical images and corresponding SAR images of each class.

Algorithm 1 Transferable Adversarial Network Training

Input: Dataset $\mathcal{X}$; batch size S; surrogate model f_s; target class C_{ta}; training loss function $\mathcal{L}_G$ of the generator; training loss function $\mathcal{L}_A$ of the attenuator; training iteration number T; learning rate η; training ratio R of the generator and attenuator; early stop condition Esc.

Output: The parameter θ_G of the well-trained generator.

Randomly initialize θ_G and θ_A
for $t = 1$ to T **do**
    According to S, randomly divide $\mathcal{X}$ into M batches $\{b_1, b_2, \cdots, b_M\}$
    for $m = 1$ to M **do**
        Calculate $\mathcal{L}_G(\theta_G, \theta_A, f_s, b_m, C_{ta})$
        Update $\theta_G = \theta_G - \eta \cdot \frac{\partial}{\partial \theta_G}\mathcal{L}_G$
        if $m\%R == 0$ **then**
            Calculate $\mathcal{L}_A(\theta_G, \theta_A, f_s, b_m)$
            Update $\theta_A = \theta_A - \eta \cdot \frac{\partial}{\partial \theta_A}\mathcal{L}_A$
        else
            $\theta_A = \theta_A$
        end if
    end for
    if $Esc == True$ **then**
        Break
    else
        Continue
    end if
end for

Figure 6. Optical images (**top**) and SAR images (**bottom**) of the MSTAR dataset.

Table 2. Details of the MSTAR dataset under SOC, including target class, serial, depression angle, and sample numbers.

Target Class	Serial	Training Data		Testing Data	
		Depression Angle	Number	Depression Angle	Number
2S1	b01	17°	299	15°	274
BMP2	9566	17°	233	15°	196
BRDM2	E-71	17°	298	15°	274
BTR60	k10yt7532	17°	256	15°	195
BTR70	c71	17°	233	15°	196
D7	92v13015	17°	299	15°	274
T62	A51	17°	299	15°	273
T72	132	17°	232	15°	196
ZIL131	E12	17°	299	15°	274
ZSU234	d08	17°	299	15°	274

4.2. Implementation Details

The proposed method is evaluated on the following six common DNN models: DenseNet121 [32], GoogLeNet [33], InceptionV3 [34], Mobilenet [35], ResNet50 [36], and Shufflenet [37]. In terms of data preprocessing, we resized all the images in MSTAR to 128×128 and uniformly sample 10% of training data to form the validation dataset. During the training phase of recognition models, the training epoch and batch size were set to 100 and 32, respectively. During the training phase of TAN, to minimize the MSE between adversarial examples and original samples, we adopted the L_2-norm to evaluate the image distortion caused by adversarial perturbations. Meanwhile, for better attack performance, the hyperparameters of TAN are fine-tuned through numerous experiments, and the following set of parameters is eventually determined to best meet our requirements. Specifically, we set the generator loss weights $[\omega_{G1}, \omega_{G2}, \omega_{G3}]$ to $[0.25, 0.25, 0.5]$, the attenuator loss weights $[\omega_{A1}, \omega_{A2}, \omega_{A3}]$ to $[0.25, 0.25, 0.5]$, the training ratio to 3, the training epoch to 50, and the batch size to 8. Due to the adversarial process involved in TAN, training can be challenging to converge. As such, we employed Adam [38], a more computationally efficient optimizer, to accelerate model convergence, which also performs better in solving non-stationary objective and sparse gradient problems. The learning rate is set to 0.001. When evaluating the transferability, we first crafted adversarial examples for each surrogate model and then assessed the transferability by testing the recognition results of victim models on corresponding adversarial examples. Detailed experiments will be given later.

Furthermore, the following six attack algorithms from the Torchattacks [39] toolbox were introduced as baseline methods for comparison with TAN: MIFGSM [19], DIFGSM [21], NIFGSM [20], SINIFGSM [20], VMIFGSM [22], and VNIFGSM [22]. All codes were written in Pytorch, and the experimental environment consisted of Windows 10 with an NVIDIA GeForce RTX 2080 Ti GPU and a 3.6 GHz Intel Core i9-9900K CPU.

4.3. Evaluation Metrics

We mainly consider two factors to comprehensively evaluate the performance of adversarial attacks: the effectiveness and stealthiness, which are directly related to the classification accuracy $\tilde{Acc}$ of victim models on adversarial examples and the norm value $\tilde{L}_p$ of adversarial perturbations, respectively. For the $\tilde{Acc}$ metric, the formula is as follows:

$$\tilde{Acc} = \begin{cases} \frac{1}{N}\sum_{n=1}^{N} D(\arg\max_i (f_v(\tilde{x}_n)_i) == C_{tr}) & \text{for the non-targeted attack} \\ \frac{1}{K \times N}\sum_{C_{ta}=1}^{K}\sum_{n=1}^{N} D(\arg\max_i (f_v(\tilde{x}_n)_i) == C_{ta}) & \text{for the targeted attack} \end{cases} \quad (18)$$

where C_{tr} and C_{ta} represent the true and target classes of the input data, K is the number of target classes, and $D(\cdot)$ is a discriminant function. In the non-targeted attack, the $\tilde{Acc}$ metric reflects the probability that the victim model $f_v(\cdot)$ identifies the adversarial example $\tilde{x}_n$ as C_{tr}, while in the targeted attack it indicates the probability that $f_v(\cdot)$ recognizes $\tilde{x}_n$ as C_{ta}. Obviously, in the non-targeted attack, the lower the $\tilde{Acc}$ metric, the better the attack. Conversely, in the targeted attack, a higher $\tilde{Acc}$ metric represents $f_v(\cdot)$ is more likely to recognize $\tilde{x}_n$ as C_{ta}, and thus the attack is more effective. In conclusion, the effectiveness of non-targeted attacks is inversely proportional to the $\tilde{Acc}$ metric, and the effectiveness of targeted attacks is proportional to this metric. Additionally, there are other three similar indicators, Acc, Acc^*, and $\tilde{Acc}^*$, that represent the classification accuracy of $f_v(\cdot)$ for the original sample x_n, the attenuated sample x_n^*, and the attenuated adversarial example $\tilde{x}_n^*$, respectively. Note that whether it is a non-targeted or targeted attack, Acc^* always represents the accuracy with which $f_v(\cdot)$ identifies x_n^* as C_{tr}, while the other three accuracy indicators need to be calculated via (18) based on the attack mode. In particular, $\tilde{Acc}^*$ represents the recognition result of $f_v(\cdot)$ on $\tilde{x}_n^*$, which indirectly reflects the strength of the transferability possessed by $\tilde{x}_n$.

Meanwhile, we applied the following L_p-norm values to measure the attack stealthiness:

$$\begin{cases} \tilde{L}_p = \frac{1}{N}\sum_{n=1}^{N}\|\tilde{x}_n - x_n\|_p & \text{for the generator} \\ L_p^* = \frac{1}{N}\sum_{n=1}^{N}\|x_n^* - x_n\|_p & \text{for the attenuator} \end{cases} \tag{19}$$

where $\tilde{L}_p$ and L_p^* represent the image distortion caused by the generator and attenuator, respectively. In our experiments, the L_p-norm defaults to L_2-norm. In summary, we can set the early stop condition Esc mentioned in Section 3.4 with the above indicators, as follows:

$$Esc = \begin{cases} \tilde{Acc} \leq 0.05, Acc^* \geq 0.9, \tilde{Acc}^* \leq 0.1, \tilde{L}_2 \leq 4, L_2^* \leq 4 & \text{for the non-targeted attack} \\ \tilde{Acc} \geq 0.95, Acc^* \geq 0.9, \tilde{Acc}^* \geq 0.9, \tilde{L}_2 \leq 4, L_2^* \leq 4 & \text{for the targeted attack} \end{cases} \tag{20}$$

Furthermore, to evaluate the real-time performance of adversarial attacks, we introduced the Tc metric to denote the time cost of generating a single adversarial example, as follows:

$$Tc = \frac{Time}{N} \tag{21}$$

where $Time$ is the total time consumed to generate N adversarial examples.

4.4. DNN-Based SAR-ATR Models

A well-trained recognition model is a prerequisite for effective adversarial attacks, so we have trained six SAR-ATR models on the MSTAR dataset: DenseNet121, GoogLeNet, InceptionV3, Mobilenet, ResNet50, and Shufflenet. All of them achieve outstanding recognition performance, with the classification accuracy of 98.72%, 98.06%, 96.17%, 96.91%, 97.98%, and 96.66% on the testing dataset, respectively. In addition, we show the confusion matrix of each model in Figure 7.

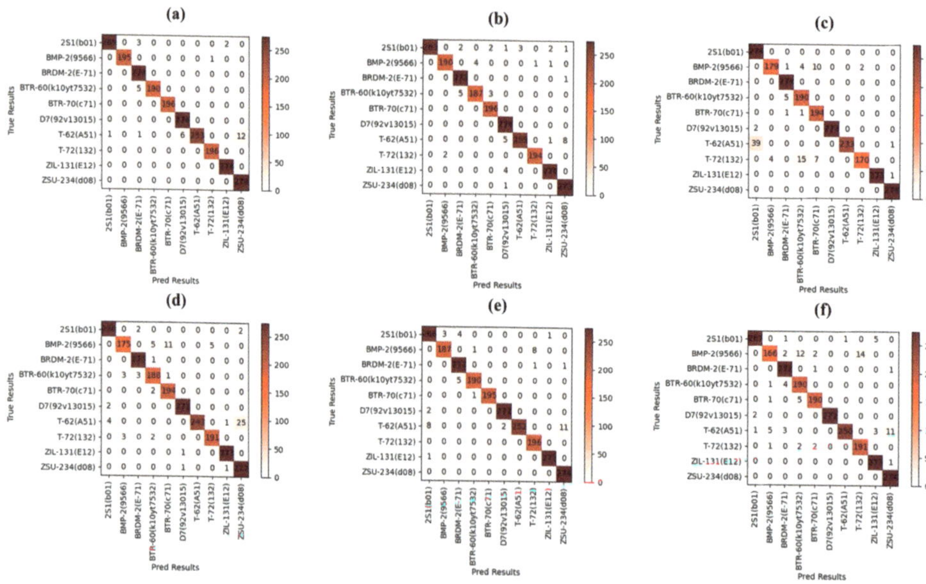

Figure 7. Confusion matrixes of DNN-based SAR-ATR models on the MSTAR dataset. (**a**) DenseNet121. (**b**) GoogLeNet. (**c**) InceptionV3. (**d**) Mobilenet. (**e**) ResNet50. (**f**) Shufflenet.

4.5. Comparison of Attack Performance

In this section, we first evaluated the attack performance of the proposed method against DNN-based SAR-ATR models on the MSTAR dataset. Specifically, during the training phase of TAN, we took each network as the surrogate model in turn and assessed the recognition results of corresponding models on the outputs of TAN at each stage. The results of non-targeted and targeted attacks are detailed in Tables 3 and 4, respectively.

Table 3. Non-targeted attack results of our method against DNN-based SAR-ATR models on the MSTAR dataset.

Surrogate	Acc	$\tilde{Acc}$	Acc^*	$\tilde{Acc}^*$	$\tilde{L}_2$	L_2^*
DenseNet121	98.72%	1.90%	81.53%	24.03%	3.595	4.959
GoogLeNet	98.06%	3.83%	89.78%	36.11%	2.884	3.305
InceptionV3	96.17%	0.82%	89.41%	19.62%	3.552	4.181
Mobilenet	96.91%	2.72%	87.88%	36.81%	3.218	4.083
ResNet50	97.98%	3.34%	83.80%	28.65%	3.684	4.568
Shufflenet	96.66%	3.46%	84.30%	23.66%	3.331	3.286
Mean	97.42%	2.68%	86.12%	28.15%	3.377	4.064

Table 4. Targeted attack results of our method against DNN-based SAR-ATR models on the MSTAR dataset.

Surrogate	Acc	$\tilde{Acc}$	Acc^*	$\tilde{Acc}^*$	$\tilde{L}_2$	L_2^*
DenseNet121	10.00%	98.08%	88.47%	78.09%	3.086	3.587
GoogLeNet	10.00%	99.09%	89.25%	85.90%	3.377	4.289
InceptionV3	10.00%	98.81%	86.87%	78.97%	3.453	3.495
Mobilenet	10.00%	97.40%	88.38%	81.37%	3.257	3.553
ResNet50	10.00%	97.69%	87.29%	82.10%	3.408	3.490
Shufflenet	10.00%	98.36%	86.85%	83.11%	3.345	3.874
Mean	10.00%	98.24%	87.85%	81.59%	3.321	3.714

In non-targeted attacks, the Acc metric of each model on the MSTAR dataset exceeds 95%. However, after the non-targeted attack, the classification accuracy of all models on the generated adversarial examples, i.e., the $\tilde{Acc}$ metric, is below 5%, and the $\tilde{L}_2$ indicator is less than 3.7. It means that adversarial examples deteriorate the recognition performance of models rapidly through minor adversarial perturbations. Meanwhile, during the training phase of TAN, we evaluate the performance of the attenuator. According to the $\tilde{Acc}^*$ metric, the attenuator leads to an average improvement of about 25% in the classification accuracy of models on adversarial examples, that is, it indeed weakens the effectiveness of adversarial examples. We also should pay attention to the metrics Acc^* and L_2^*, i.e., the recognition accuracy of models on the attenuated samples, and the deformation distortion caused by the attenuator. The fact is that the Acc^* indicator of each model exceeds 80%, and the average value of the L_2^* metric is about 4. It means that the attenuator retains most semantic information of original samples without causing excessive deformation distortion, which is in line with our requirements.

In targeted attacks, the Acc metric represents the probability that models identify original samples as target classes, so it can reflect the dataset distribution, i.e., each category accounts for about 10% of the total dataset. After the targeted attack, the probability of each model recognizing adversarial examples as target classes, i.e., the $\tilde{Acc}$ metric, is over 97%, and the $\tilde{L}_2$ indicator shows that the image distortion caused by adversarial perturbations is less than 3.5. It means that the adversarial examples crafted by the generator can induce models to output specified results with high probability through minor perturbations. As with the non-targeted attack, we evaluate the performance of the attenuator. The $\tilde{Acc}^*$ metric shows that the attenuator results in an average decrease of about 17% in the probability of adversarial examples being identified as target classes. Meanwhile, the Acc^*

metric of each model exceeds 85%, and the average value of the L_2^* indicator is about 3.7. That is, the attenuator weakens the effectiveness of adversarial examples through slight deformations, while preserving the semantic meaning of original samples well.

In summary, for both non-targeted and targeted attacks, the adversarial examples crafted by the generator can fool models with high success rates, and the attenuator is able to weaken the effectiveness of adversarial examples with slight deformations while retaining the semantic meaning of original samples. Moreover, we ensure that the generator always outperforms the attenuator by adjusting the training ratio between the two models. To visualize the attack results of TAN, we took ResNet50 as the surrogate model and display the outputs of TAN at each stage in Figure 8.

Finally, we compared the non-targeted and targeted attack performance of different methods against DNN-based SAR-ATR models on the MSTAR dataset, as detailed in Table 5. Obviously, for the same image distortion, the attack effectiveness of the proposed method against a single model may not be the best. Nevertheless, we focused more on the transferability of adversarial examples, which will be the main topic of the following section.

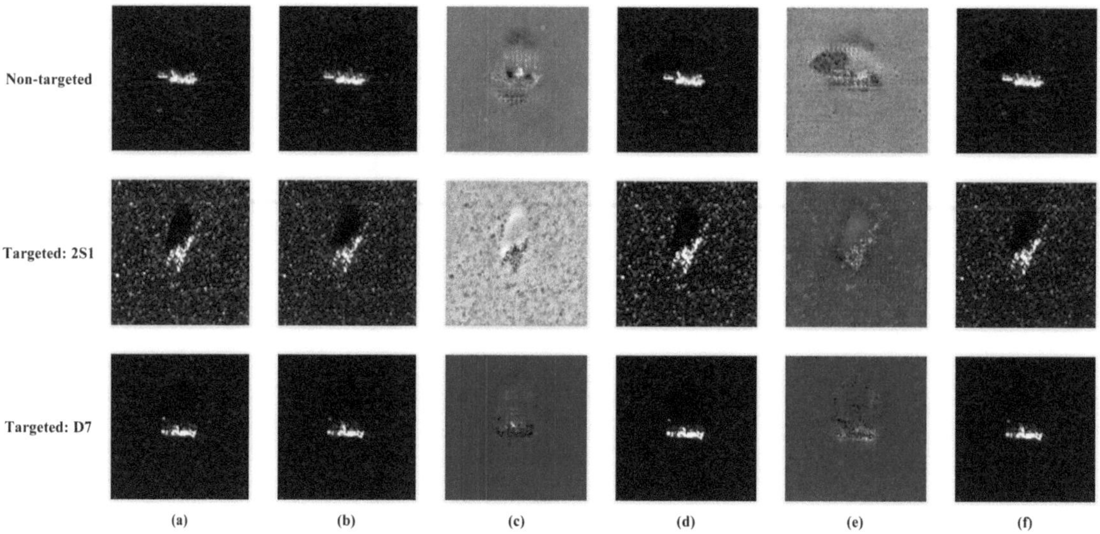

Figure 8. Visualization of attack results against ResNet50. (**a**) Original samples. (**b**) Adversarial examples. (**c**) Adversarial perturbations. (**d**) Attenuated samples. (**e**) Deformation distortion. (**f**) Attenuated adversarial examples. From top to bottom, the corresponding target classes are None, 2S1, and D7, respectively.

Table 5. Attack performance of different methods against DNN-based SAR-ATR models on the MSTAR dataset.

Surrogate	Method	Non-Targeted		Targeted	
		$\tilde{Acc}$	$\tilde{L}_2$	$\tilde{Acc}$	$\tilde{L}_2$
DenseNet121	TAN	1.90%	3.595	98.08%	3.086
	MIFGSM	0.00%	3.555	98.61%	3.613
	DIFGSM	0.00%	3.116	95.39%	2.816
	NIFGSM	0.21%	3.719	68.72%	3.550
	SINIFGSM	1.15%	3.676	82.32%	3.648
	VMIFGSM	0.00%	3.665	98.14%	3.602
	VNIFGSM	0.08%	3.691	96.89%	3.635

Table 5. *Cont.*

Surrogate	Method	Non-Targeted		Targeted	
		$\tilde{Acc}$	$\tilde{L}_2$	$\tilde{Acc}$	$\tilde{L}_2$
GoogLeNet	TAN	3.83%	2.884	99.09%	3.377
	MIFGSM	0.04%	3.615	98.36%	3.601
	DIFGSM	0.04%	3.090	94.47%	2.830
	NIFGSM	0.41%	3.674	64.32%	3.520
	SINIFGSM	4.04%	3.647	69.79%	3.615
	VMIFGSM	0.04%	3.587	97.84%	3.601
	VNIFGSM	0.37%	3.588	95.74%	3.636
InceptionV3	TAN	0.82%	3.552	98.81%	3.453
	MIFGSM	0.00%	3.599	96.00%	3.563
	DIFGSM	0.04%	3.010	86.72%	2.811
	NIFGSM	0.21%	3.671	51.66%	3.397
	SINIFGSM	2.93%	3.689	62.46%	3.593
	VMIFGSM	0.00%	3.614	91.54%	3.577
	VNIFGSM	0.00%	3.632	84.02%	3.605
Mobilenet	TAN	2.72%	3.218	97.40%	3.257
	MIFGSM	8.29%	3.557	99.86%	3.538
	DIFGSM	6.64%	2.821	91.64%	2.610
	NIFGSM	6.88%	3.575	80.05%	3.519
	SINIFGSM	1.77%	3.664	85.14%	3.662
	VMIFGSM	2.35%	3.572	99.40%	3.499
	VNIFGSM	1.32%	3.635	95.58%	3.582
ResNet50	TAN	3.34%	3.684	97.69%	3.408
	MIFGSM	0.95%	3.659	97.08%	3.613
	DIFGSM	0.33%	3.141	90.35%	2.824
	NIFGSM	0.33%	3.710	45.34%	3.501
	SINIFGSM	3.96%	3.720	71.64%	3.652
	VMIFGSM	0.87%	3.644	96.17%	3.618
	VNIFGSM	0.25%	3.692	94.17%	3.632
Shufflenet	TAN	3.46%	3.331	98.36%	3.345
	MIFGSM	0.00%	3.567	100.00%	3.518
	DIFGSM	0.00%	2.790	97.54%	2.599
	NIFGSM	0.16%	3.632	91.77%	3.455
	SINIFGSM	0.00%	3.660	95.79%	3.568
	VMIFGSM	0.00%	3.617	100.00%	3.511
	VNIFGSM	0.04%	3.654	99.73%	3.568

4.6. Comparison of Transferability

In this section, we evaluated the transferability of adversarial examples among DNN-based SAR-ATR models on the MSTAR dataset. Specifically, we first took each network as the surrogate model in turn and crafted adversarial examples for them, respectively. Then, we assessed the transferability by testing the recognition results of victim models on corresponding adversarial examples. The transferability in non-targeted and targeted attacks are shown in Tables 6 and 7, respectively.

In non-targeted attacks, when the proposed method sequentially takes DenseNet121, GoogLeNet, InceptionV3, Mobilenet, ResNet50, and Shufflenet as the surrogate model, the highest recognition accuracy of victim models on the generated adversarial examples are 12.90%, 26.88%, 23.45%, 18.59%, 11.01%, and 23.54%, respectively. Equivalently, the highest recognition accuracy of victim models on the adversarial examples produced by baseline methods are 36.11%, 44.44%, 56.06%, 65.99%, 33.84%, and 68.51%, respectively. Meanwhile, for each surrogate model, victim models always have the lowest recognition accuracy on the adversarial examples crafted by our approach. Obviously, compared with baseline methods, the proposed method slightly sacrifices the performance on attacking surrogate models, but achieves state-of-the-art transferability among victim models in non-targeted attacks. Detailed results are shown in Table 6.

Table 6. Transferability of adversarial examples generated by different attack algorithms in non-targeted attacks.

Surrogate	Method	DenseNet121	GoogLeNet	InceptionV3	Mobilenet	ResNet50	Shufflenet
DenseNet121	TAN	1.90%	4.25%	7.46%	9.93%	9.11%	12.90%
	MIFGSM	0.00%	10.10%	12.82%	26.46%	16.32%	28.65%
	DIFGSM	0.00%	8.16%	11.46%	26.01%	19.17%	30.83%
	NIFGSM	0.21%	14.67%	14.67%	26.75%	20.07%	30.67%
	SINIFGSM	1.15%	16.69%	19.29%	35.66%	17.64%	36.11%
	VMIFGSM	0.00%	8.86%	11.62%	24.40%	15.13%	25.89%
	VNIFGSM	0.08%	8.04%	11.62%	22.38%	13.60%	23.54%
GoogLeNet	TAN	6.88%	3.83%	8.16%	23.62%	10.51%	26.88%
	MIFGSM	10.18%	0.04%	17.72%	32.36%	27.66%	42.13%
	DIFGSM	8.33%	0.04%	14.47%	32.52%	24.73%	38.66%
	NIFGSM	22.88%	0.41%	24.28%	32.32%	35.16%	44.44%
	SINIFGSM	7.96%	4.04%	13.15%	33.22%	15.09%	28.07%
	VMIFGSM	8.57%	0.04%	16.32%	29.72%	25.64%	38.58%
	VNIFGSM	10.02%	0.37%	15.50%	27.99%	26.30%	36.93%
InceptionV3	TAN	8.20%	9.60%	0.82%	21.43%	14.67%	23.45%
	MIFGSM	19.25%	35.00%	0.00%	39.45%	33.14%	42.54%
	DIFGSM	16.86%	33.22%	0.04%	43.69%	33.76%	47.07%
	NIFGSM	32.11%	34.46%	0.21%	42.09%	43.08%	44.89%
	SINIFGSM	27.37%	38.05%	2.93%	49.22%	41.18%	56.06%
	VMIFGSM	18.51%	26.92%	0.00%	34.46%	31.04%	37.18%
	VNIFGSM	21.68%	26.38%	0.00%	33.80%	34.50%	37.63%
Mobilenet	TAN	14.34%	15.83%	13.56%	2.72%	14.18%	18.59%
	MIFGSM	65.99%	59.32%	53.59%	8.29%	55.56%	59.77%
	DIFGSM	51.28%	53.34%	49.34%	6.64%	49.34%	52.18%
	NIFGSM	65.75%	58.66%	51.85%	6.88%	52.31%	55.56%
	SINIFGSM	64.67%	45.14%	49.01%	1.77%	51.81%	58.37%
	VMIFGSM	62.49%	52.10%	50.45%	2.35%	49.63%	52.84%
	VNIFGSM	56.27%	50.04%	43.61%	1.32%	43.82%	48.19%
ResNet50	TAN	5.94%	9.27%	10.14%	12.94%	3.34%	11.01%
	MIFGSM	14.59%	24.15%	17.72%	16.90%	0.95%	26.42%
	DIFGSM	11.13%	17.07%	15.09%	20.45%	0.33%	26.59%
	NIFGSM	21.72%	28.19%	20.28%	19.74%	0.33%	29.43%
	SINIFGSM	26.50%	24.15%	22.59%	30.50%	3.96%	33.84%
	VMIFGSM	13.31%	22.42%	16.36%	15.95%	0.87%	23.33%
	VNIFGSM	15.00%	22.67%	16.45%	14.47%	0.25%	22.63%
Shufflenet	TAN	17.72%	23.54%	16.49%	22.22%	17.85%	3.46%
	MIFGSM	66.69%	70.03%	65.00%	55.81%	65.00%	0.00%
	DIFGSM	53.46%	57.58%	55.32%	51.44%	55.44%	0.00%
	NIFGSM	67.23%	61.58%	58.62%	48.35%	61.62%	0.16%
	SINIFGSM	68.51%	58.33%	60.92%	50.41%	56.64%	0.00%
	VMIFGSM	57.25%	55.32%	54.29%	40.23%	53.34%	0.00%
	VNIFGSM	56.68%	54.25%	51.57%	37.30%	52.14%	0.04%

In targeted attacks, the proposed method still takes DenseNet121, GoogLeNet, InceptionV3, Mobilenet, ResNet50, and Shufflenet as the surrogate model in turn, and the minimum probability that victim models identify the generated adversarial examples as target classes are 52.39%, 55.02%, 54.57%, 57.66%, 66.26%, and 47.78%, respectively. Correspondingly, the minimum probability that victim models recognize the adversarial examples produced by baseline methods as target classes are 22.18%, 19.63%, 19.49%, 15.52%, 19.36%, and 13.06%, respectively. Moreover, for each surrogate model, victim models always identify the adversarial examples crafted by our approach as target classes with the maximum probability. Thus, the proposed method also achieves state-of-the-art transferability among victim models in targeted attacks. Detailed results are shown in Table 7.

Table 7. Transferability of adversarial examples generated by different attack algorithms in targeted attacks.

Surrogate	Method	DenseNet121	GoogLeNet	InceptionV3	Mobilenet	ResNet50	Shufflenet
DenseNet121	TAN	98.08%	79.12%	70.71%	59.03%	62.31%	52.39%
	MIFGSM	98.61%	52.47%	49.05%	39.47%	43.78%	37.62%
	DIFGSM	95.39%	51.08%	46.62%	35.02%	39.51%	32.29%
	NIFGSM	68.72%	33.06%	27.61%	22.18%	25.78%	22.92%
	SINIFGSM	82.32%	40.62%	33.17%	29.95%	31.93%	30.59%
	VMIFGSM	98.14%	48.94%	44.10%	33.56%	39.29%	34.06%
	VNIFGSM	96.89%	48.78%	46.03%	34.70%	39.80%	35.52%
GoogLeNet	TAN	81.04%	99.09%	66.59%	56.72%	63.86%	55.02%
	MIFGSM	61.56%	98.36%	47.57%	34.16%	37.57%	29.75%
	DIFGSM	58.81%	94.47%	47.91%	32.17%	36.20%	26.88%
	NIFGSM	31.46%	64.32%	25.34%	19.85%	23.14%	19.63%
	SINIFGSM	41.97%	69.79%	34.39%	28.21%	29.77%	25.48%
	VMIFGSM	53.37%	97.84%	42.19%	30.67%	34.94%	26.36%
	VNIFGSM	56.26%	95.74%	43.96%	32.31%	36.11%	29.49%
InceptionV3	TAN	75.11%	71.56%	98.81%	67.23%	63.62%	54.57%
	MIFGSM	42.64%	35.92%	96.00%	32.49%	35.00%	29.51%
	DIFGSM	42.99%	33.70%	86.72%	31.16%	34.13%	28.20%
	NIFGSM	27.12%	24.67%	51.66%	19.49%	23.76%	22.45%
	SINIFGSM	26.76%	25.23%	62.46%	21.90%	24.36%	22.59%
	VMIFGSM	36.38%	34.05%	91.54%	30.15%	31.43%	28.52%
	VNIFGSM	37.82%	33.55%	84.02%	31.44%	32.28%	28.58%
Mobilenet	TAN	61.30%	57.66%	61.53%	97.40%	60.97%	63.11%
	MIFGSM	19.98%	18.66%	22.87%	99.86%	23.55%	20.31%
	DIFGSM	23.96%	21.92%	23.79%	91.64%	24.51%	22.65%
	NIFGSM	15.76%	15.58%	16.85%	80.05%	18.06%	15.91%
	SINIFGSM	16.81%	15.52%	18.96%	85.14%	21.20%	16.63%
	VMIFGSM	18.46%	17.84%	18.70%	99.40%	21.49%	19.61%
	VNIFGSM	21.60%	18.41%	22.34%	95.58%	24.67%	21.96%
ResNet50	TAN	71.39%	71.54%	71.02%	73.68%	97.69%	66.26%
	MIFGSM	43.23%	30.51%	41.57%	42.41%	97.08%	36.29%
	DIFGSM	45.18%	34.25%	42.37%	39.40%	90.35%	34.36%
	NIFGSM	22.07%	20.45%	20.33%	19.36%	45.34%	19.75%
	SINIFGSM	25.81%	21.38%	27.15%	31.01%	71.64%	26.02%
	VMIFGSM	36.44%	26.33%	35.75%	38.61%	96.17%	32.79%
	VNIFGSM	40.80%	27.10%	38.26%	38.87%	94.17%	36.49%
Shufflenet	TAN	53.91%	47.78%	51.69%	60.35%	58.78%	98.36%
	MIFGSM	18.29%	16.43%	17.06%	19.46%	17.20%	100.00%
	DIFGSM	23.55%	20.36%	20.80%	22.55%	21.35%	97.54%
	NIFGSM	13.96%	13.06%	13.14%	14.47%	13.66%	91.77%
	SINIFGSM	15.83%	15.23%	15.34%	19.42%	16.05%	95.79%
	VMIFGSM	17.58%	16.34%	17.09%	21.65%	18.46%	99.94%
	VNIFGSM	19.43%	17.97%	18.68%	22.87%	19.98%	99.73%

In conclusion, for both non-targeted and targeted attacks, our approach generates adversarial examples with the strongest transferability. In other words, it performs better on exploring the common vulnerability of DNN models. We attribute this to the adversarial training between the generator and attenuator. Figuratively speaking, it is because of the attenuator constantly creating obstacles for the generator that the attack capability of the generator is continuously enhanced and completed.

4.7. Comparison of Real-Time Performance

According to (4), compared to traditional iterative methods, the generator in our approach is capable of one-step mapping original samples to adversarial examples. It acts like a function that takes inputs and outputs results based on the mapping relationship. To evaluate the real-time performance of adversarial attacks, we compared the time cost of generating a single adversarial example through different attack algorithms. The time consumption of non-targeted and targeted attacks is shown in Tables 8 and 9, respectively.

As we can see, there is almost no difference in the time cost of crafting a single adversarial example in non-targeted and targeted attacks. Meanwhile, for all the victim models, the time cost of generating a single adversarial example through our method is stable around 2 ms. As for baseline methods, it depends on the complexity of victim models, the more complex the model, the longer the time cost. However, even for the simplest victim model, the minimum time cost of baseline methods is about 4.5 ms, consuming twice as much time as our approach. Thus, there is no doubt that the proposed method achieves the most superior and stable real-time performance.

Table 8. Time cost of generating a single adversarial example through different attack algorithms in non-targeted attacks.

Method	DenseNet121	GoogLeNet	InceptionV3	Mobilenet	ResNet50	Shufflenet	Mean
TAN	0.002029 s	0.002201 s	0.002039 s	0.002218 s	0.002031 s	0.002045 s	0.002094 s
MIFGSM	0.018285 s	0.006351 s	0.012636 s	0.005093 s	0.013445 s	0.004451 s	0.010044 s
DIFGSM	0.018276 s	0.006363 s	0.012653 s	0.005103 s	0.013468 s	0.004488 s	0.010059 s
NIFGSM	0.018312 s	0.006354 s	0.012646 s	0.005111 s	0.013477 s	0.004456 s	0.010059 s
SINIFGSM	0.091032 s	0.031499 s	0.063015 s	0.024865 s	0.067202 s	0.021676 s	0.049882 s
VMIFGSM	0.109252 s	0.037827 s	0.075580 s	0.029803 s	0.080479 s	0.025968 s	0.059818 s
VNIFGSM	0.109184 s	0.037804 s	0.075483 s	0.029776 s	0.080560 s	0.025907 s	0.059786 s

Table 9. Time cost of generating a single adversarial example through different attack algorithms in targeted attacks.

Method	DenseNet121	GoogLeNet	InceptionV3	Mobilenet	ResNet50	Shufflenet	Mean
TAN	0.002070 s	0.002069 s	0.002036 s	0.002055 s	0.002087 s	0.002097 s	0.002069 s
MIFGSM	0.018281 s	0.006353 s	0.012634 s	0.005088 s	0.013451 s	0.004446 s	0.010042 s
DIFGSM	0.018291 s	0.006369 s	0.012652 s	0.005104 s	0.013490 s	0.004488 s	0.010065 s
NIFGSM	0.018306 s	0.006358 s	0.012661 s	0.005105 s	0.013486 s	0.004460 s	0.010063 s
SINIFGSM	0.091064 s	0.031539 s	0.063066 s	0.024871 s	0.067216 s	0.021664 s	0.049903 s
VMIFGSM	0.109262 s	0.037860 s	0.075579 s	0.029776 s	0.080481 s	0.025984 s	0.059823 s
VNIFGSM	0.109176 s	0.037819 s	0.075502 s	0.029798 s	0.080546 s	0.025923 s	0.059794 s

4.8. Visualization of Adversarial Examples

In this section, we took ResNet50 as the surrogate model and visualized the adversarial examples crafted by different methods in Figure 9. Obviously, the adversarial perturbations generated by our method are continuous, and mainly focus on the target region of SAR images. In contrast, the perturbations produced by baseline methods are quite discrete, and almost cover the global area of SAR images. First, from the perspective of feature extraction, since the features that have a greater impact on recognition results are mainly concentrated in the target region rather than the background clutter area, a focused disruption of key features is certainly a more efficient attack strategy. Second, from the perspective of physical feasibility, the fewer pixels modified in adversarial examples, the smaller range perturbed in reality, so localized perturbations are more feasible than global ones. In summary, the proposed method improves the efficiency and feasibility of adversarial attacks by focusing perturbations on the target region of SAR images.

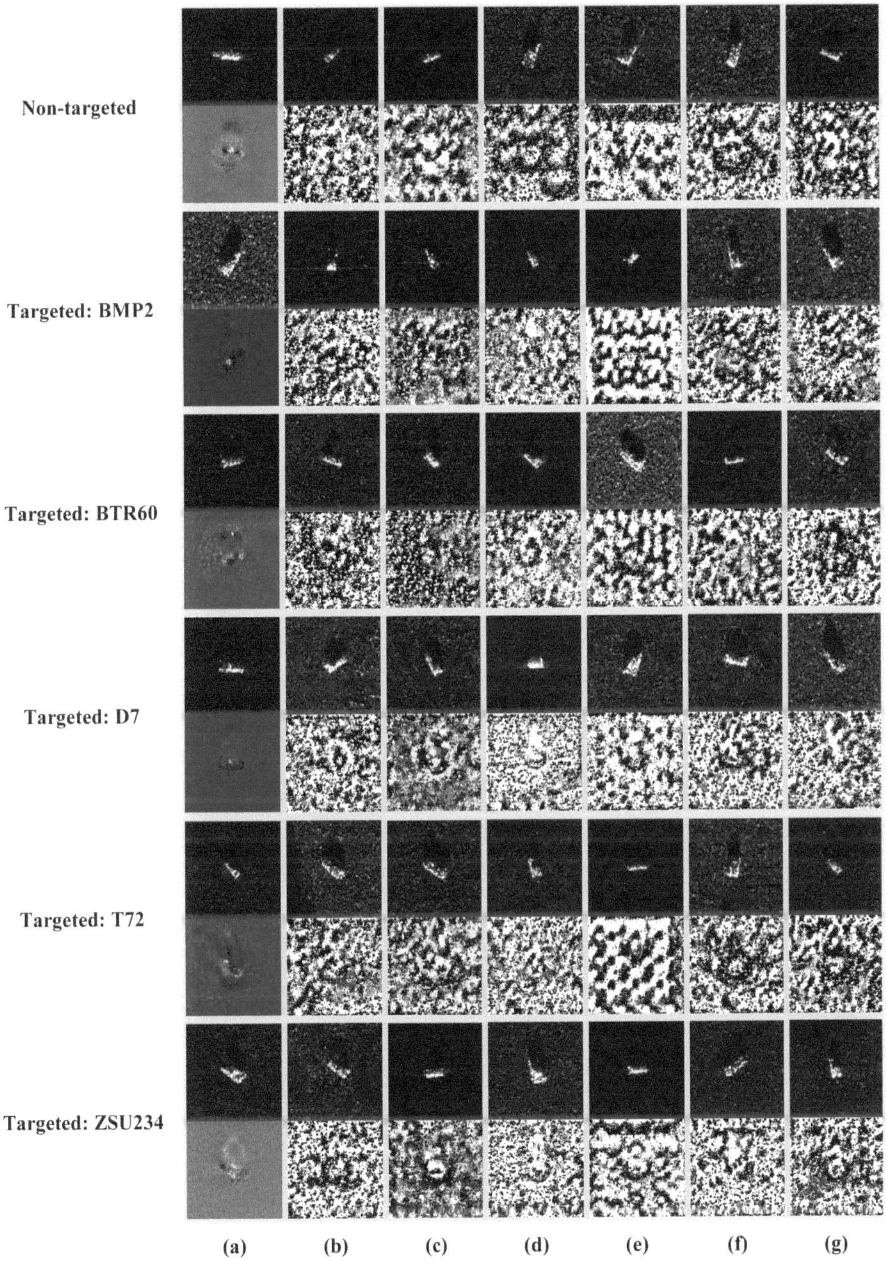

Figure 9. Visualization of adversarial examples against ResNet50. (**a**) TAN. (**b**) MIFGSM. (**c**) DIFGSM. (**d**) NIFGSM. (**e**) SINIFGSM. (**f**) VMIFGSM. (**g**) VNIFGSM. From top to bottom, the corresponding target classes are None, BMP2, BTR60, D7, T72, and ZSU234, respectively. For each attack, the first row shows adversarial examples, and the second row shows corresponding adversarial perturbations.

5. Discussions

So far, the proposed method has been proven to be effective for SAR images. Further studies should verify its effectiveness in other fields, such as optical [40,41], infrared [42,43], and synthetic aperture sonar (SAS) [44–47] images, etc. Although different imaging principles lead to huge differences in the resolution, dimension, and target type of images, we argue that TAN can be well-suitable to these fields. The reason is that adversarial examples essentially attack the inherent vulnerability of DNN models, independent of the input data. However, the non-negligible challenge is how to realize these adversarial examples in the real world. Specifically, the physical implementation depends on the imaging principle, e.g., crafting adversarial patches against optical cameras, changing temperature against infrared devices, and emitting acoustic signals against SAS, etc. This is a worthwhile topic in the future.

6. Conclusions

This paper proposed a transferable adversarial network (TAN) to attack DNN-based SAR-ATR models, with the benefit that the transferability and the real-time performance of adversarial examples is significantly improved, which is of great significance for real-world black-box attacks. In the proposed method, we simultaneously trained two encoder–decoder models: a generator that learns the one-step forward mapping from original data to adversarial examples, and an attenuator that captures the most harmful deformations to malicious samples. It is motivated by enabling real-time attacks by one-step mapping original data to adversarial examples, and enhancing the transferability through a two-player game between the generator and attenuator. Experimental results demonstrated that our approach achieves state-of-the-art transferability with acceptable adversarial perturbations and minimum time costs compared to existing attack methods, making real-time black-box attacks without any prior knowledge a reality. Potential future work could consider attacking DNN-based SAR-ATR models under small sample conditions. In addition to improving the performance of attack algorithms, it makes sense to implement adversarial examples in the real world.

Author Contributions: Conceptualization, M.D. (Meng Du) and D.B.; methodology, M.D. (Meng Du); software, M.D. (Meng Du); validation, D.B., Y.S., B.S. and Z.W.; formal analysis, D.B. and M.D. (Mingyang Du); investigation, M.D. (Mingyang Du); resources, D.B.; data curation, M.D. (Meng Du); writing—original draft preparation, M.D. (Meng Du); writing—review and editing, M.D. (Meng Du), L.L. and D.B.; visualization, M.D. (Meng Du); supervision, D.B.; project administration, D.B.; funding acquisition, D.B. All authors have read and agreed to the published version of the manuscript.

Funding: This work was supported by the National Natural Science Foundation of China under Grant 62071476.

Institutional Review Board Statement: The study does not involve humans or animals.

Informed Consent Statement: The study does not involve humans.

Data Availability Statement: The experiments in this paper use public datasets, so no data are reported in this work.

Conflicts of Interest: The authors declare that they have no conflict of interest to report regarding the present study.

References

1. Li, D.; Kuai, Y.; Wen, G.; Liu, L. Robust Visual Tracking via Collaborative and Reinforced Convolutional Feature Learning. In Proceedings of the IEEE/CVF Conference on Computer Vision and Pattern Recognition Workshops, Long Beach, CA, USA, 15–19 June 2019. [CrossRef]
2. Kuai, Y.; Wen, G.; Li, D. Masked and dynamic Siamese network for robust visual tracking. *Inf. Sci.* **2019**, *503*, 169–182. [CrossRef]
3. Cong, R.; Yang, N.; Li, C.; Fu, H.; Zhao, Y.; Huang, Q.; Kwong, S. Global-and-local collaborative learning for co-salient object detection. *IEEE Trans. Cybern.* **2022**, *53*, 1920–1931. [CrossRef] [PubMed]
4. Tang, J.; Xiang, D.; Zhang, F.; Ma, F.; Zhou, Y.; Li, H. Incremental SAR Automatic Target Recognition With Error Correction and High Plasticity. *IEEE J. Sel. Top. Appl. Earth Obs. Remote Sens.* **2022**, *15*, 1327–1339. [CrossRef]
5. Wang, L.; Yang, X.; Tan, H.; Bai, X.; Zhou, F. Few-Shot Class-Incremental SAR Target Recognition Based on Hierarchical Embedding and Incremental Evolutionary Network. *IEEE Trans. Geosci. Remote Sens.* **2023**, *2023*, 3248040. [CrossRef]

6. Kwak, Y.; Song, W.J.; Kim, S.E. Speckle-Noise-Invariant Convolutional Neural Network for SAR Target Recognition. *IEEE Geosci. Remote Sens. Lett.* **2019**, *16*, 549–553. [CrossRef]
7. Du, C.; Chen, B.; Xu, B.; Guo, D.; Liu, H. Factorized discriminative conditional variational auto-encoder for radar HRRP target recognition. *Signal Process.* **2019**, *158*, 176–189. [CrossRef]
8. Vint, D.; Anderson, M.; Yang, Y.; Ilioudis, C.; Di Caterina, G.; Clemente, C. Automatic Target Recognition for Low Resolution Foliage Penetrating SAR Images Using CNNs and GANs. *Remote Sens.* **2021**, *13*, 596. [CrossRef]
9. Huang, T.; Zhang, Q.; Liu, J.; Hou, R.; Wang, X.; Li, Y. Adversarial attacks on deep-learning-based SAR image target recognition. *J. Netw. Comput. Appl.* **2020**, *162*, 102632. [CrossRef]
10. Szegedy, C.; Zaremba, W.; Sutskever, I.; Bruna, J.; Erhan, D.; Goodfellow, I.; Fergus, R. Intriguing properties of neural networks. *arXiv* **2013**, arXiv:1312.6199.
11. Goodfellow, I.J.; Shlens, J.; Szegedy, C. Explaining and harnessing adversarial examples. *arXiv* **2014**, arXiv:1412.6572. [CrossRef]
12. Kurakin, A.; Goodfellow, I.J.; Bengio, S. Adversarial examples in the physical world. In *Artificial Intelligence Safety and Security*; Chapman and Hall/CRC: London, UK, 2018; pp. 99–112.
13. Moosavi-Dezfooli, S.M.; Fawzi, A.; Frossard, P. Deepfool: A simple and accurate method to fool deep neural networks. In Proceedings of the IEEE Conference on Computer Vision and Pattern Recognition, Las Vegas, NV, USA, 26 June–1 July 2016; pp. 2574–2582. [CrossRef]
14. Papernot, N.; McDaniel, P.; Jha, S.; Fredrikson, M.; Celik, Z.B.; Swami, A. The limitations of deep learning in adversarial settings. In Proceedings of the 2016 IEEE European Symposium on Security and Privacy (EuroS&P), Saarbrücken, Germany, 21–24 March 2016; pp. 372–387. [CrossRef]
15. Su, J.; Vargas, D.V.; Sakurai, K. One pixel attack for fooling deep neural networks. *IEEE Trans. Evol. Comput.* **2019**, *23*, 828–841. [CrossRef]
16. Chen, P.Y.; Zhang, H.; Sharma, Y.; Yi, J.; Hsieh, C.J. Zoo: Zeroth order optimization based black-box attacks to deep neural networks without training substitute models. In Proceedings of the 10th ACM Workshop on Artificial Intelligence and Security, Dallas, TX, USA, 3 November 2017; pp. 15–26. [CrossRef]
17. Chen, J.; Jordan, M.I.; Wainwright, M.J. Hopskipjumpattack: A query-efficient decision-based attack. In Proceedings of the 2020 IEEE Symposium on Security and Privacy (SP), San Francisco, CA, USA, 18–21 May 2020; pp. 1277–1294. [CrossRef]
18. Papernot, N.; McDaniel, P.; Goodfellow, I. Transferability in machine learning: From phenomena to black-box attacks using adversarial samples. *arXiv* **2016**, arXiv:1605.07277.
19. Dong, Y.; Liao, F.; Pang, T.; Su, H.; Zhu, J.; Hu, X.; Li, J. Boosting adversarial attacks with momentum. In Proceedings of the IEEE Conference on Computer Vision and Pattern Recognition, Salt Lake City, UT, USA, 18–23 June 2018; pp. 9185–9193. [CrossRef]
20. Lin, J.; Song, C.; He, K.; Wang, L.; Hopcroft, J.E. Nesterov accelerated gradient and scale invariance for adversarial attacks. *arXiv* **2019**, arXiv:1908.06281.
21. Xie, C.; Zhang, Z.; Zhou, Y.; Bai, S.; Wang, J.; Ren, Z.; Yuille, A.L. Improving transferability of adversarial examples with input diversity. In Proceedings of the IEEE/CVF Conference on Computer Vision and Pattern Recognition, Long Beach, CA, USA, 15–20 June 2019; pp. 2730–2739. [CrossRef]
22. Wang, X.; He, K. Enhancing the transferability of adversarial attacks through variance tuning. In Proceedings of the IEEE/CVF Conference on Computer Vision and Pattern Recognition, Nashville, TN, USA, 20–25 June 2021; pp. 1924–1933. [CrossRef]
23. Xu, Y.; Du, B.; Zhang, L. Assessing the threat of adversarial examples on deep neural networks for remote sensing scene classification: Attacks and defenses. *IEEE Trans. Geosci. Remote Sens.* **2020**, *59*, 1604–1617. [CrossRef]
24. Xu, Y.; Ghamisi, P. Universal Adversarial Examples in Remote Sensing: Methodology and Benchmark. *IEEE Trans. Geosci. Remote Sens.* **2022**, *60*, 1–15. [CrossRef]
25. Li, H.; Huang, H.; Chen, L.; Peng, J.; Huang, H.; Cui, Z.; Mei, X.; Wu, G. Adversarial examples for CNN-based SAR image classification: An experience study. *IEEE J. Sel. Top. Appl. Earth Obs. Remote Sens.* **2020**, *14*, 1333–1347. [CrossRef]
26. Du, C.; Huo, C.; Zhang, L.; Chen, B.; Yuan, Y. Fast C&W: A Fast Adversarial Attack Algorithm to Fool SAR Target Recognition with Deep Convolutional Neural Networks. *IEEE Geosci. Remote Sens. Lett.* **2021**, *19*, 1–5. [CrossRef]
27. Du, M.; Bi, D.; Du, M.; Xu, X.; Wu, Z. ULAN: A Universal Local Adversarial Network for SAR Target Recognition Based on Layer-Wise Relevance Propagation. *Remote Sens.* **2022**, *15*, 21. [CrossRef]
28. Xia, W.; Liu, Z.; Li, Y. SAR-PeGA: A Generation Method of Adversarial Examples for SAR Image Target Recognition Network. *IEEE Trans. Aerosp. Electron. Syst.* **2022**, *2022*, 3206261. [CrossRef]
29. Johnson, J.; Alahi, A.; Fei-Fei, L. Perceptual losses for real-time style transfer and super-resolution. In *Computer Vision–ECCV 2016, Proceedings of the 14th European Conference, Amsterdam, The Netherlands, 11–14 October 2016*; Proceedings, Part II 14; Springer: Cham, Switzerland, 2016; pp. 694–711. [CrossRef]
30. Goodfellow, I.; Pouget-Abadie, J.; Mirza, M.; Xu, B.; Warde-Farley, D.; Ozair, S.; Courville, A.; Bengio, Y. Generative adversarial networks. *Commun. ACM* **2020**, *63*, 139–144. [CrossRef]
31. Keydel, E.R.; Lee, S.W.; Moore, J.T. MSTAR extended operating conditions: A tutorial. *Algorithms Synth. Aperture Radar Imag. III* **1996**, *2757*, 228–242. [CrossRef]
32. Huang, G.; Liu, Z.; Van Der Maaten, L.; Weinberger, K.Q. Densely connected convolutional networks. In Proceedings of the IEEE Conference on Computer Vision and Pattern recognition, Honolulu, HI, USA, 21–26 July 2017; pp. 4700–4708.

33. Szegedy, C.; Liu, W.; Jia, Y.; Sermanet, P.; Reed, S.; Anguelov, D.; Erhan, D.; Vanhoucke, V.; Rabinovich, A. Going deeper with convolutions. In Proceedings of the IEEE Conference on Computer Vision and Pattern Recognition, Boston, MA, USA, 7–12 June 2015; pp. 1–9. [CrossRef]
34. Szegedy, C.; Vanhoucke, V.; Ioffe, S.; Shlens, J.; Wojna, Z. Rethinking the inception architecture for computer vision. In Proceedings of the IEEE Conference on Computer Vision and Pattern Recognition, Las Vegas, NV, USA, 26 June–1 June 2016; pp. 2818–2826. [CrossRef]
35. Howard, A.G.; Zhu, M.; Chen, B.; Kalenichenko, D.; Wang, W.; Weyand, T.; Andreetto, M.; Adam, H. Mobilenets: Efficient convolutional neural networks for mobile vision applications. *arXiv* **2017**, arXiv:1704.04861.
36. He, K.; Zhang, X.; Ren, S.; Sun, J. Deep residual learning for image recognition. In Proceedings of the IEEE Conference on Computer Vision and Pattern Recognition, Las Vegas, NV, USA, 27–30 June 2016; pp. 770–778.
37. Zhang, X.; Zhou, X.; Lin, M.; Sun, J. Shufflenet: An extremely efficient convolutional neural network for mobile devices. In Proceedings of the IEEE Conference on Computer Vision and Pattern Recognition, Salt Lake City, UT, USA, 18–22 June 2018; pp. 6848–6856.
38. Kingma, D.P.; Ba, J. Adam: A method for stochastic optimization. *arXiv* **2014**, arXiv:1412.6980.
39. Kim, H. Torchattacks: A pytorch repository for adversarial attacks. *arXiv* **2020**, arXiv:2010.01950.
40. Kang, J.; Wang, Z.; Zhu, R.; Xia, J.; Sun, X.; Fernandez-Beltran, R.; Plaza, A. DisOptNet: Distilling Semantic Knowledge From Optical Images for Weather-Independent Building Segmentation. *IEEE Trans. Geosci. Remote Sens.* **2022**, *60*, 1–15. [CrossRef]
41. Liu, K.; Liang, Y. Underwater optical image enhancement based on super-resolution convolutional neural network and perceptual fusion. *Opt. Express* **2023**, *31*, 9688–9712. [CrossRef]
42. Tang, L.; Yuan, J.; Ma, J. Image fusion in the loop of high-level vision tasks: A semantic-aware real-time infrared and visible image fusion network. *Inf. Fusion* **2022**, *82*, 28–42. [CrossRef]
43. Liu, J.; Fan, X.; Huang, Z.; Wu, G.; Liu, R.; Zhong, W.; Luo, Z. Target-aware dual adversarial learning and a multi-scenario multi-modality benchmark to fuse infrared and visible for object detection. In Proceedings of the IEEE/CVF Conference on Computer Vision and Pattern Recognition, New Orleans, LA, USA, 18–24 June 2022; pp. 5802–5811.
44. Kiang, C.W.; Kiang, J.F. Imaging on Underwater Moving Targets With Multistatic Synthetic Aperture Sonar. *IEEE Trans. Geosci. Remote Sens.* **2022**, *60*, 1–18. [CrossRef]
45. Zhang, X.; Wu, H.; Sun, H.; Ying, W. Multireceiver SAS imagery based on monostatic conversion. *IEEE J. Sel. Top. Appl. Earth Obs. Remote Sens.* **2021**, *14*, 10835–10853. [CrossRef]
46. Choi, H.m.; Yang, H.s.; Seong, W.J. Compressive underwater sonar imaging with synthetic aperture processing. *Remote Sens.* **2021**, *13*, 1924. [CrossRef]
47. Pate, D.J.; Cook, D.A.; O'Donnell, B.N. Estimation of Synthetic Aperture Resolution by Measuring Point Scatterer Responses. *IEEE J. Ocean. Eng.* **2021**, *47*, 457–471. [CrossRef]

drones MDPI

Article

Special Vehicle Detection from UAV Perspective via YOLO-GNS Based Deep Learning Network

Zifeng Qiu [1,2,3], Huihui Bai [1] and Taoyi Chen [3,*]

1 Institute of Information Science, Beijing Jiaotong University, Beijing 100044, China
2 Key Laboratory of Aerospace Information Applications of CETC, Shijiazhuang 050081, China
3 The 54th Research Institute of CETC, Shijiazhuang 050081, China
* Correspondence: 20112012@bjtu.edu.cn

Abstract: At this moment, many special vehicles are engaged in illegal activities such as illegal mining, oil and gas theft, the destruction of green spaces, and illegal construction, which have serious negative impacts on the environment and the economy. The illegal activities of these special vehicles are becoming more and more rampant because of the limited number of inspectors and the high cost required for surveillance. The development of drone remote sensing is playing an important role in allowing efficient and intelligent monitoring of special vehicles. Due to limited onboard computing resources, special vehicle object detection still faces challenges in practical applications. In order to achieve the balance between detection accuracy and computational cost, we propose a novel algorithm named YOLO-GNS for special vehicle detection from the UAV perspective. Firstly, the Single Stage Headless (SSH) context structure is introduced to improve the feature extraction and facilitate the detection of small or obscured objects. Meanwhile, the computational cost of the algorithm is reduced in view of GhostNet by replacing the complex convolution with a linear transform by simple operation. To illustrate the performance of the algorithm, thousands of images are dedicated to sculpting in a variety of scenes and weather, each with a UAV view of special vehicles. Quantitative and comparative experiments have also been performed. Compared to other derivatives, the algorithm shows a 4.4% increase in average detection accuracy and a 1.6 increase in detection frame rate. These improvements are considered to be useful for UAV applications, especially for special vehicle detection in a variety of scenarios.

Keywords: drone; special vehicle; object detection; YOLO; SSH; GhostNet

Citation: Qiu, Z.; Bai, H.; Chen, T. Special Vehicle Detection from UAV Perspective via YOLO-GNS Based Deep Learning Network. *Drones* **2023**, 7, 117. https://doi.org/10.3390/drones7020117

Academic Editor: Anastasios Dimou

Received: 16 January 2023
Revised: 31 January 2023
Accepted: 6 February 2023
Published: 8 February 2023

1. Introduction

Special vehicles refer to motorized machines that are distinct from conventional automobiles in terms of their physical characteristics, such as shape, size, and weight. Those vehicles are typically used for a variety of purposes, including traction, obstacle removal, cleaning, lifting, loading and unloading, mixing, excavation, bulldozing, and road rolling, etc.

The detection of special vehicles in oil and gas pipelines [1], transmission lines [2], urban illegal construction [3], theft, and excavation scenarios is of great importance in order to ensure the security of these areas. This is because in the above scenarios, the presence of special vehicles often represents a high risk that these scenarios will occur, and the nature of special vehicles may cause damage to important property. The use of unmanned aerial vehicles to patrol and search for special vehicles in these scenarios has gradually become a mainstream application trend [4]. However, due to the particular shape of special vehicles, manual interpretation has low efficiency, high misjudgment, and omission. The application of a deep neural network in the automatic detection of special vehicles has been applied to some extent, but it is not mature yet, and the accuracy of existing methods is relatively poor.

Experts and scholars have proposed a variety of depth neural network methods for target detection in UAV aerial images including various vehicles. Various techniques including CNNs, RNNs, autoencoders, and GANs have been used in vehicle detection and have yielded interesting results for many tasks [5]. To detect small objects, some techniques divide the last layer of the neural network into multiple variable-sized chunks to extract features at different scales, while other approaches remove the deeper layers of the CNN, allowing the number of feature points of the target to increase [6]. Liu W et al. proposed the YOLOV5-Tassel network, which combines CSPDarknet53 and BiFPN to efficiently extract minute features and introduces the SimAM attention mechanism in the neck module to extract the features of interest before each detection head [7]. Zhou H et al. designed a data augmentation method including background replacement and noise increase in order to solve the detection of tiny targets such as cars and planes, and constructed the ADCSPDarkent53 backbone network based on YOLO, which was used to modify the loss of localization function and improve the detection accuracy [8]. In order to solve the problems of low contrast, dense distribution, and weak features of small targets, Wang J et al. constructed corresponding feature mapping relations, solved the level of adjacency between misaligned features, adjusted and fused shallow spatial features and deep semantic features, and finally improved the recognition ability of small objects [9]. Li Q et al. proposed a "rotatable region-based residual network (R3-Net)" to distinguish vehicles with different directions from aerial images and used VGG16 or ResNet101 as the backbone of R3-Net [10]. Li et al. presented an algorithm for detecting sea targets based on UAV. This algorithm optimizes feature fusion calculation and enhances feature extraction at the same time, but the computational load is too large [11]. Wang et al. used the Initial Horizontal Connection Network to enhance the Feature Pyramid Network. In addition, the use of the Semantic Attention Network to provide semantic features helps to distinguish interesting objects from cluttered backgrounds, but how the algorithm performs as expected in complex and variable aerial images needs further study [12]. Mantau et al. used visible light and thermal infrared data taken from drones to find poachers. They used YOLOv5 as their basic network and optimized it using migration learning, but this method did not work well with the fusion of different data sources [13]. Deng et al. proposed a network for detecting small objects in aerial images. They designed a Vehicle Proposal Network, which proposed areas similar to vehicles [14]. Tian et al. proposed a bineural network review method, which classifies the secondary characteristics of the suspicious target area in the unmanned aerial vehicle image, quickly filters the missing targets in one-stage detection, and achieves high-quality detection of small targets [15].

In terms of drone inspection of vehicles, Jianghuan Xie et al. proposed an anchor-free detector, called residual feature enhanced pyramid network (RFEPNet), for vehicle detection from the UAV perspective. RFEPNet contains a cross-layer context fusion network (CLCFNet) and a residual feature enhancement module (RFEM) based on pyramid convolution to achieve small target vehicle detection [16]. Wan Y et al. proposed an adaptive region selection detection framework for the retrieval of targets, such as vehicles in the field of search and rescue, adding a new detection head to achieve better detection of small targets [17]. Liu Mingjie et al. developed a detection method for small-sized vehicles in drone view, specifically optimized by connecting two ResNet units with the same width and height and adding convolutional operations in the early layers to enrich the spatial information [18]. Zhongyu Zhang et al. proposed a YOLOv3-based Deeply Separable attention-guided network (DAGN) that combines feature cascading and attention blocks and improves the loss function and candidate merging algorithm of YOLOv3. With these strategies, the performance of vehicle detection is improved while sacrificing some detection speed [19]. Wang Zhang et al. proposed a novel multiscale and occlusion-aware network (MSOA-Net) for UAV-based vehicle segmentation, which consists of two parts, including a multiscale feature adaptive fusion network (MSFAF-Net) and a region-attention-based three-headed network (RATH-Net) [20]. Xin Luo et al. developed a fast automatic vehicle detection method for UAV images, constructed a vehicle dataset for target recognition, and proposed a YOLOv3 vehicle detection framework for relatively small and dense vehicle targets [21]. Navaneeth Balamuralidhar proposed MultEYE that can detect, track, and estimate the velocity of a vehicle in a sequence of aerial images using a multi-task learning approach with a segmentation head added to the backbone of the object detector to form the MultEYE object detection architecture [22].

When drones patrol oil and gas pipelines, power transmission lines, urban violations and other fields, the size of special vehicles in the images change greatly, and there are many small targets. The feature information carried by camera overhead is limited and changeable, which increases the difficulty of detection. Secondly, the UAV cruises across complex and changeable scenes such as cities, wilderness, green areas, bare soil, and so on. Some areas contain dense targets, which makes it difficult to distinguish some similar objects. Finally, the shooting angle also brings more noise interference, and the special vehicle will be weakened, obscured, or even camouflaged, unable to expose the characteristics of the target. Due to the characteristics of variable target scale, a number of small targets, and the complex background of special vehicles, it is difficult to meet the requirements of speed and accuracy for patrol tasks if the above research methods are directly applied to special vehicle detection from a UAV perspective.

In order to solve the problem of special vehicle detection in complex backgrounds from the perspective of drones, we propose a deep neural network algorithm (YOLO-GNS) based on YOLO and optimized by GhostNet (GN) and Single Stage Headless (SSH), which can be used to detect special vehicles effectively. Firstly, the SSH network structure is added behind the FPN network to parallel several convolution layers, which enhances the convolution layer perception field and extracts the high semantic features of the special vehicle targets. Secondly, in order to improve the detection speed to meet the requirements of UAV, the GPU version of GN (G-GN) is used to reduce the computational consumption of the network. Finally, we have searched for a large number of rare places to take aerial photos and created a dataset containing a large number of special vehicle targets. We have experimented with YOLO-GS on the special vehicle (SEVE) dataset and public dataset to verify the effectiveness of the proposed method.

The rest of this paper is arranged as follows. Section 2 describes the proposed target detection method YOLO-GNS and the necessary theoretical information. Section 3 introduces the special data sets, evaluation methods, and detailed experimental results. In Section 4, we draw conclusions and determine the direction of future research.

2. Materials and Methods

2.1. Principle of YOLOv7 Network Structure

YOLO (You Only Look Once) is a one-stage target detection algorithm based on regression method proposed by Redmon et al. It has been developed into several versions [23–29]. As the latest upgrade of YOLO series, YOLOv7 has been improved from data enhancement, backbone network, activation function, and loss function, so that it has higher detection accuracy and faster detection speed.

The YOLOv7 algorithm employs strategies such as extended efficient long-range attention network (E-ELAN), Concatenation-Based models, and convolution parameterization to achieve a good balance between detection efficiency and accuracy.

As shown in Figure 1, YOLOv7 network is composed of four parts: Input, Backbone, Neck, and Head.

The Input section scales the input image to a uniform pixel size to meet the input size requirements of the backbone network. The Backbone part is composed of several CBS modules, E-ELAN modules, and MP1 modules. The CBS module is composed of convolution layer, batch normalization layer (BN), sigmoid-weighted linear unit activation function to extract image features at different scales, as shown in Figure 2.

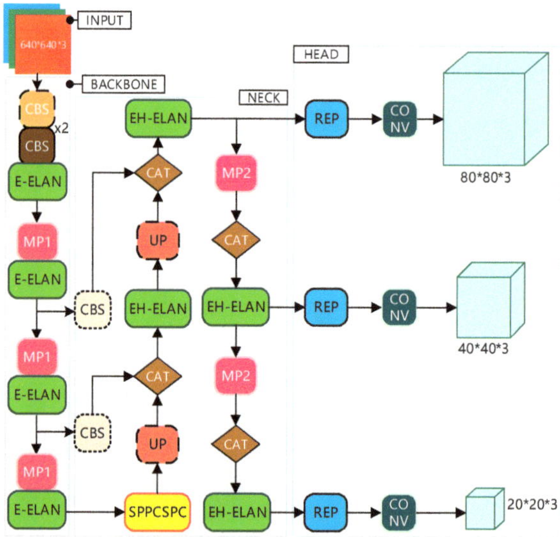

Figure 1. The original structure of yolov7.

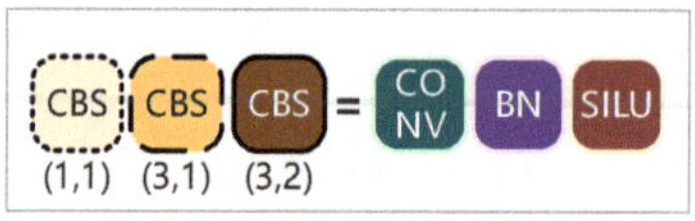

Figure 2. The structure of CBS module.

ELAN module consists of several CBS modules, whose input and output feature sizes remain the same. By guiding the computing blocks of different feature groups to learn more diverse features, the learning ability of the network is improved without destroying the original gradient path, as shown in Figure 3.

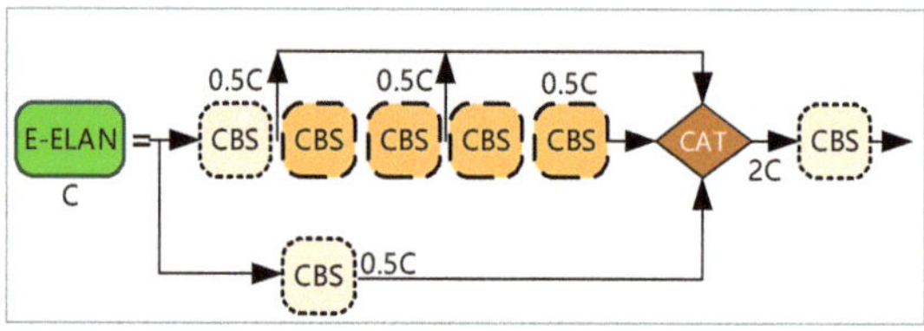

Figure 3. The structure of E-ELAN module.

MP1 module adds Maxpool layer on the basis of CBS module, which constitutes the upper and lower branches. The upper branch halves the image length and width through Maxpool and the image channel through CBS module. The lower branch halves the image channel through the first CBS module; the second CBS layer halves the image length and width and finally uses the Cat operation to fuse the features extracted from the upper and lower branches, which improves the feature extraction ability of the network, as shown in Figure 4.

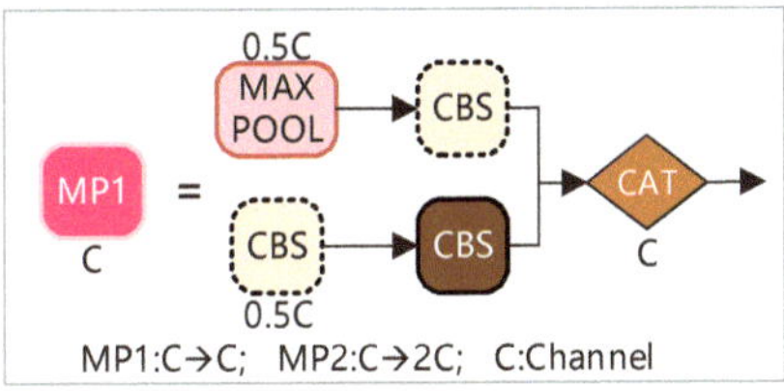

Figure 4. The structure of MP1module.

The Neck part is composed of Path Aggregation Feature Pyramid Network (PAFPN) structure, mainly including SPPCSPC module, ELAN-H module, and UP module. By introducing the bottom-up path, the bottom-level information can be easily transferred to the top level, which enables the efficient fusion of different hierarchical features.

The SPPCSPC module is mainly composed of CBS module, CAT module, and Maxpool module, which get different perception fields through maximum pooling, as shown in Figure 5.

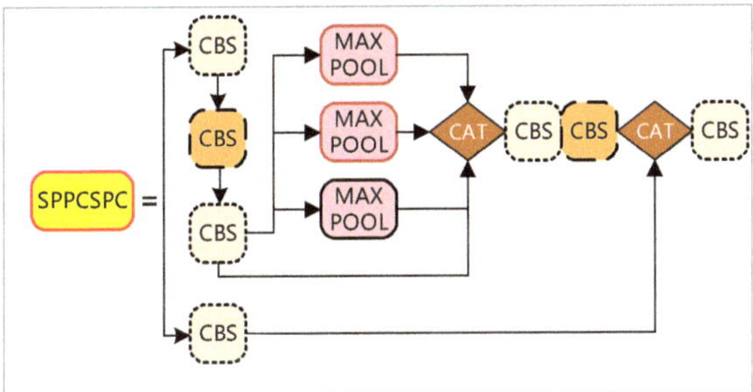

Figure 5. The structure of SPPCSPC module.

EH-ELAN module is similar to E-ELAN module but slightly different in that it selects five branches to add up with different number of outputs, as shown in Figure 6.

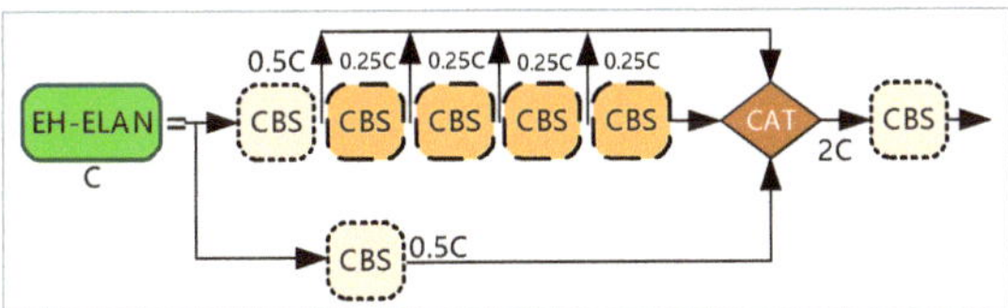

Figure 6. The structure of EH-ELAN module.

The UP module is composed of CBS and up sampling modules, as shown in Figure 7.

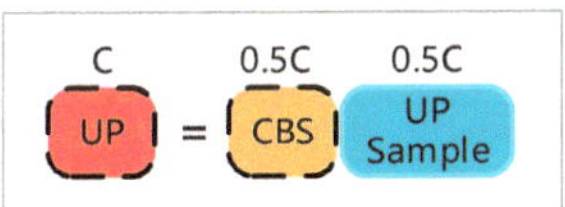

Figure 7. The structure of UP module.

Head adjusts the number of image channels for three different scales of Neck output through RepVGG Block (REP) structure, and then passes through 1×1 Convolution is used for predicting confidence, category, and anchor frame.

The REP structure is divided into train and deploy versions, as shown in Figure 8. The train version has three branches. The top branch is 3×3 convolution, which is used for feature extraction; the middle branch is 1×1 convolution, which is used for smoothing features; and the bottom branch is an Identity, which is moved without convolution and finally added together. The deploy version contains a 3×3 convolution with a stride of 1, which is converted from the training module parameterization.

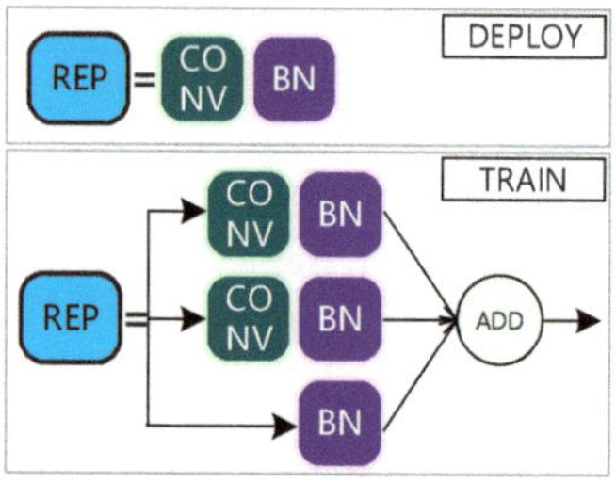

Figure 8. The structure of REP module.

Although the YOLOv7 algorithm framework performs well in common task scenarios, such as pedestrian and normal vehicle detection, there are still many problems when it is applied directly to the detection of special vehicles from the perspective of unmanned aerial vehicles: (1) Compared with common scenarios, the target scale in unmanned aerial vehicle images changes more, and there are more small targets, which further increases the difficulty of special vehicle detection; (2) The background of the scene in which the special vehicle is located is complex, and there is no corresponding context mechanism to handle the complex background, which results in the inaccurate detection of the special vehicle in the complex background; (3) UAV images require higher detection speed, but conventional YOLOv7 does not have the detection acceleration function for UAV. To solve the above problems, the algorithm in this paper is based on YOLOv7 and improved.

2.2. *YOLO-GNS Algorithm*

This section introduces the special vehicle target detection algorithm from the perspective of UAV, as shown in Figure 9. With YOLOv7 as the framework, the Backbone is improved based on GhostNet to enhance the feature extraction ability and improve the detection speed; in the view of UAV, it is beneficial to detect the weakened or occluded special vehicles from the complex scene. In order to improve the ability to detect small targets, SSH modules are added behind the pafpn structure of yolov7 to merge context information. Therefore, the algorithm is named YOLO-GNS. Compared with YOLOv7 and other derivatives, YOLO-GNS can achieve the best balance between detection accuracy and calculation cost.

2.2.1. Improvement of Backbone Network Based on GhostNet

In the backbone network of the original YOLOv7, due to the high redundancy of the intermediate feature map calculated by a large number of conventional convolutional CBS modules, the computing cost will increase. YOLO-GNS built GhostMP and GhostELAN modules to form a backbone network to extract UAV image features by drawing on the ideas of GhostNet [30]. GhostNet has the advantages of maintaining the recognition performance of similarity and reducing the convolution operation at the same time, which can greatly reduce the number of model parameters while maintaining high performance.

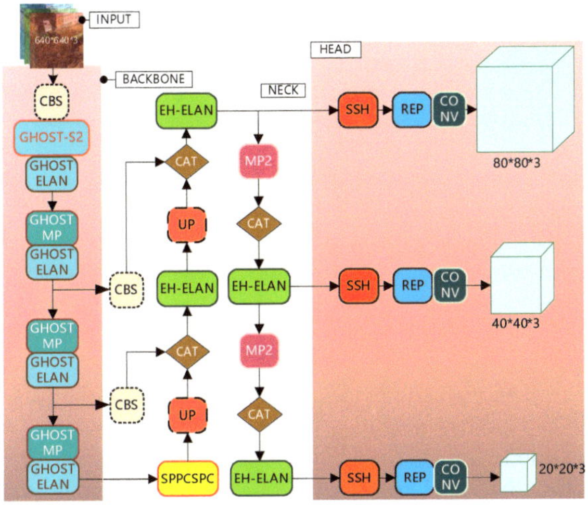

Figure 9. The structure of YOLO-GNS algorithm.

The GhostMP module is composed of Maxpool, GhostS2, CBS, CAT. The GhostELAN module is composed of GhostS1, GhostS2, CBS, and CAT. GhostS1 consists of two stacked Ghost convolutions (Ghost Conv), the first Ghost Conv increasing the number of channels and the second Ghost Conv reduces the number of channels to match the shortcut path, making the number of channels for the input signature map the same as which in the output signature map for the second Ghost Conv. The shortcut path of GhostS2 is implemented by depth-wise convolution (DW Conv) with a downsampling layer and a stride = 2 to reduce the number of channels. Add represents a signature graph addition operation where the number of channels does not change.

The implementation of GhostConv is divided into three steps: the first step is to use ordinary convolution calculation to get a feature map with less channel information, the second step is to use inexpensive operation to generate more feature maps, and the last step is to connect different feature maps to form a new output.

In ordinary convolution, given input data $X \in R^{c \times h \times w}$, c denotes the number of input channels; h and w denote the height and width of the input data, respectively, and are used to generate any convolution layer of N feature map, as shown in Equation (1):

$$Y \in X^{*} f + B \tag{1}$$

where: * is a convolution operator, B is a deviation term, $Y \in R^{h' \times w' \times n}$ represents the output feature map of N channels, $f \in R^{c \times k \times k \times n}$ is the convolution kernel size in a convolution layer, h' and w' represent the height and width of the output data, respectively, $k \times k$ denotes the size of the convolution kernel f. In ordinary convolution operations, because the number of convolution cores n and channel c is very large, the number of FLOPs required is $n \cdot h' \cdot w' \cdot c \cdot k \cdot k$.

Thus, the parameters to be optimized for operation (f and B) are determined by the size of the input and output feature maps. Since the output feature maps of ordinary convolution layers are usually redundant and may have similar redundancy to each other, it is not necessary to use a large number of parameters FLOP to generate redundant feature maps, which are "Ghost" converted from a few original feature maps by some inexpensive linear operations. These original feature maps are usually generated by ordinary convolution kernels and have less channel information. Generally, m original feature map $Y' \in R^{h' \times w' \times m}$ is generated by once convolution:

$$Y' = X^{*} f' \tag{2}$$

where: $f' \in R^{c \times k \times k \times m}$ is a convolution kernel, $m \leqslant n$. To maintain the same spatial size as the output feature map, the hyperparametric (convolution size, stride, padding) is the same as the ordinary convolution. To further obtain the required n feature maps, a series of inexpensive linear operations are used for each original feature in Y', resulting in s Ghost feature maps, as shown in Formula (3):

$$y_{ij} = \phi_{i,j}\left(y_i'\right); \forall i = 1, 2, \cdots, m, j = 1, 2, \cdots, s \tag{3}$$

where:y_{ij} represents the first primitive feature map in Y'. $\phi_{i,j}$ represents the jth linear operation used to generate the jth Ghost feature graph. By using inexpensive linear operations, we can get $n = m \cdot s$ feature maps as output of the Ghost module, as shown in Formula (4):

$$Y = [y_{11}, y_{12}, \cdots, y_{ms}] \tag{4}$$

The Ghost module divides the original convolution layer into two phases, as shown in Figure 10. The first phase uses a small number of convolution cores to generate the original feature map, and the second phase uses inexpensive transformation to generate more Ghost feature maps. Linear operations are used on each channel to reduce computational effort.

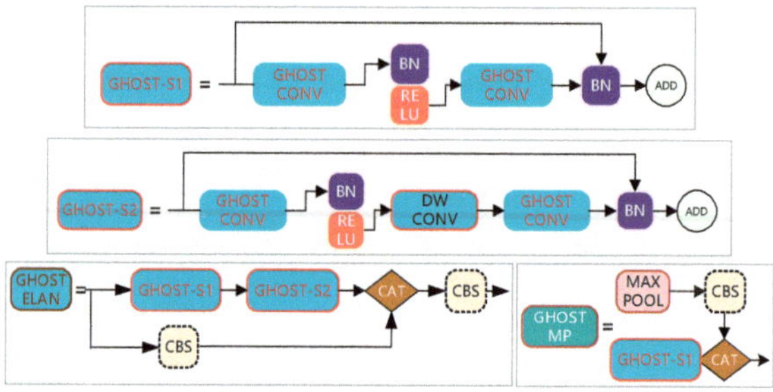

Figure 10. The structure of GhostNet in YOLO-GNS.

2.2.2. Prediction Optimization Based on SSH Structure

In order to improve the small target detection ability and further shorten the inference time, Single Stage Headless (SSH) algorithm [31] is introduced into the network, which is a single-stage context network structure. The two-stage context network structure combines more context information by increasing the size of the candidate box. Nevertheless, SSH combines context information through a single convolution layer, where the Context-Network structure of the SSH detection module is shown in Figure 11, which requires less memory to detect and locate more accurately.

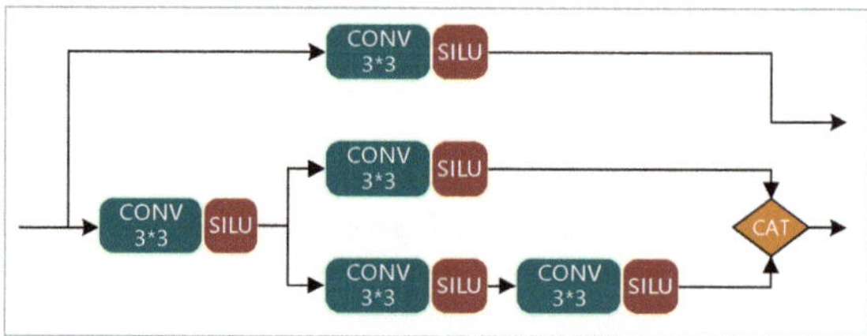

Figure 11. The structure of Context-Network in SSH.

In YOLO-GNS, add the SSH context network structure before the REP structure. First, reduce the number of channels to X/2 through 3×3 convolution layer and SILU activation function (3C-SILU), and then send this result to two branches. One branch contains only one 3C-SILU operation, which results in the feature that the channel is X/2. The other branch contains two consecutive 3C-SILU operations, which also results in the feature that the channel is X/2. Finally, concatenate the two feature maps to get the final output of the SSH context network structure.

The SSH context network structure incorporates more context information and is approximated by increasing the sensory field of the feature maps. For example, a small field can only see the special vehicle itself, while a larger field can see the excavator head, caterpillar, and other places.

Generally, deeper feature layers contain more abstract semantic information to facilitate classification, while shallow features have more specific information, such as edges, angles, and so on, to facilitate the positioning of bounding box.

Therefore, the SSH context network structure integrates the current and high-level feature information, effectively improves the detection ability of the weakened and obstructed special vehicles in complex environments, helps to improve the accuracy of the algorithm, and does not significantly increase the additional computational load.

3. Results

In order to evaluate the special vehicle detection performance of YOLO-GNS algorithm in this paper, this experiment conducts training and testing on special vehicle (SEVE) dataset. Additionally, to evaluate the general performance of the algorithm, this experiment adds training and testing on the Microsoft COCO dataset.

3.1. Special Vehicle Dataset

Heretofore, there is no public data set of special vehicles from the perspective of drones. Therefore, from January 2021 to June 2022, we used UAV to shoot a large number of videos at multiple heights and angles over construction areas, wilderness, building sites, and other areas. After that, frames are extracted and labeled from these videos to form a special vehicle dataset. This dataset contains 17,992 pairs of images and labels, including 14,392 training sets, 1800 validation sets, and 1800 test sets. The image resolution in SEVE dataset is 1920 × 1080. The types of special vehicles include cranes, traction vehicles, tank trucks, obstacle removal vehicles, cleaning vehicles, lifting vehicles, loading and unloading vehicles, mixing vehicles, excavators, bulldozers, and road rollers. The different scene types include urban, rural, arable, woodland, grassland, construction land, roads, etc. Some examples of the dataset are shown in Figure 12.

3.2. Experimental Environment and Settings

The experiment is based on 64-bit operating system Windows 10, the CPU is Intel Xeon Gold 6246R, the GPU uses NVIDIA GeForce RTX3090, and the deep learning framework is Pytorch v1.7.0. We use Frames Per Second (FPS) to measure the detection speed, which indicates the number of images processed by the specified hardware per second by the detection model. In the experiment, the FPS for each method is tested on a single GPU device. IOU is set to 0.5, The mAP (mean Average Precision), an index related to the IOU threshold, was used as the standard of detection accuracy. In multi-category target detection, the curve drawn by each category based on its accuracy (Precision) and recall (Recall) is called a P-R curve, in which the average recognition accuracy of a category is equal. AP@0.5 (Average Precision, IoU threshold greater than 0.5) is the size of the area below the P-R curve of this category. mAP@0.5 Average recognition accuracy by all categories AP@0.5 add up to get the average.

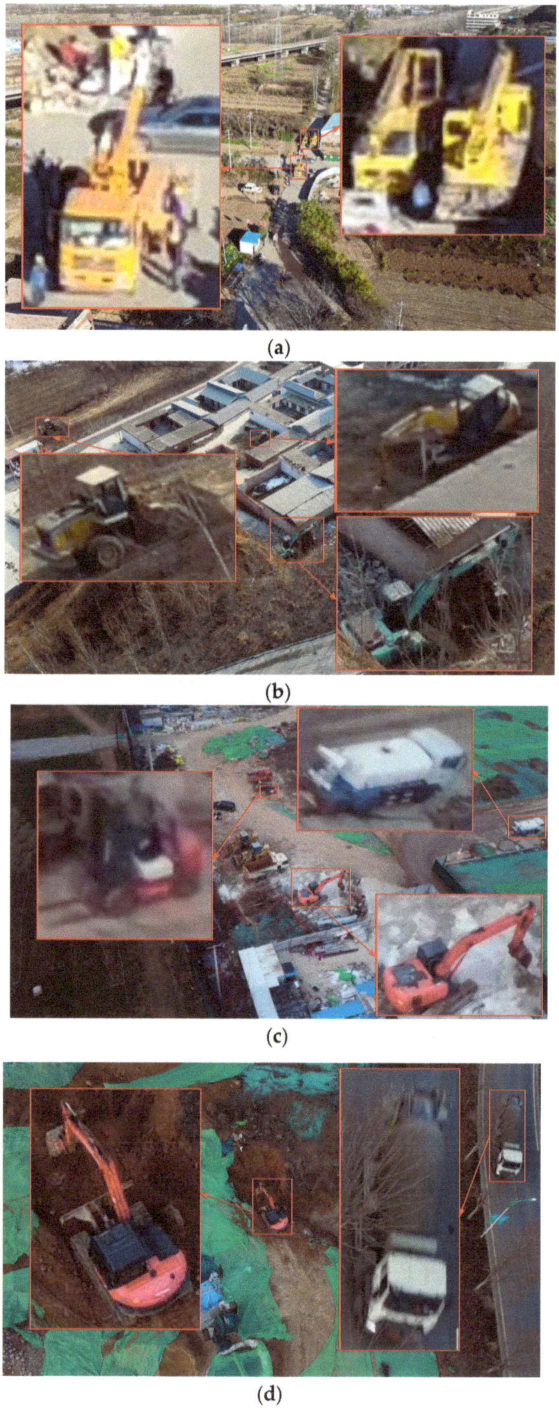

Figure 12. Sample Images of SEVE dataset. (**a**) Cranes in construction areas; (**b**) Excavator and loaders in building sites; (**c**) Forklifts in construction areas; (**d**) Excavator and tank trucks in wilderness.

Precision and recall are defined as:

$$Recall = \frac{TP}{TP + FN} \tag{5}$$

$$Precision = \frac{TP}{TP + FP} \tag{6}$$

$$AP = \int_0^1 P(R)dR \tag{7}$$

$$mAP = \frac{1}{c}\sum_{i=1}^{c} AP_i \tag{8}$$

Among them, *TP* was the real case, *FP* was the false positive case, *FN* was the false negative case, and *C* was the total number of categories detected for the target.

Due to the limitation of the experimental device, the input image size is scaled to 800×800 pixels. The optimizer uses SGD; the learning rate is 1×10^{-2}; the momentum is 0.9; the weight decay is 5×10^{-4}, using the Cosine Annealing algorithm to adjust the learning rate; the batch size is 8; and the training durations are 300 epochs, 10 training epochs, and 1 test epochs alternately.

3.3. Experimental Results and Analysis

This paper conducts experiments on the open dataset COCO and the SEVE dataset created in this paper to verify the validity of the proposed methods. The experiment is divided into three parts:

(1) Experiments are carried out on the SEVE dataset to verify the feasibility of the proposed method, and to compare the results with those of other target detection methods on this dataset to illustrate the advantages of this method;

(2) Verify the universality of this method on COCO datasets;

(3) Designing an ablation experiment further demonstrates the validity of the method.

3.3.1. Experiments on SEVE Dataset

In this experiment, the YOLO-GNS algorithm is compared with the prevailing target detection algorithms in the SEVE dataset created in this paper. The experimental results are shown in Table 1. Table 1 contains nine categories: C, L, T, M, F, P, R, EL, and EX, corresponding to the SEVE dataset and referring to cranes, loader cars, tank cars, mixer cars, forklifts, piling machines, road rollers, elevate cars, and excavators. The resulting data AP@0.5 represent the average recognition accuracy of this category under different methods, while data in column mAP@0.5 represents the average recognition accuracy of all categories. Params represent the size of the paraments of each method. The resulting data represent the average recognition accuracy for all categories for different datasets under different methods.

Table 1. Comparison of Detection Accuracy of Different Target Detection Algorithms on SEVE dataset.

Methods	AP@0.5(%)									mAP@0.5 (%)	Params(M)	FPS
	C	L	T	M	F	P	R	EL	EX			
Faster-RCNN	73.2	75.5	76.1	80.2	78.1	81.3	56.3	45.5	21.3	65.3	186.3	16.8
RetinaNet	77.5	78.6	85.1	82.3	81.5	80.6	57.6	49.1	23.5	68.4	28.5	19.5
YOLOV4	78.7	80.1	82.3	83.5	82.6	78.3	60.5	55.8	30.3	70.2	64.4	25.6
YOLOV5-X	79.8	78.1	85.6	83.9	83.1	82.5	59.1	58.3	32.5	71.4	86.7	29.2
YOLOV7	80.5	82.3	86.4	88.6	85.3	86.4	65.3	60.8	45.8	75.7	36.9	31.5
YOLO-GNS	85.9	86.9	89.4	91.3	90.1	89.6	69.5	67.3	50.8	80.1	30.7	33.1

In the SEVE dataset, special vehicle targets vary greatly in scale and there are mostly small targets. The image background is complex and volatile, and it is difficult to distinguish the targets into the background, and some targets are also obscured, which brings some difficulty to the detection. The improved network in this paper has significant accuracy advantages compared with other mainstream target detection algorithms. The method in this paper achieves the best results on the SEVE dataset with 80.1%, which is 4.4% higher accuracy compared to YOLOV7; meanwhile, the mAP is 14.8%, 11.7%, 9.9%, and 8.7% higher compared to four target detection algorithms, namely Faster R-CNN, RetinaNet, YOLOV4, and YOLOV5, respectively; although the YOLOv7 and YOLOv5 detection speeds are close to that of YOLO-GNS, the mAPs are all lower than the methods in this paper. Owing to GhostNet applied in the backbone section, the parameters of YOLO-GNS are reduced by 6.2M. In the case of low differentiation of YOLO series backbone networks, the mAP of this paper's method is higher and the detection speed is faster, which indicates that this paper's method makes up for the difference of backbone networks and reflects greater advantages. Due to the reconstructed backbone network and the parallel SSH context network that makes the network structure of this paper in the case of increasing complexity, the detection speed is not reduced and can meet the needs of engineering applications.

The detection results of YOLOV7 and this paper's method YOLO-GNS are shown in Figures 13–15. Column (a) shows the recognition results of the YOLO-GNS network, and column (b) shows the recognition results of the original YOLOV7 network. A comparison of the results of the two networks shows that the YOLO-GNS network in this paper has improved accuracy in terms of bounding box and category probabilities. On the other hand, the recognition of special vehicles, such as cranes, loader cars, tank cars, mixer cars, forklifts, and excavators, and their differences from ordinary vehicles are improved in the proposed model.

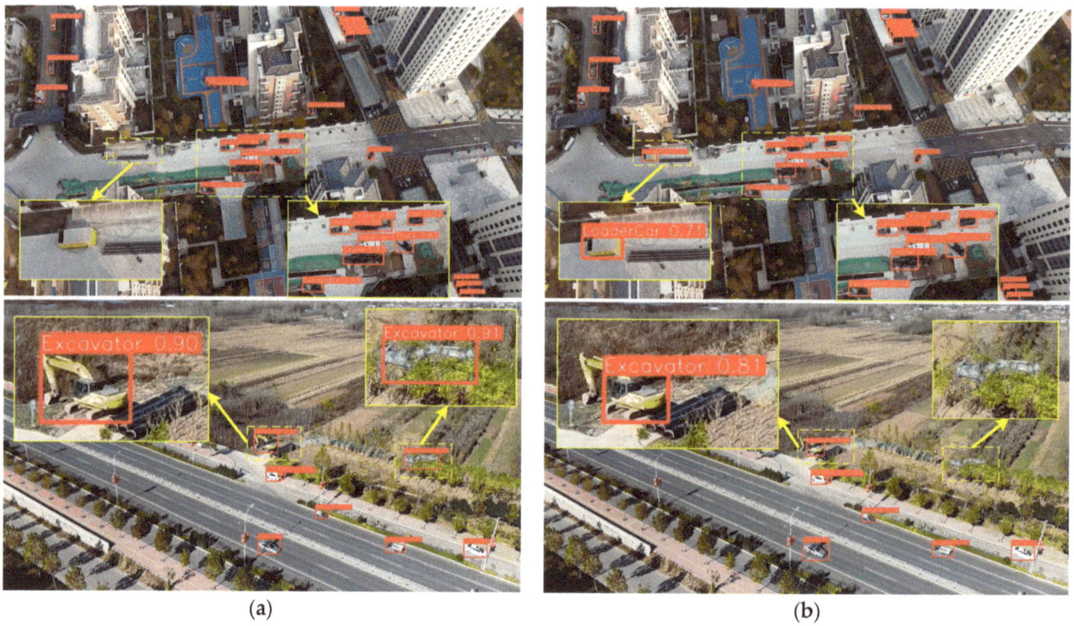

Figure 13. Recognition results in crowded environments of SEVE Dataset. (**a**) Recognition results of the YOLO-GNS network; (**b**) Recognition results of the YOLO-V7 network.

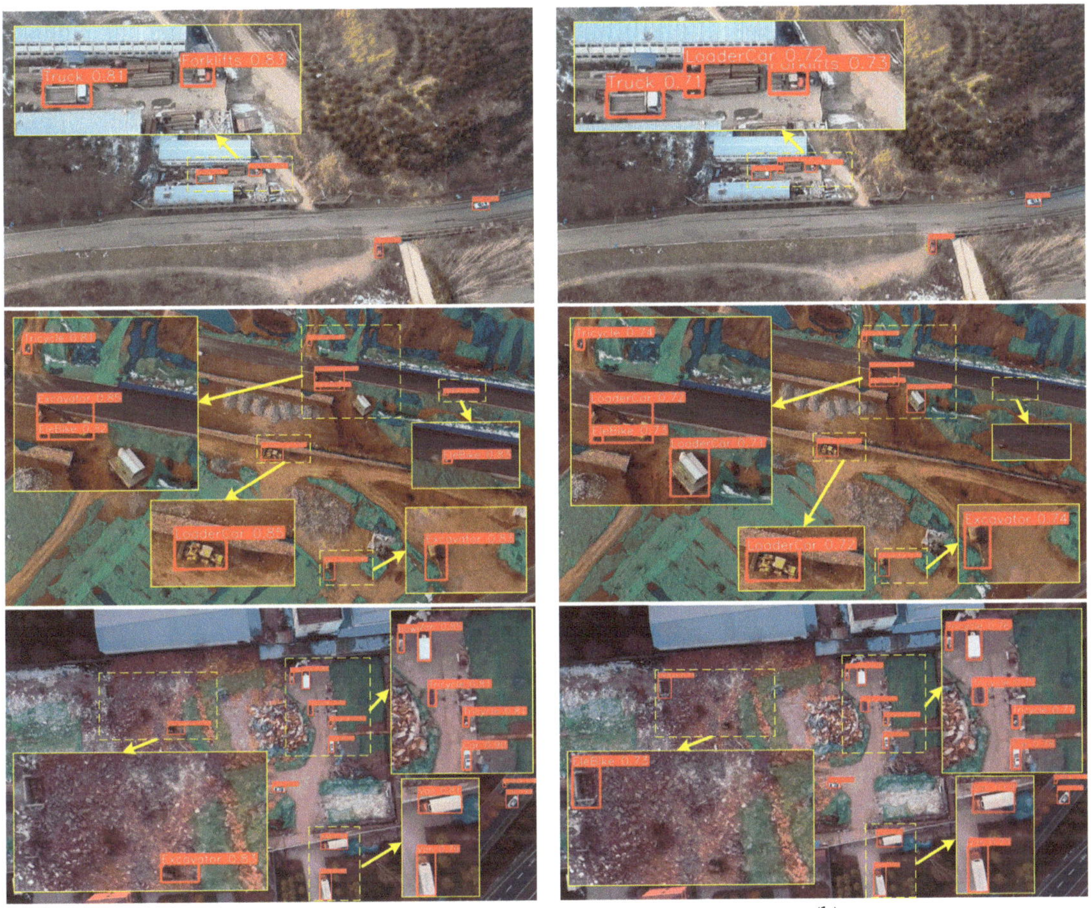

(**a**) (**b**)

Figure 14. Recognition results in complex background of SEVE Dataset. (**a**) Recognition results of the YOLO-GNS network; (**b**) Recognition results of the YOLO-V7 network.

In Figure 13, it is shown that in crowded environments such as cities and roads, YOLO-GNS can identify obscured special vehicles and does not cause false detections, while YOLOV7 produces false detections and missed detections and has lower class probability values than the modified model. In Figure 14, it is shown that YOLO-GNS distinguishes special vehicles from ordinary vehicles by extracting smaller and more accurate features in environments with camouflage characteristics, such as construction sites, and can identify special vehicles that are highly similar to the background. In Figure 15, it is shown that the YOLO-GNS network is able to identify different special vehicle types in complex and challenging conditions under poor lighting conditions and bad weather, while the original YOLOV7 model would show quite a few missed and false detections. In conclusion, the YOLO-GNS proposed in this paper is able to identify targets with a high prediction probability under a variety of complex scenarios. In some cases, the base model YOLOV7 cannot accurately identify special vehicles, or it has a lower probability than YOLO-GNS.

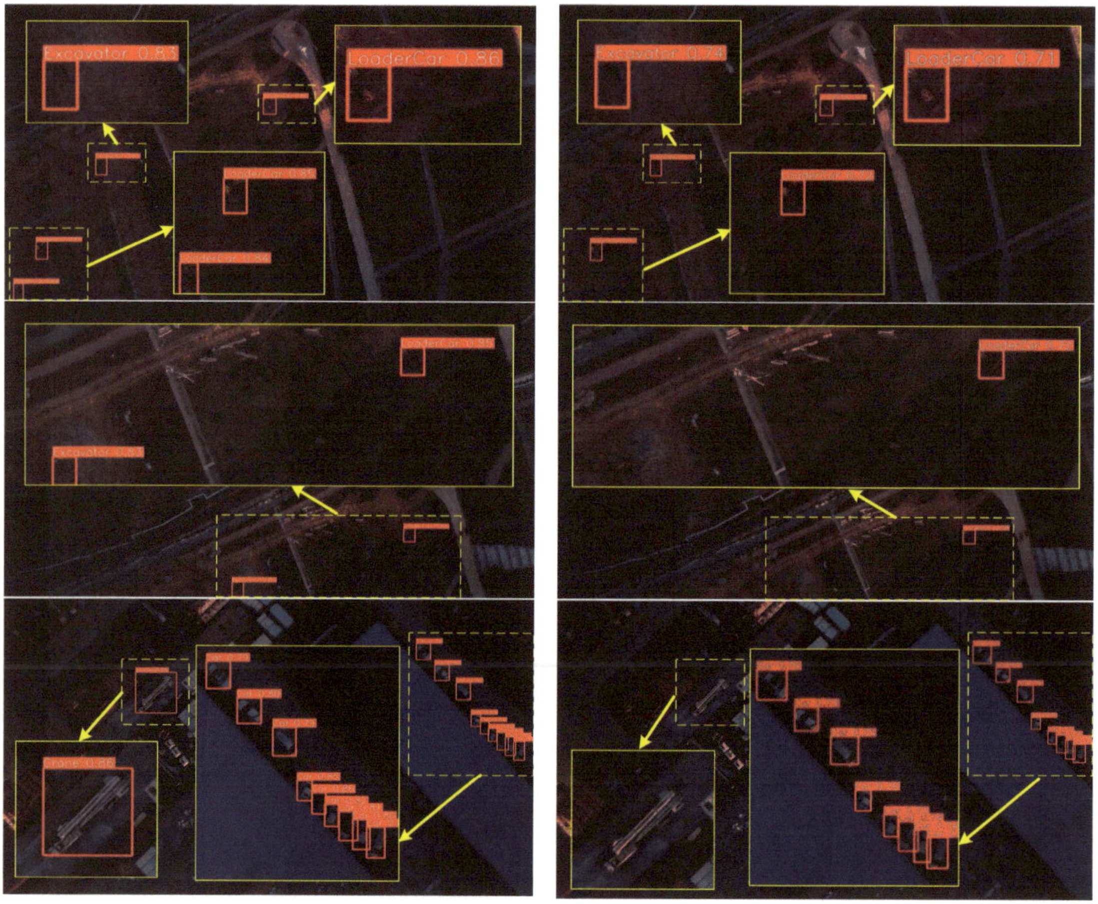

(**a**) (**b**)

Figure 15. Recognition results in adverse light environment of SEVE Dataset. (**a**) Recognition results of the YOLO-GNS network; (**b**) Recognition results of the YOLO-V7 network.

3.3.2. Experiments on COCO Datasets

The evaluation metrics are mAP0.5, mAP0.75, and mAP0.5:0.95. mAP0.5 and mAP0.75 are the average accuracy of all target categories calculated at IOU thresholds of 0.5 and 0.75. mAP0.5:0.95 is the average accuracy of 0.5 to 0.95 at 0.05 intervals of 10. mAP0.5:0.95 is the average accuracy at 10 threshold values from 0.5 to 0.95 at 0.05 intervals.

As shown in Table 2, the experimental data show that the method in this paper also works well on the COCO dataset. The mAP0.5:0.95 is improved by 0.1% for YOLO-GNS compared to the original method with a similar speed. The mAP0.5 of YOLOV4 reaches 65.7% under this dataset; the mAP0.5 of YOLOV5-X is 68.8% under this dataset, but both networks are based on Darknet and its improvements with complex structures, and the detection speed is slightly lower than that of the present method. YOLO-GNS has 0.2% lower mAP0.75 than YOLOV7 on the COCO dataset but 0.1% higher mAP0.5; YOLO-GNS has improved detection speed and higher mAP than YOLOV4 and YOLOV5-X methods, indicating that the method in this paper is still effective on the public dataset COCO.

Table 2. Experimental results on coco dataset.

Methods	Backbone	mAP0.5:0.95	mAP0.5	mAP0.75
Faster-RCNN	ResNet50	36.2	59.2	39.1
RetinaNet	ResNet50	36.9	56.3	39.3
YOLOV4	CSPDarknet-53	43.5	65.7	47.3
YOLOV5-X	Modified CSP v5	50.4	68.8	-
YOLOV7	E-ELAN	51.4	69.7	55.9
YOLO-GNS	GhostELAN	51.5	69.8	55.7

3.3.3. Ablation Experiment

Ablation experiments were conducted on the SEVE dataset to verify the effect of different network structures on the final detection results, and the experimental results are shown in Table 3.

Table 3. Results of ablation experiments.

Methods	Backbone	GhostNet	SSH	mAP@0.5(%)
YOLOV7	E-ELAN	×	×	75.7
YOLOV7	E-ELAN	×	√	78.9
YOLOV7	E-ELAN	√	×	79.2
YOLOV7	E-ELAN	√	√	80.1

"×" means no addition, "√" means addition.

With the addition of GhostNet in YOLOV7, the mAP value is improved by 3.5%. GhostNet forms the backbone network by forming GhostMP and GhostELAN modules, which has the advantages of maintaining the recognition performance of similarity and reducing the convolution operation at the same time and continuing to effectively increase the exploitation of feature maps, which is beneficial to the recognition of small targets. The addition of SSH structure in YOLOV7 improves the mAP value by 3.2%. SSH contextual network structure incorporates more concrete information and enhances the recognition of multiple details of special vehicles by increasing the perceptual field of the features, thus improving the detection performance. After adding both GhostNet and SSH structures in YOLOV7, the AP increases by 4.4%, further demonstrating that GhostNet and SSH can improve detection accuracy.

4. Discussion

The evaluation metrics examined in this study were AP and mAP. In the modified network, the values obtained from these criteria were as follows. The AP of cranes was 85.9%, the AP of loader cars was 86.9%, the AP of tank cars was 89.4%, the AP of mixer cars was 91.3%, the AP of forklifts was 90.1%, the AP for piling machines is 89.6%, the AP for road rollers is 69.5%, the AP for elevate cars is 67.3%, and the AP for excavators is 50.8%. Based on the basic results of the YOLOv7 network, it can be said that the proposed network has improved on average by 4.4% in accuracy and 1.6 in FPS, indicating that the improved network has improved speed to some extent with improved accuracy.

In recent years, the employment of artificial intelligence and deep learning methods has become one of the most popular and useful approaches in object recognition. Scholars have made many efforts to better detect vehicles in the context of UAV observations. Jianghuan Xie et al. proposed the residual feature enhanced pyramid network (RFEPNet), which uses pyramidal convolution and residual connectivity structure to enhance the semantic information of vehicle features [16]. One of the problems of these studies is the inability to detect small vehicles over long distances. Zhongyu Zhang et al. used a YOLOv3-based deep separable attention-guided network (DAGN), improved the loss function of YOLOv3, and combined feature tandem and attention blocks to enable the model to distinguish between important and unimportant vehicle features [19]. One of the limitations of this study is the lack of types of vehicles and the lack of challenging images. Wang Zhang et al. helped the feature pyramid network (FPN) to handle the scale variation of vehicles by using the multi-scale feature adaptive fusion network (MSFAF-Net) and the region attention-based three-headed network (RATH-Net) [20]. However, the study did not address the crowded background images, hidden regions, and vehicle target-sensor distance, etc. Xin Luo et al. constructed a vehicle dataset for target recognition and used it for vehicle detection by an improved YOLO [21], but the dataset did not include special vehicles.

Previous research has focused on general vehicle detection, with a few studies examining the identification of different types of vehicles. In addition, the challenges of specialty vehicle identification, such as the small size of vehicles, crowded environments, hidden areas, and confusion with contexts such as construction sites, have not been comprehensively addressed in these studies. Thus, it can be argued that the unauthorized presence of specialty vehicles in challenging environments and the inaccurate identification of sensitive infrastructures remain some of the most important issues in ensuring public safety. The main goal of this study is to identify multiple types of specialty vehicles and distinguish them from ordinary vehicles at a distance, despite challenges such as the small size of specialty vehicles, crowded backgrounds, and the presence of occlusions.

In this study, the YOLOV7 network was modified to improve the challenges of specialty vehicle identification. A large number of visible images of different types of special vehicles and ordinary vehicles at close and long distances in different environments were collected and labeled to identify multiple types of special vehicles and distinguish them from ordinary vehicles. Considering the limited computational power of the airborne system, GhostNet is introduced to reduce the computational cost of the proposed algorithm. The proposed algorithm facilitates the deployment of airborne systems by using linear transformation to generate feature maps in GhostNet instead of the usual convolutional computation and requires less FLOP. On the other hand, the SSH structure is shown to have the ability to improve the detection accuracy of the algorithm. The context network is able to compute the contexts of pixels at different locations from multiple subspaces, which facilitates YOLO-GNS to extract important features from large-scale scenes. For example, in Figure 13, there are examples of special vehicles that the basic model cannot recognize in some cases. However, the modified model is able to recognize them; moreover, in other cases, they operate with lower accuracy than the modified network. This result indicates that the current network has improved in identifying special vehicles compared to the basic network. By applying these changes in the network structure and using a wide range of data sets, the proposed method is able to identify all specialty vehicle types in challenging environments. In Figures 14 and 15, examples of difficult images and poor lighting conditions are provided, all of which have higher recognition accuracy in the modified network than in the basic network.

5. Conclusions

As already pointed out, specialty vehicle recognition in various scenarios is a complex process; the usual approaches and even traditional deep learning network methods do not work well in some cases. When using UAVs to detect small or obscured specialty vehicles from large-scale scenes, both detection accuracy and computational consumption need to be considered. In this work, we propose a novel UAV-based algorithm for special vehicle target detection that enhances feature extraction while optimizing the feature fusion computation. A dedicated dataset of 17,992 UAV image datasets including multiple types of special vehicles is introduced, and extensive comparative experiments are conducted to illustrate the effectiveness of the proposed algorithm. The results show that the AP and FPS are improved by 4.4% and 1.6, respectively, compared to the primary YOLOv7. It can be demonstrated that the algorithm provides a single optimal solution for UAV-based target detection in the field of special vehicle identification. In the next work, the special vehicle detection method with visible and infrared fusion will be investigated.

Author Contributions: Conceptualization, Z.Q. and H.B.; Data curation, H.B.; Formal analysis, Z.Q. and H.B.; Funding acquisition, T.C.; Investigation, H.B.; Methodology, Z.Q.; Project administration, T.C.; Resources, T.C.; Software, Z.Q.; Supervision, H.B. and T.C.; Validation, Z.Q.; Visualization, Z.Q.; Writing—original draft, Z.Q.; Writing—review & editing, H.B. and T.C. All authors have read and agreed to the published version of the manuscript.

Funding: This work was supported in part by the National Key R&D Program of China (No.2022YFE0200300), the National Natural Science Foundation of China (No. 61972023), and the Beijing Natural Science Foundation (L223022).

Data Availability Statement: Not applicable.

Conflicts of Interest: The authors declare no conflict of interest.

References

1. Li, Q.; Ban, X.; Wu, H. Design of Informationized Operation and Maintenance System for Long-Distance Oil and Gas Pipelines. In Proceedings of the International Conference on Computer Science and Application Engineering, Sanya, China, 22–24 October 2019. [CrossRef]
2. Bao, W.; Ren, Y.; Wang, N.; Hu, G.; Yang, X. Detection of Abnormal Vibration Dampers on Transmission Lines in UAV Remote Sensing Images with PMA-YOLO. *Remote Sens.* **2021**, *13*, 4134. [CrossRef]
3. Jiang, Y.; Huang, Y.; Liu, J.; Li, D.; Li, S.; Nie, W.; Chung, I.-H. Automatic Volume Calculation and Mapping of Construction and Demolition Debris Using Drones, Deep Learning, and GIS. *Drones* **2022**, *6*, 279. [CrossRef]
4. Mittal, P.; Singh, R.; Sharma, A. Deep Learning-Based Object Detection in Low-Altitude UAV Datasets: A Survey. *Image Vis. Comput.* **2020**, *104*, 104046. [CrossRef]
5. Bouguettaya, A.; Zarzour, H.; Kechida, A.; Taberkit, A.M. Vehicle Detection From UAV Imagery with Deep Learning: A Review. *IEEE Trans. Neural Netw. Learn. Syst.* **2022**, *33*, 6047–6067. [CrossRef]
6. Srivastava, S.; Narayan, S.; Mittal, S. A Survey of Deep Learning Techniques for Vehicle Detection from UAV Images. *J. Syst. Archit.* **2021**, *117*, 102152. [CrossRef]
7. Liu, W.; Quijano, K.; Crawford, M.M. YOLOv5-Tassel: Detecting Tassels in RGB UAV Imagery with Improved YOLOv5 Based on Transfer Learning. *IEEE J. Sel. Top. Appl. Earth Obs. Remote Sens.* **2022**, *15*, 8085–8094. [CrossRef]
8. Zhou, H.; Ma, A.; Niu, Y.; Ma, Z. Small-Object Detection for UAV-Based Images Using a Distance Metric Method. *Drones* **2022**, *6*, 308. [CrossRef]
9. Wang, J.; Shao, F.; He, X.; Lu, G. A Novel Method of Small Object Detection in UAV Remote Sensing Images Based on Feature Alignment of Candidate Regions. *Drones* **2022**, *6*, 292. [CrossRef]
10. Li, Q.; Mou, L.; Xu, Q.; Zhang, Y.; Zhu, X.X. R^3-Net: A Deep Network for Multioriented Vehicle Detection in Aerial Images and Videos. *IEEE Trans. Geosci. Remote Sens.* **2019**, *57*, 5028–5042. [CrossRef]
11. Li, Y.; Yuan, H.; Wang, Y.; Xiao, C. GGT-YOLO: A Novel Object Detection Algorithm for Drone-Based Maritime Cruising. *Drones* **2022**, *6*, 335. [CrossRef]
12. Wang, J.; Ding, J.; Guo, H.; Cheng, W.; Pan, T.; Yang, W. Mask OBB: A Semantic Attention-Based Mask Oriented Bounding Box Representation for Multi-Category Object Detection in Aerial Images. *Remote Sens.* **2019**, *11*, 2930. [CrossRef]
13. Mantau, A.J.; Widayat, I.W.; Leu, J.-S.; Köppen, M. A Human-Detection Method Based on YOLOv5 and Transfer Learning Using Thermal Image Data from UAV Perspective for Surveillance System. *Drones* **2022**, *6*, 290. [CrossRef]
14. Deng, Z.; Sun, H.; Zhou, S.; Zhao, J.; Zou, H. Toward Fast and Accurate Vehicle Detection in Aerial Images Using Coupled Region-Based Convolutional Neural Networks. *IEEE J. Sel. Top. Appl. Earth Obs. Remote Sens.* **2017**, *10*, 3652–3664. [CrossRef]
15. Tian, G.; Liu, J.; Yang, W. A Dual Neural Network for Object Detection in UAV Images. *Neurocomputing* **2021**, *443*, 292–301. [CrossRef]
16. Xie, J.; Wang, D.; Guo, J.; Han, P.; Fang, J.; Xu, Z. An Anchor-Free Detector Based on Residual Feature Enhancement Pyramid Network for UAV Vehicle Detection. In Proceedings of the 2021 4th International Conference on Artificial Intelligence and Pattern Recognition, Xiamen, China, 24–26 September 2021; ACM: New York, NY, USA, 2021; pp. 287–294.
17. Wan, Y.; Zhong, Y.; Huang, Y.; Han, Y.; Cui, Y.; Yang, Q.; Li, Z.; Yuan, Z.; Li, Q. ARSD: An Adaptive Region Selection Object Detection Framework for UAV Images. *Drones* **2022**, *6*, 228. [CrossRef]
18. Liu, M.; Wang, X.; Zhou, A.; Fu, X.; Ma, Y.; Piao, C. UAV-YOLO: Small Object Detection on Unmanned Aerial Vehicle Perspective. *Sensors* **2020**, *20*, 2238. [CrossRef]
19. Zhang, Z.; Liu, Y.; Liu, T.; Lin, Z.; Wang, S. DAGN: A Real-Time UAV Remote Sensing Image Vehicle Detection Framework. *IEEE Geosci. Remote Sens. Lett.* **2020**, *17*, 1884–1888. [CrossRef]
20. Zhang, W.; Liu, C.; Chang, F.; Song, Y. Multi-Scale and Occlusion Aware Network for Vehicle Detection and Segmentation on UAV Aerial Images. *Remote Sens.* **2020**, *12*, 1760. [CrossRef]
21. Luo, X.; Tian, X.; Zhang, H.; Hou, W.; Leng, G.; Xu, W.; Jia, H.; He, X.; Wang, M.; Zhang, J. Fast Automatic Vehicle Detection in UAV Images Using Convolutional Neural Networks. *Remote Sens.* **2020**, *12*, 1994. [CrossRef]

22. Balamuralidhar, N.; Tilon, S.; Nex, F. MultEYE: Monitoring System for Real-Time Vehicle Detection, Tracking and Speed Estimation from UAV Imagery on Edge-Computing Platforms. *Remote Sens.* **2021**, *13*, 573. [CrossRef]
23. Redmon, J.; Divvala, S.; Girshick, R.; Farhadi, A. You only look once: Unified, real-time object detection. In Proceedings of the IEEE Conference on Computer Vision and Pattern Recognition, Las Vegas, NV, USA, 27–30 June 2016; pp. 779–788.
24. Redmon, J.; Farhadi, A. YOLO9000: Better, faster, stronger. In Proceedings of the IEEE Conference on Computer Vision and Pattern Recognition, Honolulu, HI, USA, 21–26 July 2017; pp. 7263–7271.
25. Redmon, J.; Farhadi, A. Yolov3: An incremental improvement. *arXiv* **2018**, arXiv:1804.02767.
26. Bochkovskiy, A.; Wang, C.Y.; Liao, H.Y.M. Yolov4: Optimal speed and accuracy of object detection. *arXiv* **2020**, arXiv:2004.10934.
27. Ultralytics. YOLOv5. Available online: https://github.com/ultralytics/yolov5 (accessed on 1 November 2020).
28. Li, C.; Li, L.; Jiang, H.; Weng, K.; Geng, Y.; Li, L.; Ke, Z.; Li, Q.; Cheng, M.; Nie, W.; et al. YOLOv6: A Single-Stage Object Detection Framework for Industrial Applications. *arXiv* **2022**, arXiv:2209.02976.
29. Wang, C.-Y.; Bochkovskiy, A.; Liao, H.-Y.M. YOLOv7: Trainable Bag-of-Freebies Sets New State-of-the-Art for Real-Time Object Detectors. *arXiv* **2022**, arXiv:2207.02696.
30. Han, K.; Wang, Y.; Xu, C.; Guo, J.; Xu, C.; Wu, E.; Tian, Q. GhostNets on Heterogeneous Devices via Cheap Operations. *Int. J. Comput. Vis.* **2022**, *130*, 1050–1069. [CrossRef]
31. Najibi, M.; Samangouei, P.; Chellappa, R.; Davis, L. SSH: Single Stage Headless Face Detector. *arXiv* **2017**, arXiv:1708.03979.

MDPI AG
Grosspeteranlage 5
4052 Basel
Switzerland
Tel.: +41 61 683 77 34

Drones Editorial Office
E-mail: drones@mdpi.com
www.mdpi.com/journal/drones

www.ingramcontent.com/pod-product-compliance
Lightning Source LLC
Chambersburg PA
CBHW041358210926
51850CB00013B/3112

* 9 7 8 3 7 2 5 8 2 5 0 7 3 *